Ondes Martenot with Tubes

Ondes Martenot with Tubes

Thierry Courrier
Laurent Quartier

iSTE

WILEY

First published 2023 in Great Britain and the United States by ISTE Ltd and John Wiley & Sons, Inc.

ISTE Ltd
27-37 St George's Road
London SW19 4EU
UK

www.iste.co.uk

John Wiley & Sons, Inc.
111 River Street
Hoboken, NJ 07030
USA

www.wiley.com

Library of Congress Control Number: 2022950708

British Library Cataloguing-in-Publication Data
A CIP record for this book is available from the British Library
ISBN 978-1-78630-905-1

Contents

Foreword

The 20th century saw the birth of new forms of lutherie, which no longer relied on the mechanical properties of materials to produce sounds, but used electricity as a raw material for various components, active and passive, making it possible to modulate electric flow to achieve a rich palette of timbres and control the intensity and pitch of musical sounds.

However, while a large number of electronic instruments have emerged, only a few have succeeded in establishing themselves in the long term and, among these, the ondes musicales invented by Maurice Martenot has occupied a special place for almost a century!

Few electronic instruments can claim to have found a place in such varied musical fields. Thus, many composers and artists, from Olivier Messiaen to Jonny Greenwood, from André Jolivet to Jacques Brel, have been seduced by the expressive possibilities of the ondes.

No other electronic instrument makes it possible to shape the sound with such precision, such refinement in the nuances of intensity and timbre, and this is undoubtedly one of the reasons why the ondes Martenot continues to be played, taught and still inspire many composers today.

However, there are only a few works that present the instrument in detail, and none have given an account of its evolution and analyzed in detail each major type of the ondes, from the first instrument presented to the public in 1928 to the very first transistor model in 1974. Maurice Martenot, in fact, never ceased to develop his instrument, and it is no exaggeration to say that each model from his workshop has unique characteristics.

This book is the result of considerable work over several years. I am sure it will be of interest to ondists as well as for lovers of old musical electronics.

Hugues GENEVOIS
Lutheries-Acoustics-Music team
Jean le Rond d'Alembert Institute
(Sorbonne University – CNRS – Ministry of Culture)

Photo Credits

For those without particular mention, the photographs belong to Mr. Courrier.

All photographs bearing the caption "Musée de la Musique, Paris" were taken on the premises of the Museum (Philharmonie de Paris) by Mr. Courrier, with the exception of the photographs corresponding to the Martenot Fund.

Ms. Loepper-Attia, in charge of the preventive and curative conservation of the collections at the Cité de la Musique, is the author. These are Figures 3.22, 3.23, 3.28, 3.31, 3.32, 4.5, 4.18, 5.1, 5.2, 5.5, 6.21, 6.23, as well as the radiograph (Figure 2.26).

The photographs showing Mr. Martenot playing on the first ondes models (Figures 1.1, 1.2, and 2.8) are documents belonging to the Martenot family, except Figure 1.2 (© Jacques Boyer/Roger-Viollet).

The photographs of the diffusers (Figures 2.16, 2.17, 2.18 and 2.19) come from the site: https://tsf.pagesperso-orange.fr/tsf/hp.html.

The photograph of the diffuser for model '37 (Figure 3.34) belongs to a private collection.

Introduction

This book is written for those who, out of simple curiosity, would like to know how a particular type of musical instrument works: the ondes musicales Martenot.

We summarize in what follows the main stages that mark the life of its inventor: Mr. Martenot, and for more details it is advisable to consult the work of Jean Laurendeau (1990).

Maurice Martenot was born in Paris on October 14, 1898. He was the third child of Juliette and François Martenot. A cellist by training, it was in the army that he discovered wireless radios and the existence of a particular device: the triode tube. Demobilized in 1918, he made his first attempts at musical expression using an electronic instrument of his own design using the triode tube.

On May 3, 1928, he presented a first prototype at the Paris Opera, which he named "ondes musicale", and the success was immediate. The device, equipped with triode tubes, generates a sound whose frequency is determined by the position of a metal wire that moves parallel in front of an alignment of flat screwheads, and whose amplitude is regulated by the pressure exerted on a key. This key, called the intensity key, is made from a small glass bottle containing mercury into which a glass slide drawn in pencil can be plunged. The sound is immediately distinguished from the sounds specific to classical musical instruments. The device covers seven octaves.

Mr. Martenot devoted all of his life to the development of his invention. In parallel, he invented and developed diffusers that enabled a diversification of the sounds of his instrument like the Palme and the Metallic. The last ondes model was presented in 1974 and was equipped with transistors.

Mr. Martenot died in Clichy on October 8, 1980.

The peculiarities of this instrument, which we will analyze and explain, have greatly contributed to its success: the possibility of performing a glissando over seven octaves, the vibrato controlled by the keyboard, the new sounds obtained by using the Palme or the Metal Gong.

The different prototypes will be studied in chronological order, and we will present a possible new approach to classify these instruments. According to Antoine Cuvelier (Loriod 1987), the different ondes are classified into seven different groups. The criteria used concern the dates of presentation, the external appearance, the position of the musician, and the control methods: wire, ribbon and keyboard. We propose a classification into three groups defined by the musical ondes generation technique and which will be explained in detail in the following chapters: additive synthesis, multiplicative synthesis, relaxation.

The works presented in what follows were carried out over a fairly long period (8 years) and in different places: the laboratory of the Cité de la Musique in Paris, the LAM (Laboratory of Musical Acoustics, in Jussieu[1]), as well as restoration work carried out on instruments at the request of ondists.

Our approach is based on the analyses of the functioning of the ondes on which we could proceed: ondes no. 15, no. 159, no. 169, no. 195, no. 208, no. 222, no. 257, no. 270, no. 275, no. 278[2].

The other study medium is of course the diagram for the oldest devices recorded in situ. For the latest models with electronic tubes, we have a diagram.

In order to fully understand the operation of the different sub-assemblies that make up the instrument, and also to highlight the different techniques of manufacture most likely that its inventor used, we have made mock-ups (ondes 1928, ondes 15-169, ondes 1947). Then we decided to undertake the manufacture of a copy of ondes no. 208. It seemed to us that this model was the most representative in terms of the design approaches and know-how of the inventor.

1 The laboratory which has now become the "Lutherie – Acoustics – Music Team" of the Jean Le Rond D'Alembert Institute.

2 Without detailing the classification developed by Antoine Cuvelier (Loriod 1987), ondes no. 15 corresponds to the "3rd model" (between 1929 and 1936), ondes no. 159 and 169 correspond to the "5th model" (or "model 1937"), the other ondes studied correspond to the "6th model" (between 1955 and 1974) and the "7th model" (and last model) corresponds to the transistorized version.

Finally, we have written, for the attention of our fellow ondists, an article about the manufacture of the intensity key and the ribbon, which will make it possible to maintain in working order some ondes with tubes for a long time to come.

1

Ondes 1928

1.1. Presentation

This was the model presented for the first time at the Paris Opera in 1928. We have no material trace of this version; on the contrary, we have five important pieces of information, two photographs showing Mr. Martenot in front of his instrument, a film with a sound recording (Pathé 1934), a technical description by Weiss (1930) and finally, the various patents that Mr. Martenot filed to protect his invention (Martenot 1931).

Figure 1.1. *Excerpt from the advertising brochure for the recital-demonstration of May 3, 1928*

For the color versions of the figures in this chapter, see www.iste.co.uk/courrier/ondesmartenot.zip.

We also have another photograph where we see the same ondes model at Mr. Martenot's house.

Figure 1.2. *The ondes prototype. Musée de la Musique, Paris (© Jacques Boyer/Roger-Viollet)*

A detailed examination of the two photographs proves to us that they were the same prototype.

The shape of the wooden box with its elliptical cutout of the front face and its two large control buttons shows that it is, in fact, a radio receiver, the model of which was very common in the years 1927–1928. The manufacturer L.G.M. built wireless receivers with a casing design very similar to that shown in the photograph.

Old radios with a switch were generally used prior to 1928 and operated most of the time using the principle of positive feedback. This switch was used to select circuit fractions on a multi-tapped inductor. Both of the large buttons on the front panel each controlled a large capacity variable capacitor (~ 600 pF). They had two to five tubes, depending on the complexity of the LF amplifier associated with the detector. Thus, in these receivers there are the elements necessary to build an ondes.

The other devices are the wooden box on which Mr. Martenot placed one or more (?) control levers for the intensity keys, the console with the tuning capacitor controlled by a wire, the graduated mark on the ground to locate the musical note played, the rubber band, not visible in the photograph, which makes it possible to keep the wire taut, the wooden box that probably contains the accumulators and finally, not visible either in this photograph, but observable in the film, the diffuser.

1.2. Principle of operation

The principle of operation of this ondes, is the additive mixture of two inaudible frequencies, one fixed and the other variable. The resulting signal has an audible frequency. To vary this musical frequency, we use the properties of resonant circuits; they consist of an inductance of self-induction coefficient L placed in parallel with a capacitor of capacitance C. The resonant frequency is determined by the equation:

$$F = \frac{1}{2\pi\sqrt{LC}}.$$

It is also the frequency of the heterodyne in which the LC circuit is placed. To vary the frequency of the oscillator, we can act either on the value of L or on the value of C. Mr. Martenot chose to vary the value of the capacitance C of the resonant circuit. For this, he used a conductive wire that passed at a very short distance in front of a row of flat heads of nails or screws, themselves made of electrically conductive metal.

In order to operate an oscillator, an electronic device, the triode, is essential. It was invented and developed by De Forest in 1906 in the USA. It was only available in Europe, in particular for radio enthusiasts, after the 1914–1918 conflict.

1.3. The diagram

We do not currently have a diagram by M. Martenot concerning this prototype, on the contrary, in an old work (Weiss 1930), we found a diagram that we have reproduced in Figure 1.3.

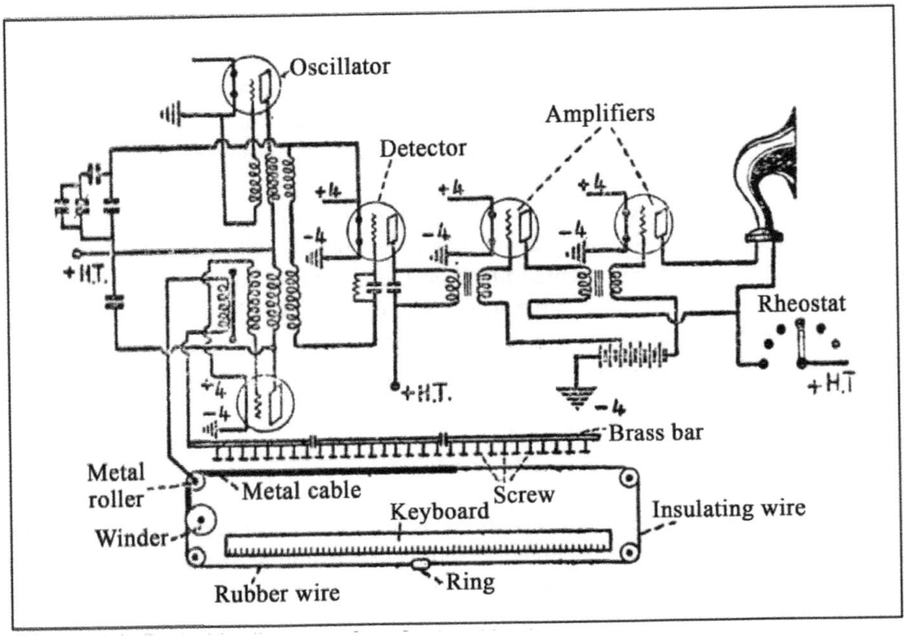

Figure 1.3. *Probable diagram of an ondes with wire*

This diagram is probably that of the second or third ondes models according to the classification proposed by Antoine Cuvelier (Loriod 1987). It presents inaccuracies, the absence of the intensity key, the polarization of the HF triodes, the splitting of the brass bar which supports the screws, the role of the winder, the intensity key and so on. But if we observe the different ondes models that Mr. Martenot built over the period 1928–1947, we observe a great inertia to change at the level of the diagram. Indeed up to and including the '37 model, the basic 5-triodes diagram is used.

The tuning capacitor controlled by the wire is produced by an alignment of screw heads, in front of which the conductive wire driven by the performer's finger moves. The detailed operation of this device, which is identical to that of the ribbon capacitor, is explained in the chapter dealing with ondes no. 208 (Chapter 5).

The screw heads constitute the fixed armature of the capacitor and the wire the mobile armature, the assembly constitutes the variable capacitor which makes it possible to vary the frequency of the oscillator. With a fixed frequency of 100 kHz, over seven octaves, we must obtain a frequency variation close to 4 kHz, the total variation in capacitance of the tuning capacitor is given by the following equation:

$$1 - \frac{\Delta F}{F_0} = \frac{1}{\sqrt{1 + {}^{\Delta C}/_{C_0}}}$$

with: F_0 = 100 kHz, ΔF = 4 kHz and C_0 = 500 pF, we obtain: ΔC = 42 pF, approximately.

The ratio between the increase in capacity at the end of the high register and the increase in capacity at the beginning of the low register is approximately 136. Whatever the solution used to make the fixed walls of this tuning capacitor, for constant dimensions of the facing surfaces, the ratio of the extreme walls/wire distances is also equal to 136. If we have a distance e = 10 mm at the input of the capacitor, its output value will be e = 10:136 ~ 0.074 mm. These are orders of magnitude, but it is enough to show that it is technically difficult to make. It is therefore necessary to determine means which make it possible to circumvent this difficulty.

It can be seen in the diagram that the HF transformer of the variable oscillator has four windings. Two windings are used for the oscillator function, a winding in series with the plate circuit of the oscillating triode, and a feedback winding connected to the grid of this same triode. The third winding is the one that picks up the HF signal for the additive mix. Finally, the fourth winding, which is connected to the tuning capacitor, constitutes with the winding of the plate circuit, on the one hand, an HF circuit tuned by the tuning capacitor, and on the other hand, a transformer of ratio m. On this transformer with two separate windings, if an impedance Zs is connected to the secondary, this can be brought back to the primary. The value of the equivalent impedance is $Z_p = \frac{Z_s}{m^2}$. If this impedance is a capacitor, the capacitance brought back to the primary is divided by m^2. To divide the impedance by 2, the number of turns in the secondary must be multiplied by approximately 1.4. This configuration is possible, but the maximum value of the capacitance remains high 42/2 = 21 pF. If we double the number of secondary winding turns, then m = 4 and the maximum value of the capacitance of the wire capacitor is 42/4 = 10.5 pF. This capacitance value is small enough to consider building a tuning capacitor controlled by a wire. But this solution leads to the manufacture of a large HF transformer. Several documents (Willems 1980) confirm this configuration, but we have no trace of material realization of this HF transformer.

It is also possible to carry out the assembly as an auto-transformer. In this case, the transformation ratio has the value m' = m + 1. This structure notably reduces the dimensions of the device, and it is also more efficient in terms of operation in HF, because an inductance is constructed with an intermediate access point. It also corresponds to the structure of the windings of the HF transformer observed on the ondes no. 15. Indeed, if the number of turns in the primary and secondary are identical, then m = 1 and m' = 2, the value of the capacitance brought back to the secondary is divided by 4. This is probably the solution adopted for the very first ondes. It is enough to wind the windings simultaneously with two wires in hand. Optionally, it is possible to reduce the final size of the windings by taking a conductor section of the wire to be wound that is smaller. In the realization of our model, taking into account the available material, we chose a compromise between the final size of the inductor, the section of conductor available, and the initial value of the increase in capacity at the input of the low register. We adopt a coefficient m' = 1.75. The diagram is as follows.

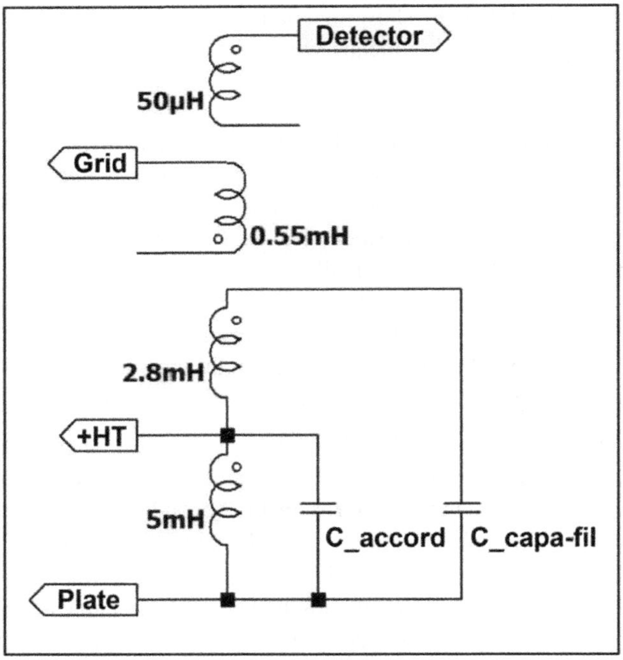

Figure 1.4. *Partial diagram of the oscillating circuit*

This mode of connection of the windings associated with the values mentioned in the diagram makes it possible to divide by approximately 3 the amplitude of variation of the tuning capacity. The initial choice of the value of the capacitance C_T (the frequency control capacitor) associated with an operating frequency of 100 kHz is probably the result of reusing existing components. In the years 1925–1930, the electronic components available to amateurs were all intended to be used for the production of radio receivers. And, if you were a little dexterous with your hands, you could quite easily make your own inductors. The C_T capacitor is one of the two capacitors on the front of the prototype; its maximum value is between 600 pF and 700 pF depending on the model. The C_W is the tuning capacitor, controlled by the wire, and is located away from the oscillator block.

Finally, after modifications, we propose the following two diagrams.

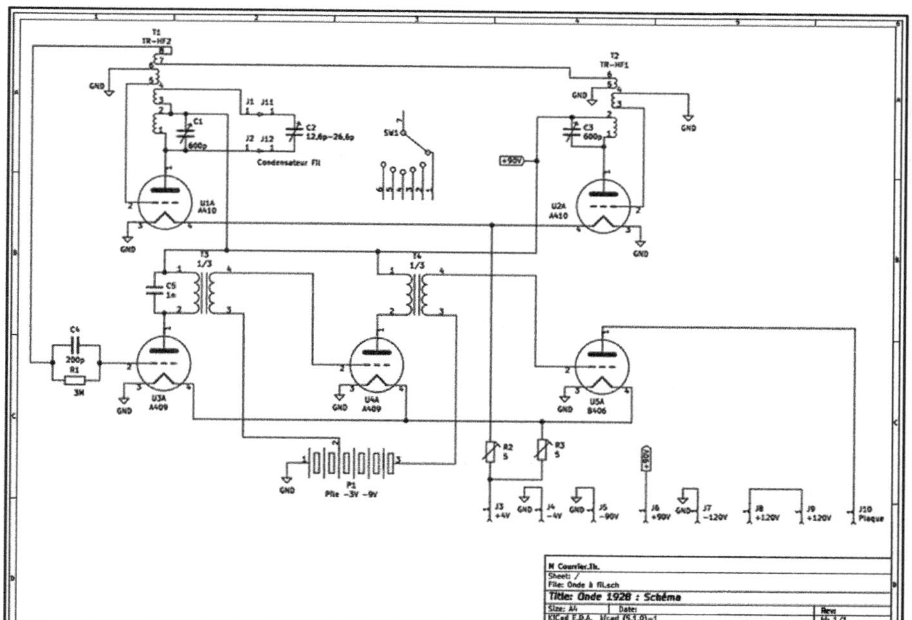

Figure 1.5. *Ondes 1928, principle diagram*

At the level of the diagram of the box containing the intensity key, we have generally taken over the diagram of the drawer noted on ondes no. 15; note that it is a mercury key.

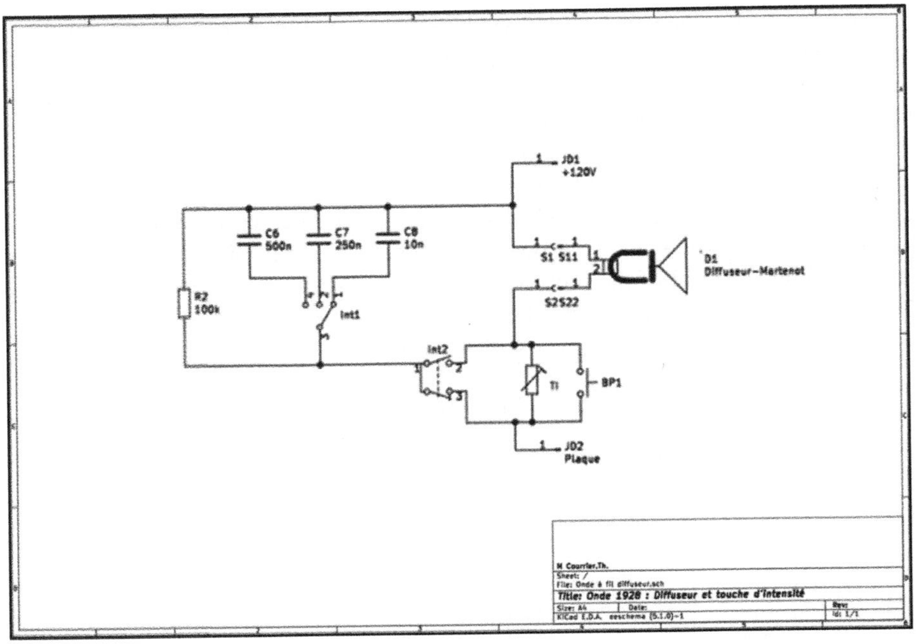

Figure 1.6. *Diffuser and intensity key diagram*

1.4. Construction of a model of the "1928" ondes

As we have seen previously, Mr. Martenot built the ondes from a radio receiver of the time. It is therefore quite easy to gather the information necessary for the realization of a model, by combining the techniques used in radio technology and based on existing documentation, in particular patents. This will allow us to highlight the difficulties that Mr. Martenot had to overcome to build a functional instrument.

1.4.1. *The box*

In these external aspects, it takes up the characters visible in the two previous photographs. The dimensions are those of the L.G.M. receiver (L = 70 cm, H = 27 cm, D = 24 cm). This case is constructed from fir wood panels, the thickness of which is 18 mm. The front face, which supports the adjustment elements, is cut from a 6-mm thick solid-colored Plexiglas plate. At the back, a 5-mm thick plywood board is removable and allows access to the interior of the cabinet or to connections to external devices.

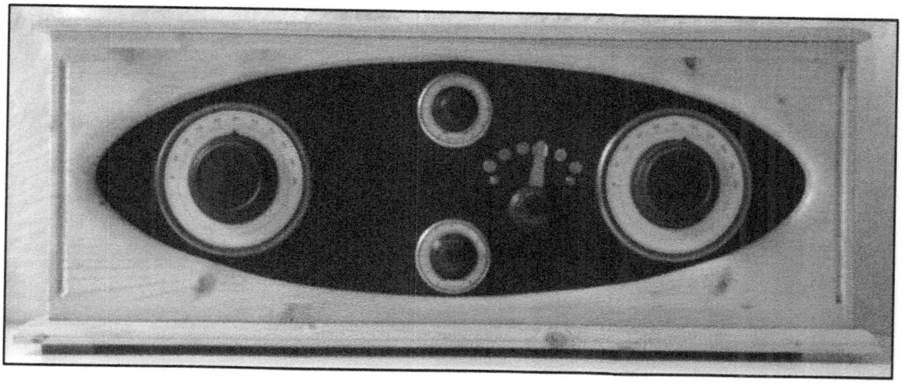

Figure 1.7. *The 1928 ondes model*

1.4.2. *Construction*

In general, for all the components necessary for the operation of the instrument, we will use period equipment. Only the HF transformers and the tuning capacitor will be manufactured by us.

At the fixed oscillator level, the primary of the HF transformer is also the inductance of the tuning circuit. Its value is 5 mH, the first secondary winding (0.55 mH) is connected to the grid, and it will have a number of turns equal to 1/3 of the total number of turns in the primary. The second secondary (50 μH) is connected to the detector; it will have a number of turns equal to 1/10 of the total number of turns in the primary. The winding technique used is that of the "honeycomb"; it makes it possible to reduce the parasitic capacitances generated by the contiguous turns of insulated conductors. Each coil is spaced from the next by a distance equal to 12 mm, the height of a coil is 22 mm, the diameters are 62 mm, 74 mm and 86 mm (three windings) or 90 mm (four windings), and the assembly height is 110 mm.

For the fixed frequency oscillator, the value of the capacity to obtain the resonance is approximately C = 506 pF. We use a variable tuning capacitor from the old radio receivers (C~ 600 pF). For the variable frequency oscillator, we use a capacitor of the same type. We will not reuse the variometer switch.

The layout of the equipment in the box is determined by the location of the variable capacitors arranged on the front panel, and also by the need to separate the two heterodyne circuits as far as possible. The wiring technique is the so-called "breadboard" technique. All connections are obtained by screwing.

Figure 1.8. *The chassis, partial view of the wiring*

1.4.3. *The tuning capacitor controlled by a wire*

As we specified in section 1.3, the variation of the capacity according to the frequency follows a particular law: the increases in capacity are very weak at the beginning of the low register and are high at the end of the high register. The diagram by Weiss does not indicate the presence of an LC resonant circuit connected in series with the tuning capacitor. This device, which makes it possible to reduce in large proportions the increase in the value of the capacitance of the wire capacitor in the high register, is described by Mr. Martenot in his patent no. 1.004.406 (Martenot 1947). It is unlikely that the instrument presented by Mr. Martenot in 1928 has this system. On the contrary, the tuning capacitor is represented by an alignment of screw heads in front of which the conductive wire passes. These screws are mounted on a brass bar and grouped by 8, 10 or 11 screws. Three sections are represented. In the same way as for the resonant corrector circuit, Mr. Martenot proposes in his patent (Martenot 1947) an association of capacitors in series with the variable ribbon capacitor, this association makes it possible to increase the capacitance variation of the ribbon capacitor in the high register. This solution does not seem to be used here. This is, of course, a representation intended for an uninitiated public, but the fact of specifying that the brass bar is cut in length allows us to doubt the use of this solution. Ondes no. 15, which works with a ribbon capacitor, does not have this configuration.

Note that if we have obtained a reduction in the amplitude of variation of the tuning capacity: 42.2/3.01 = 14 pf, the ratio of the difference between the wall/wire distances remains 136, so we will implement different tricks that make it possible to build a working tuning capacitor controlled by a wire.

By carefully observing the photographs, and taking into account the hardware items currently available, we have made this variable tuning capacitor as follows. To make the fixed frame, we will use M4 screws (zinc-plated steel) with flat heads, whose diameter is approximately 7 mm, and M4 screws (zinc-plated steel) with pan heads, whose diameter is around 9 mm. All the screws have their heads leveled and then tinned in order to present a flat surface facing the conductive wire. These screws will be mounted on angle elements (20 mm x 15 mm) in 2 mm thick aluminum. Each element is 142 mm long, and the holes spaced 12 mm apart are M4 threaded. To allow the adjustment of the screws, these are slotted at the other end, and a nut is provided to ensure the locking of the screw after adjustment. The useful length of wire travel is 12 x 84 = 1,008 mm, for seven octaves. For the wire, we used a cello string with a nylon core wrapped in a chrome metal spiral. Its length, when removing the synthetic thread coverings at both ends, is just sufficient. The angles are placed parallel to the wire, one after the other, leaving a space of 2 mm between them. On the contrary, from the low register to the high register, the angles are arranged at a distance from the wire that decreases when the rank of the octave increases.

An approximate calculation of the head/wire capacitance using the equation $C = \varepsilon \times \frac{S}{e}$[1] gives $C = 5.717\ 10^{-14}$F/mm, with a screw head diameter $D_v = 7$ mm, and a conductor wire diameter $d_f = 1$ mm. ε is the vacuum permittivity: $8.85\ 10^{-12}$ F.m^{-1}, the facing surface S is approximately $S = D_v * d_f$ and e designates the distance between the wire and the screw head.

To obtain the Do_0 (32.7 Hz), the increase in capacity is 1.8559×10^{-14} F. This value is divided by 3 thanks to the choice we have made at the level of the structure of the oscillating circuit. This leads to a distance e = 9.24 mm, in reality, this distance is slightly greater due to the capacity created by the vertical wall of the angle iron. However, the diameter of the screw head becomes insufficient at the level of La_3 (440 Hz) because the distance is calculated in tenths of mm, we then use pan head screws of 9 mm in diameter.

1 In several places in this work, we use the equation that defines the value of the capacity measured between two flat and parallel conductive metal walls. To obtain a capacitance value as close as possible to the measured value, the distance that separates the two plane surfaces must be less than or of the same order of magnitude as the smallest dimension of the surfaces concerned. For the first ondes studied, these conditions are not strictly respected, in particular for very low capacitance values (which are necessary for the operation of the oscillator in the low register). The equation is then used to give an order of magnitude of the dimensions which define the tuning capacitor. The final dimensioning is obtained by a frequency measurement.

Again, for the note Si_4, the distance e is still too low, and the diameter of the screw head must be doubled, but this will make the adjustment complex (two screw heads facing each side of the conductor wire). It is preferable to use a wall made from the same angle irons arranged opposite the screw heads. With a pitch of 12 mm, the total useful surface is therefore 9 + 12 = 21 mm, which leads for the note Si_6 to a distance equal to 0.1 mm. This wall/wire distance seems to be a minimum and can be used if the wire is well guided.

Figure 1.9. *Tuning capacitor controlled by the wire, overview*

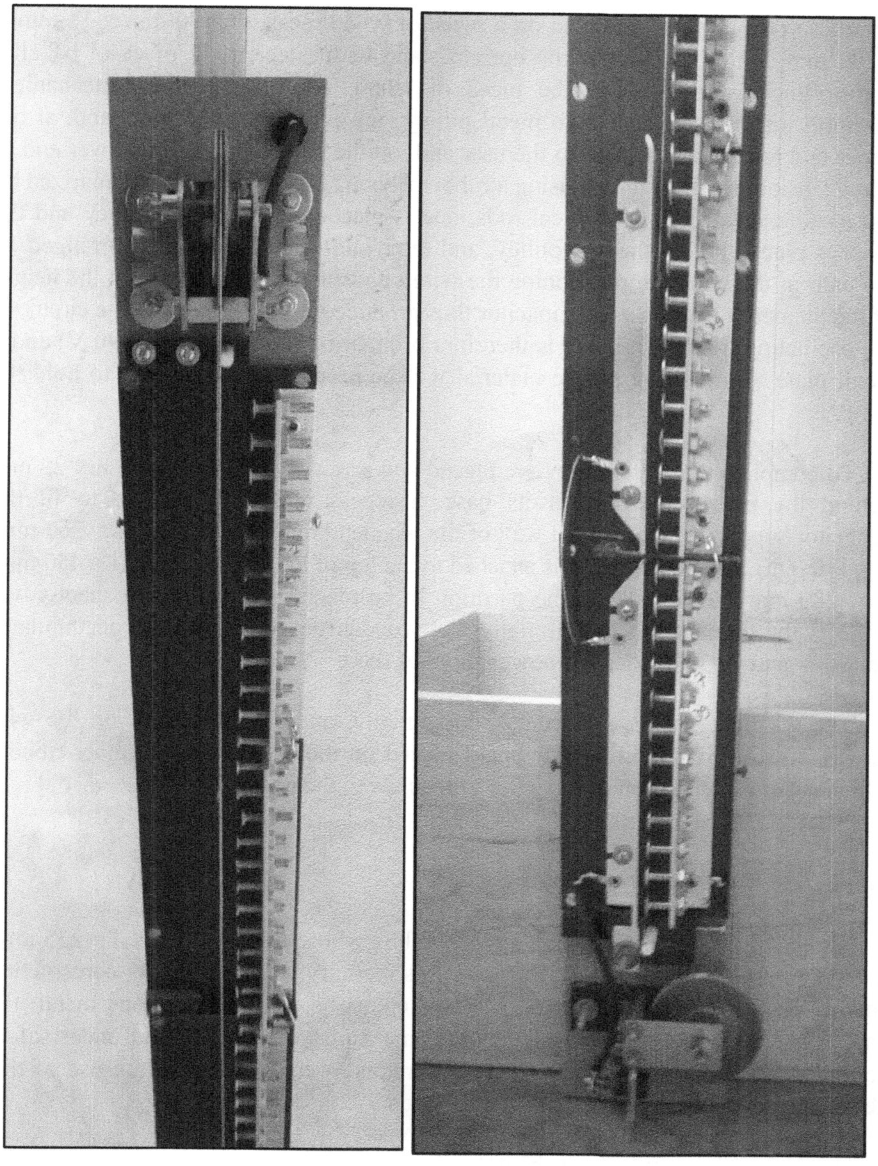

Figures 1.10 and 1.11. *Tuning capacitor, high pulley (left) and low pulley (right)*

These elements are arranged on a piece of wood measuring 70 mm x 15 mm x 1,200 mm, which constitutes the upright stand of the lectern. A piece of bakelite 3 mm thick, screwed onto the piece of wood, allows meticulous mechanical assembly of the elements. Two metal pulleys are placed, on the one hand, at the upper end to deliver the wire to the user and, on the other hand, at the lower end to allow tensioning of the thread using a rubber ribbon. The guidance is summarized by the installation of two cylindrical rods, one is placed after the high pulley and the other is placed before the low pulley, and three sliding pads, regularly arranged on the path of the wire and positioning the wire opposite the screw heads at the height of the diameter. Note that the capacitor thus produced is placed in the plate circuit of the oscillating triode. The wire is therefore brought to a high potential (90 V) and a small plate of insulating plastic material will be used to allow the user to hold the wire.

To complete this description, we placed and screwed a cleat of 15 mm x 22 mm behind the piece of wood; at its base, a recess makes it possible to fit the capacitor support on the vertical wall of the wooden box (l = 330 mm, w = 260 mm, h = 170 mm). A wooden frame is attached to the top of the upright stand (l = 450 mm, w = 280 mm), which allows the partition to be placed. An aluminum sheet cover protects the tuning capacitor. Initially, the wooden box contained the accumulator batteries, and we will place the power supplies there.

The wire extends to the ground, horizontally, and a cursor fixed on the wire moves in front of a fictitious keyboard placed on the floor. A long rubber ribbon, which ensures the tension of the wire, is attached to the end of the wire.

1.4.4. *The intensity key*

The intensity key is housed in a wooden box whose dimensions are l = 235 mm, w = 145 mm and h = 95 mm. This key uses mercury; we describe its construction and its characteristics in Chapter 7. But mercury is not an innocuous metal, the vapors are toxic and polluting. If in the context of specific tests and under safety conditions that can be obtained in the laboratory, we can consider the use of this key, we renounce it in the context of prolonged use.

We looked for a mercury substitute. A chemical manufacturer (Metallpulver24) offers Galinstan, nut unfortunately this product, which is an eutectic mixture of three metals (Gallium, Indium, Tin), is not suitable because it is reactive with most metals. We have also created a purely electronic key, which can be housed in a test tube, which operates from a linear response photodiode, but it requires the use of an external voltage source. Finally, we made this key using the leather bag and the graphite/mica powder imagined by Mr. Martenot.

It will be vertically operated and removable, so it can be replaced if necessary by a mercury key.

Figure 1.12. *The intensity key*

The detailed examination of the photographs of the box, which supports the key, shows two buttons on the front face that probably control rheostats and had to act on the output power. Between the two, we have a small button that probably had to carry out shorting of the intensity key.

On the right side, we have the handle and the two closing latches.

On the left side, we have the connections grouped in pairs, which ensured, on the one hand, the connection with the diffuser and, on the other hand, the connection with the heterodyne chassis. We also observe the presence of small aligned buttons; it is probably a "telephone" type rocker switch, which activates or not the capacitors of the timbres.

With the materials at our disposal, we have generally adopted the provisions adopted by Mr. Martenot, but on the front face: the two large buttons control switches, on the left there is a three-position switch for switching on the tone capacitors, on the right there is an inverter that allows the association of the circuit of the timbres in series or in parallel with the key. The key short-circuit button is retained in the center. The rheostats are removed. From an electro-acoustic point of view, they are not essential because the sound level can be adjusted using the intensity key; they are not present in ondes no. 15.

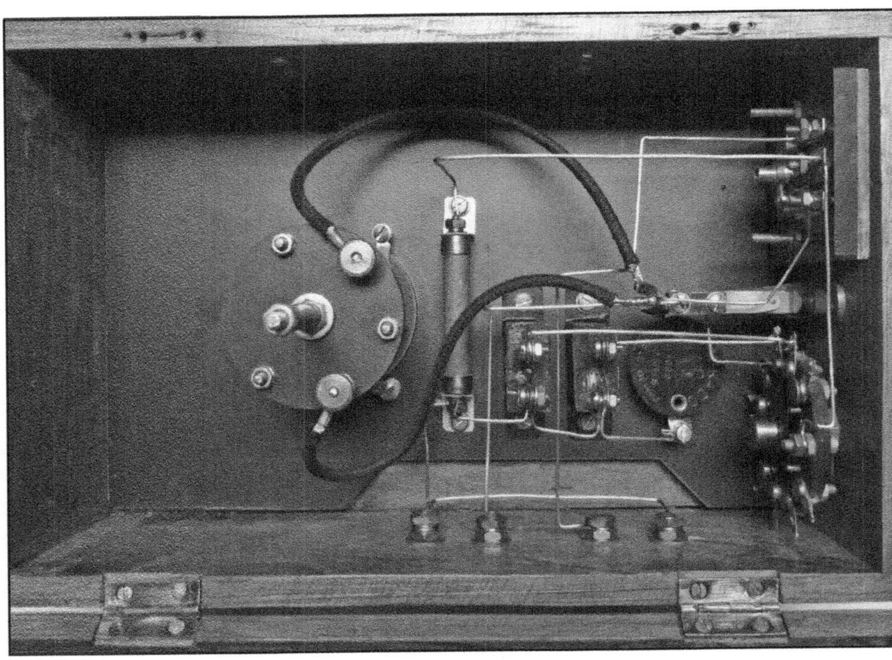

Figure 1.13. *Interior of the box*

Figure 1.14. *Box, front view*

On the left side are the banana plugs for connections to the diffuser and heterodyne chassis. The right side is true to the original. Finally, on the top of the box, a wooden lever rests on a rod which crosses the lid through the appropriate hole and acts on the bag of the intensity key.

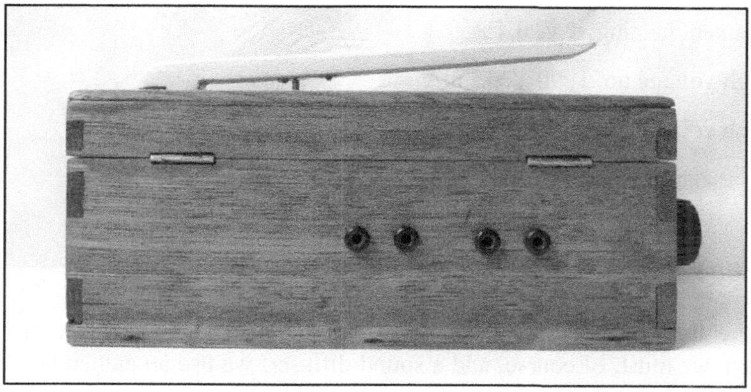

Figure 1.15. *Left side of the box*

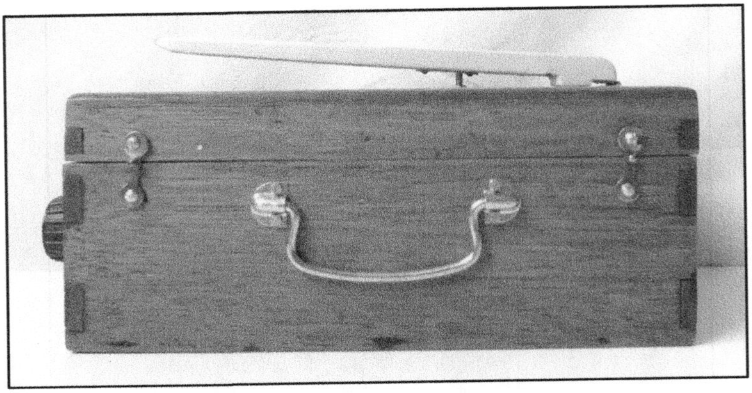

Figure 1.16. *Right side of the box*

1.5. Tests

1.5.1. *Energy sources*

To make this ondes work, we need four DC voltage sources; initially, they were made up of accumulator batteries, probably lead/acid and a dry cell. The accumulators were placed inside the wooden box placed at the foot of the lectern. The dry battery was installed on the chassis.

With the exception of the dry cell, we replaced them with conventional sources:

– Filament heating: 4 V, 0.4 A.

– High voltage no. 1: 90 V, 15 mA.

– High voltage no. 2: 120 V, 30 mA.

– Polarization of the grids: a dry battery delivering -3 V and -9 V.

1.5.2. *The wiring*

To the three devices we have built (heterodyne chassis, intensity key, wire capacitor), we must, of course, add a sound diffuser; we use an authentic cardboard cone diffuser.

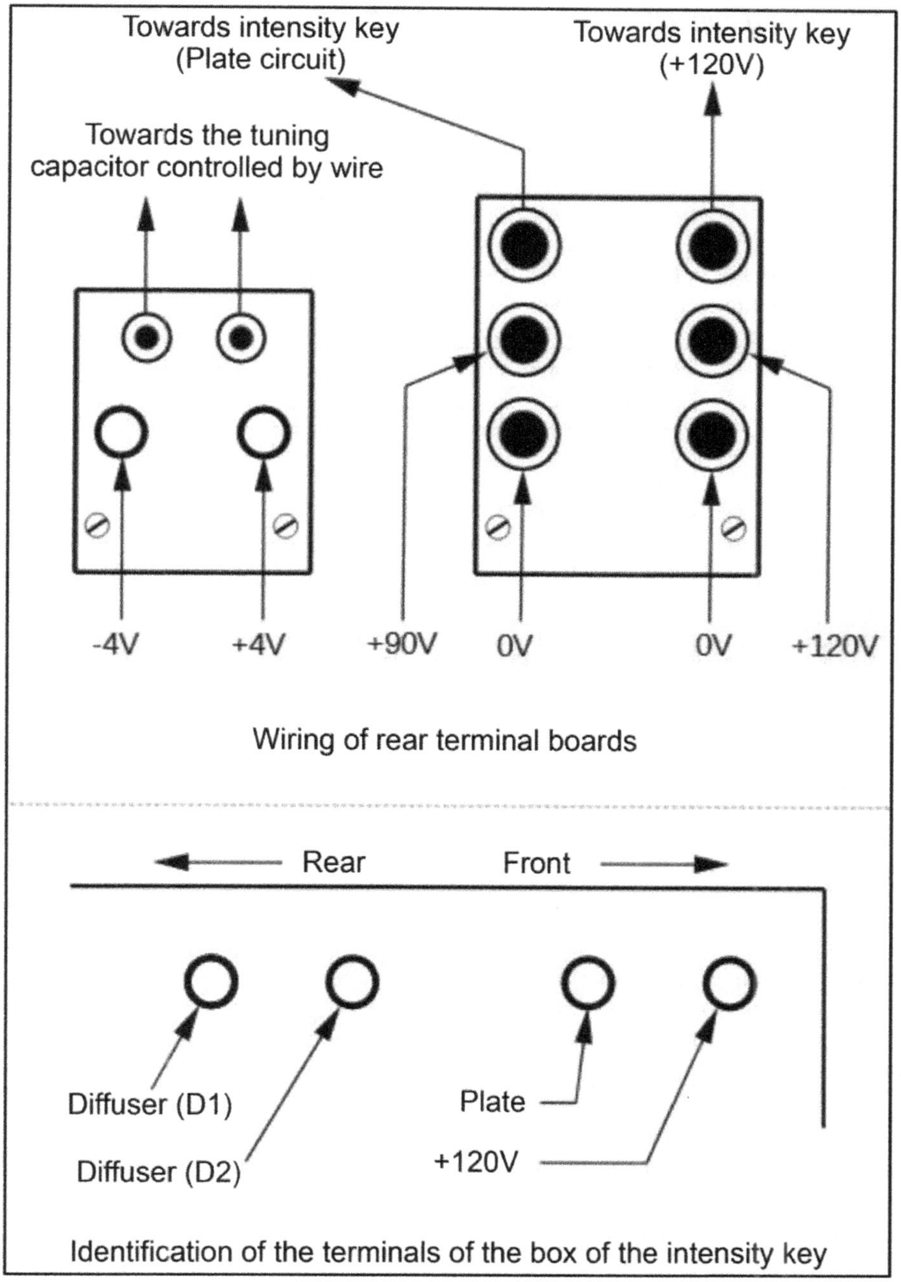

Figure 1.17. *Terminal board wiring*

1.5.3. *Tuning capacitor calibration*

This is the trickiest part of tuning the instrument. Note that the capacities necessary for the operation of the device are very low, in particular for octaves 0 and 1, for which measurements with a bridge are not possible. The presetting is then carried out by the theoretical calculation of the wall/wire distances for a frequency of 100 kHz.

For octaves 2–6, the screws are adjusted so as to obtain the following increases in capacity at the end of the octave:

Note	ΔC to pF	Comment
Do_0	x	Unmeasurable (e # 10 mm)
Do_1	x	Unmeasurable (e # 5 mm)
Do_2	0.33	
Do_3	0.77	
Do_4	1.67	
Do_5	3.48	
Do_6	7.18	
Si_6	14.00	see *

*By increasing the frequency of the fixed heterodyne to approximately 108 kHz, we obtained an overall ΔC variation of less than 12 pF, which made it possible to gain precious 1/10 mm of wall/wire distance adjustment for the high register.

Table 1.1. *Calculation of the capacitance increases of the capacitor controlled by a wire*

A bridge measurement makes it possible to obtain accuracy of the order of 0.1 pF. The values in the table are the calculated values.

In our model, the initial capacitance value due to the mounting of the metallic elements is 12.6 pF, so for each note, we have to measure 12.6 + ΔC.

For the first two octaves, the distances are positioned along a curve whose concavity is quite pronounced, it is only by ear that we can determine the final setting. For the following octaves, an alignment along a straight line is largely sufficient, and it was only in exceptional cases that we needed to make adjustments.

1.5.4. *Adjusting the instrument*

We are ready to test the overall operation of the instrument. We proceed to power up the filaments, then a few minutes later, we power up, first the heterodynes and then the power triode. The capacitor of the fixed-frequency heterodyne is adjusted so as to obtain F = 100 kHz. The adjustment capacitor of the variable frequency heterodyne is set to its maximum value, the intensity key is pressed, and this same capacitor is adjusted to obtain the musical frequency of 32.7 Hz. Then, we scan the entire course of the wire, and check that we have an overall variation $\Delta f = 4$ kHz, if this is not the case, we must then resume the adjustment of the fixed frequency, and we repeat all the operations. In our model, the final fixed frequency setting is 105 kHz. Finally, we finish the adjustment by ear, particularly for octaves 0 and 1. It should be noted that the device has no shielding, and the tuning capacitor controlled by the wire is placed on the plate circuit of the oscillator. Neither of the two poles are fixed to a ground potential, the effects of hand proximity are very important. On the contrary, fine adjustments of the tuning capacitors are possible thanks to the presence of a ball reducer. This device is provided from the outset by the manufacturer (large diameter knob = coarse adjustment, small diameter knob = fine adjustment).

In operation over long periods of time, the effects of drifts are perceived quite quickly, which oblige the user to adapt the length of the wire; the fictitious keyboard on the ground is only a reference point. The shift is of the order of ± 2 tones at the La 440 level. If the high frequencies are easy to play, on the other hand, the low frequencies are very unstable due to the drift, and the heterodynes drop out. In addition, the bandwidth of the amplification chain is quite narrow and the low frequencies are cut. We found a sound identical to that of ondes no. 15. The effectiveness of the timbres is quite similar to that of ondes no. 15, the signal delivered by the power triode is a rectangular type signal obtained by saturation, and the timbre filters act on this signal. The forms of the signals recorded on the oscilloscope are strictly identical to those recorded on ondes no. 15. It is therefore sufficient to refer to Chapter 2.

In the high register, it is the mechanical connection between the conductive thread and the nylon thread which poses a problem, indeed, at the output of the conductive thread, the space e is 1/10 mm, and the knot causes a slight lateral displacement of the conductive wire which comes into contact with the wall or the screw heads of the fixed armature of the capacitor. We observed that the contact took place at the level of Sol#$_5$. If we can extend the nylon part of the chord so that the knot is far enough away that it does not cause contact, this problem goes away.

Finally, in our model, we have a proper operation over a range of just over four octaves: from Do_1 to $Sol\#_5$.

1.6. Perspectives of evolution

The original idea of Mr. Martenot is concretized by the realization of this first prototype in 1928, two triodes, three capacitors and two HF transformers are enough to generate a musical sound. Of course, the triode has a fundamental role, but it was to Mr. Martenot's credit that he was able to exploit it effectively. This ondes diagram with five triodes was used until 1953 (the first not commercialized prototype with a pentode tube appearing in 1947).

The first improvements that Mr. Martenot will bring to this prototype are:

– the shielding of the HF parts. In fact, in use and in the absence of any form of shielding, the instrument is very sensitive to all electromagnetic or electrostatic disturbances;

– the polarization of the grids of the oscillating triodes by RC network which has the effect of stabilizing the operation of the oscillators and reducing the drift;

– the series connection of a resonant LC circuit parallel with the tuning capacitor controlled by the wire, which makes it possible to reduce the increase in capacitance of the capacitor in the high register;

– the replacement of the wire by a ribbon;

– the replacement of the screw heads by a continuous wall;

– the replacement of the mercury key by a key using a graphite/mica mixture.

These modifications are all present in ondes no. 15.

2

Ondes No. 15 (1930)

2.1. Introduction

The following study was carried out in the laboratory of the Museum of Music of Paris. Firstly, we analyzed the operation of this ondes, and then noted that it was possible to put it in operation again. Subsequently, we carried out tests and measurements that led us to an effective operation of this ondes.

2.2. Presentation

This model is part of a series whose manufacture ran from between 1929 and 1936.

Figure 2.1. *Ondes no. 15, the drawer has been removed (Musée de la Musique, Paris)*

Ondes no. 15, which belongs to a private individual, is at the Musée de la Musique.

For the color versions of the figures in this chapter, see www.iste.co.uk/courrier/ondesmartenot.zip.

It has the particularity of belonging to the first generation "3rd model" of instruments with both wire playing and ribbon playing associated with a fictitious keyboard. It is probably one of the first models working with a ribbon, this one making it possible to span seven octaves: from Do_0 to Si_6 of the French system (32 Hz – 3,951 Hz).

The instrument, from an electrical point of view, is practically complete. The wire system, the batteries and the polarization batteries are absent in this instrument. There is also no associated diffuser.

2.3. Organization of the instrument

Figure 2.2. *General provisions (Musée de la Musique, Paris)*

The general configuration of the instrument conforms to the descriptions made by Mr. Martenot in his various patents.

We present the structure of this ondes in Figure 2.3.

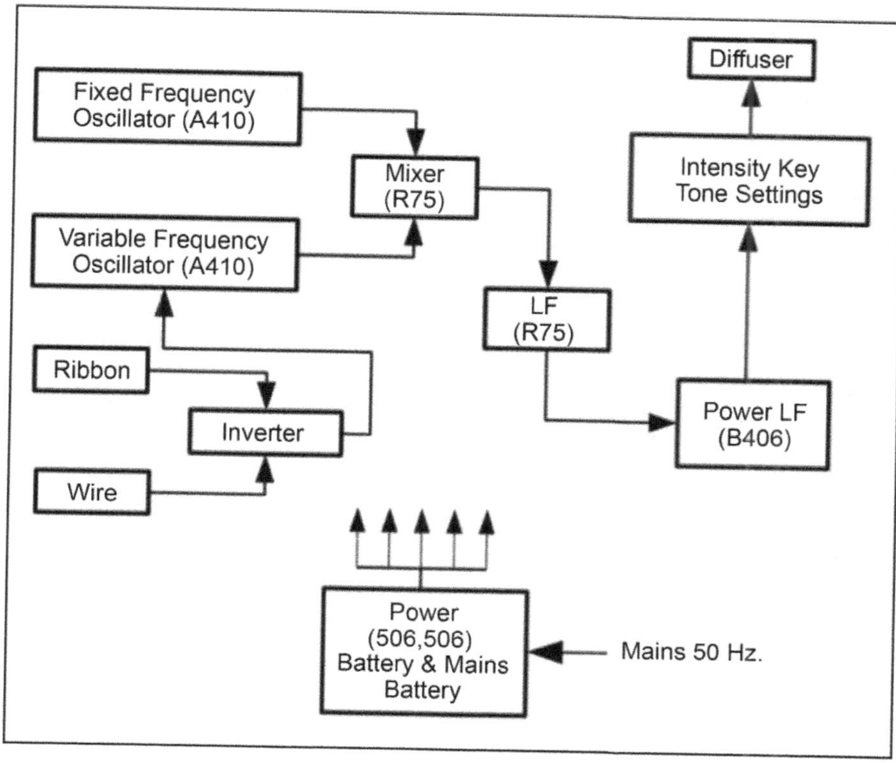

Figure 2.3. *Synoptic diagram of the operation of ondes no. 15*

The system works on the principle of frequency subtraction generated by two heterodynes one of fixed frequency (f_f) and the other of variable frequency (f_v), controlled either by the ribbon or by the wire. Neither of these two frequencies are perceptible to the human ear, but the subtraction of one by the other generates a third audible frequency ($f_{musical}$). Thanks to the variable frequency heterodyne, it is possible to cover a wide spectrum of musical frequencies.

This signal is pre-amplified by an amplifier tube, itself followed by a triode power amplifier feeding the diffuser. The particularity of the latter is the structure of the sound volume control; indeed, the intensity adjustment is made at the level of the power stage using an intensity key.

The harmonic processing of the signal is also carried out at the power level by a set of switches located on the drawer and making it possible to select the filter or filters. Similarly, a switch authorizes the operation of the diffuser.

All of these sub-assemblies work with the first industrial series of battery-type vacuum tubes, Axxx and Bxxx series from the vacuum tube manufacturer Philips. Several accumulator/charger devices, batteries and rectifier block on an 110 V network (not shown in this figure) supply the electronic tubes with the different voltages required.

NOTE.– The diffuser is, from the point of view of electro-acoustic technology, the predecessor of the loudspeaker.

To carry out the electrical functional analysis of the instrument, we will group the various devices into eight subsets:

– the "heterodynes + mixer-preamplifier + amplifier" system supported by an aluminum sheet chassis;

– the ribbon system;

– the wire system;

– the "drawer";

– the HT power supply from the 110 V network;

– the charger + accumulator system for heating electronic tubes at 4 V;

– the batteries for the grid polarization of the last LF tubes;

– the sub-assembly made up of the diffuser.

The modular design developed by Mr. Martenot for his instrument greatly facilitates all maintenance, adjustment and calibration operations. Indeed, the main plate, made on an L-shaped aluminum alloy sheet, supports all the electronic components. These are wired using the so-called "breadboard" technique: the connections are screwed. In principle at least, because we see in the case of this ondes a mixed system of connections (screwing/welding). The components are laid out on the plate in a visible way and are connected by wires of bare or insulated copper wiring. It is therefore easy to remove this plate from the cabinet of the instrument and to intervene as close as possible to a component to carry out any repairs.

2.4. The heterodyne system + mixer-preamplifier + amplifier

2.4.1. *General organization*

Figure 2.4. *The heterodyne + preamplifier + amplifier block (Musée de la Musique, Paris)*

We do not have a diagram of the "generator block" of this instrument. We have therefore identified it and presented it in Figure 2.5. Components whose values could not be identified are marked with the symbol (?).

The V_1 and V_5 electronic tubes of the A410 triode type are the oscillator tubes. The V_2 triode has, for reference: R75, the French manufacturer of electronic tubes Dario manufactured it; it is strictly equivalent to the A409 triode and performs the function of mixer and pre-amplifier. The V_3 triode of the R75 type (= A409) is also a pre-amplifier and controls the V_4 triode of the B406 type. This last triode performs the power amplification function and delivers the amplified signal to the diffuser.

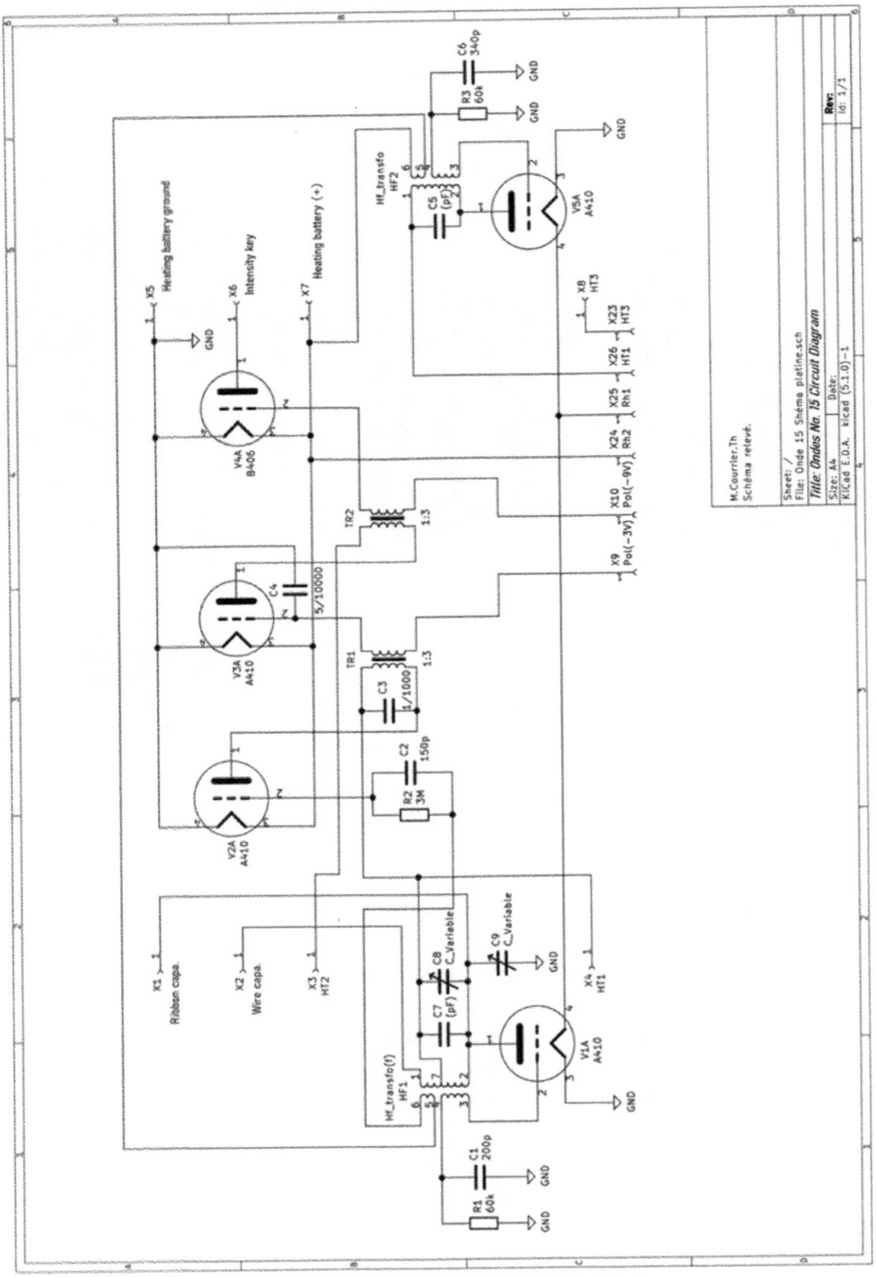

Figure 2.5. *Principle diagram of ondes no. 15*

This tube delivers electrical power (< 1 W) relatively low at the diffuser terminals. This triode ensemble will be reproduced in its structure in the following instruments, including those that will function with a keyboard (ondes no. 169).

At the level of the realization, it is to be noted that Mr. Martenot used a type of connector made by himself, and which consisted of mounting semi-rigid copper blades, on the one hand, on the plate and, on the other hand, on the bottom of the cabinet. A strong contact pressure was obtained by its tightening against the bottom of the cabinet thanks to two threaded rods. These contacts are marked X_1 to X_8.

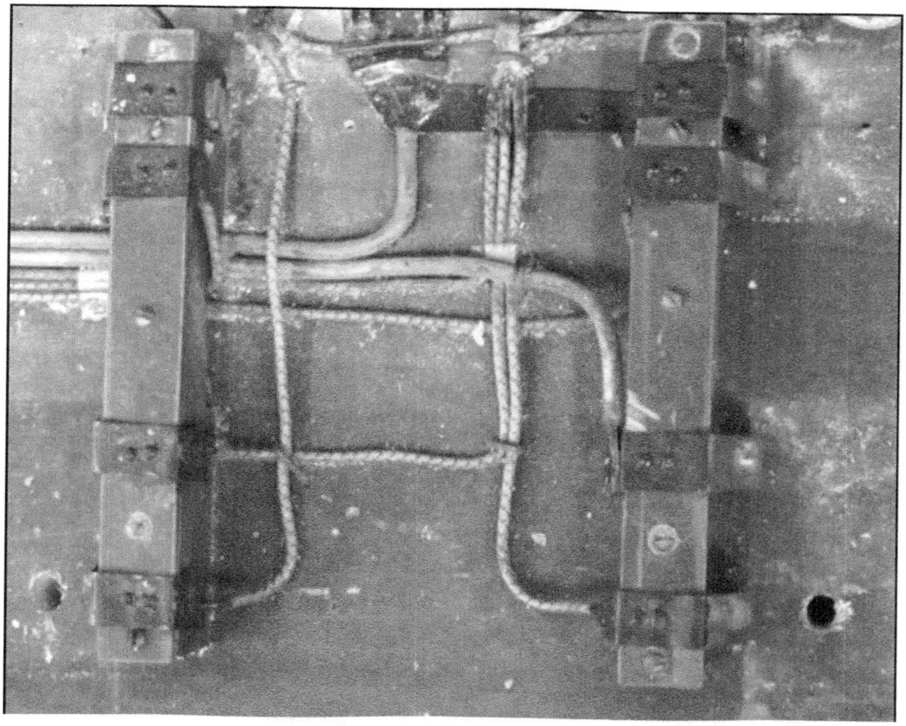

Figure 2.6. *The cabinet bottom connector (Musée de la Musique, Paris)*

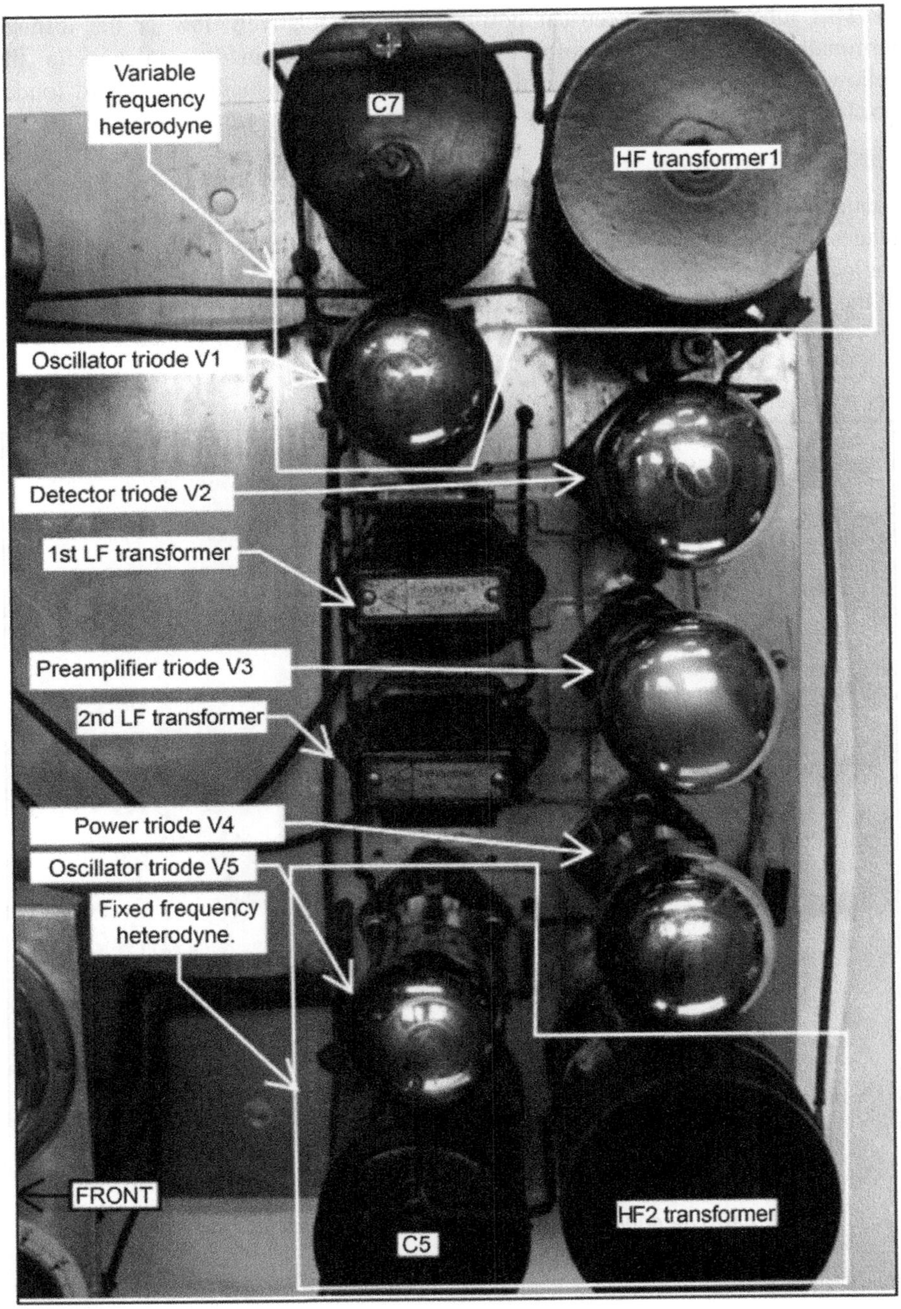

Figure 2.7. *Main plate, component positioning (Musée de la Musique, Paris)*

2.4.2. *Heterodynes*

2.4.2.1. *Diagram*

Let us now analyze in more detail the operation of the oscillators (see the principle diagram of ondes no. 15, presented in the previous section). The analysis will be identical for ribbon or wire operation.

The variable frequency oscillator is made around the triode V_1 (A410). The primary inductance of the transformer HF_1 associated with the capacitors C_7, C_8 and C_9 as well as the capacitor produced by the ribbon define the resonant frequency of the circuit. We then find the diagram as it is defined by Mr. Martenot on his patent no. 1.004.406 (Martenot 1947).

Capacitor C_7, arranged behind the plate, allows coarse adjustment of the frequency; Capacitor C_9, accessible on the front panel by the operator, is adjustable using a wheel and allows precise adjustment of the tuning. Note that the wire acts in the same way as the ribbon on the heterodyne because the capacitance obtained on the X_2 terminal is applied to an additional part of the primary winding. The secondary of winding HF_1 is in series with the grid circuit of the triode and forms a feedback circuit, which forces the triode to oscillate. This is biased by its grid current through resistor R_1, in parallel with capacitor C_1. A second secondary winding with a low number of turns will make it possible to take the voltage necessary to achieve the mixture with the signal of the fixed oscillator. This signal is a pure sinusoid.

At the level of the fixed oscillator (HF_2/V_5), the assembly is identical, apart from the capacitors. It is therefore a triode oscillator with a tuned plate circuit. The reaction by the grid is obtained by a first secondary whose number of turns is clearly lower than that of the primary. Bias is set by grid current through resistor R_3, decoupled by C_3. Capacitor C_5 and primary inductance of transformer HF_2 set the resonant frequency. This frequency is not adjustable. A second secondary winding with a low number of turns will make it possible to take the voltage necessary to achieve the mixture with the signal from the variable oscillator. This signal is also a pure sinusoid.

We will show in sections 2.8.2 and 2.8.3 that the frequencies of the oscillators are between 80 kHz and 90 kHz and comply with the values announced by Mr. Martenot in his patent.

The HF_1 and HF_2 transformers are large and wound using the so-called "honeycomb" technique. This technique was used at that time to reduce distributed capacities. Indeed, the primary inductance (>5 mH) of these transformers is high compared to the frequencies of use and requires a large number of turns. The manufacturing technique for these inductors is fully explained in the work by Alindret (1925): The C119.

If the implementation of this diagram is simple, on the contrary, the system is very sensitive to drifts. Let us recall that Mr. Martenot raises in his patent (Martenot 1939) these problems of drifts. To partially reduce them, the HT power supply circuit (HT_1) of the oscillating and detecting triodes is separated from the HT power supply circuit (HT_3) of the amplifier triodes. An adjustment of the heating circuit of the oscillating triodes is provided.

The wire technique is explained by Mr. Martenot in his patents no. 66807 (Martenot 1931) and 1004406 (Martenot 1947). On the instrument, all of the mechanical parts which support the wire variable capacitor are absent. The following can be identified: the associated side pulleys, in particular the wire exit inclined pulley, the ribbon/wire mode switch and a wooden cover to the right of the cabinet, allowing the wire to be taken out.

2.4.2.2. *Wire*

Figure 2.8. *Model similar to no. 15 (playing with the wire on the left, playing with the ribbon on the right)*

Ondes no. 15 was able to play with the wire or the ribbon. Unfortunately, the device producing the variable capacitor controlled by a wire is absent from the instrument: there remains only an empty slot, a set of pulleys and an access hatch with its cover. A connection terminal to the oscillating circuit is freely accessible; it is probably the terminal which allowed, as in the 1928 model, the connection of the variable capacitor made from a screw head alignment. The components that take part in the operation in wire mode could be located, and a diagram is noted. But their electrical functions are not formally identified. These are components C_{10}, C_{11}, C_{13} and INT2. The role of the INT2 switch controlled by the opening/closing of the cabinet lid is difficult to define. This part of the diagram is given as an indication and additional research is necessary to define exactly the functions of these components.

This mode of operation with wire does not correspond to the description made in patent no. 1004406 (Martenot 1947).

2.4.2.3. *Ribbon*

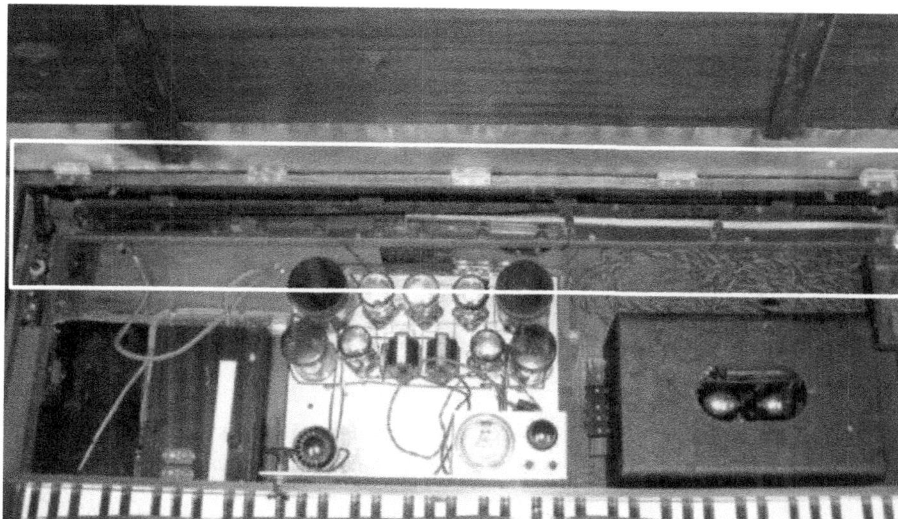

Figure 2.9. *The ribbon system (white box) (Musée de la Musique, Paris)*

The ribbon is one of the interfaces between the musician and the instrument. Mr. Martenot in his patent no. 666.807 gave the first description (Martenot 1931) while a first version, which used a conductive wire, had been produced. The ribbon is the logical development of this invention. These first capacitors are made from continuous walls of flared shape. For low musical frequencies, a single wall is used because the capacitance required is low. For high frequencies, and in order to reduce the frequency increase, Mr. Martenot uses a second oscillating circuit whose resonant frequency is constant. The armatures of these capacitors are fixed and do not allow any adjustment. The detailed operation of the ribbon system will be covered in section 5.4 of Chapter 5.

2.4.3. *The mixer-preamplifier*

At the level of the heterodynes, the mixing of the sinusoidal signals coming from the two oscillators is carried out by the series connection of two secondaries of the transformers HF_1 and HF_2. The mixture is therefore of the **additive type** (in reality a subtraction). In application of the physical principle of the beat, we thus obtain at the level of the triode V_2 a sinusoidal signal of frequency:

$$f_{inf} = {(f_f - f_v)}/{2}$$

This signal is the envelope of a second sinusoidal signal of frequency:

$$f_{sup} = {(f_f + f_v)}/{2}$$

But the resulting signal is applied to the grid of the triode V_2, which operates in detection or rectifier mode. That is, at the output of this one, we obtain a rectified sinusoid whose frequency is:

$$f_{musical} = f_f - f_v$$

Indeed, the capacitor C_2, in parallel with the high value resistor R_2, constitutes with the grid of the triode V_2 the detection cell. The musical signal is not sinusoidal.

The connection between the detection triode (V_2) and the pre-amplifier triode (V_3) is made using an LF transformer (TR1) with a 1/3 ratio. Capacitor C_3 reduces the amplitude of the frequency signal (f_f + f_v). In this way, only the musical frequency signal is applied to this transformer. The secondary of this transformer is connected to the grid of the triode (V_3), which operates as a voltage amplifier. The triode is loaded in its plate circuit by a second LF transformer (TR2) of the same 1/3 ratio. The grid bias is provided by a battery. According to the manufacturer's documents, the following operating point can be provided: plate voltage $V_a = 90$ V, grid voltage $V_g = -3$ V.

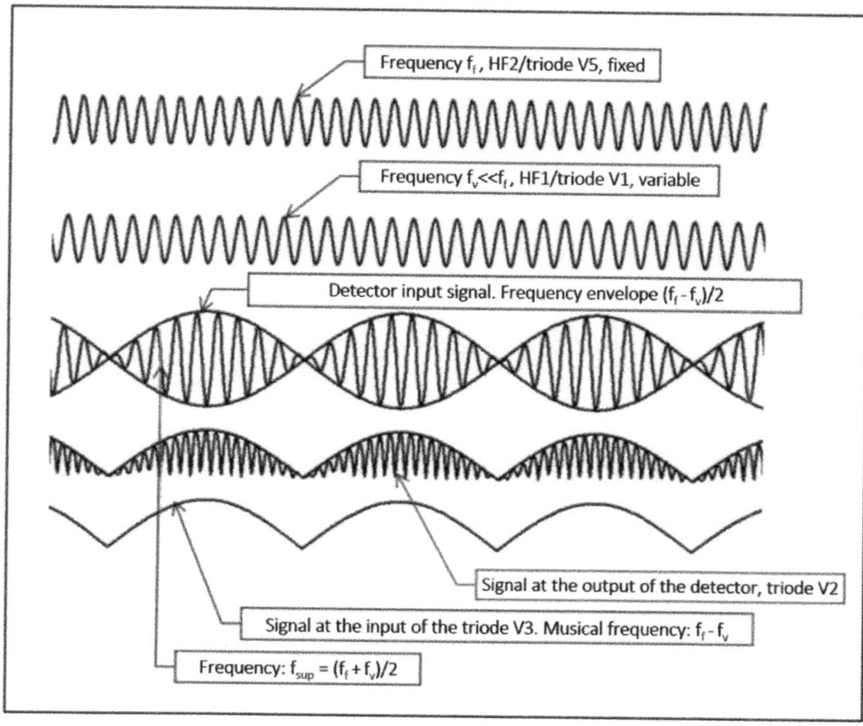

Figure 2.10. *Principle of operation of the additive mixing of signals of ondes no. 15*

2.4.4. *The amplifier*

It is built around the B406 type V_4 triode. The power delivered by it in class A is very modest (<1 W). The signal is applied to the grid by the secondary of the transformer TR2. A battery polarizes the tube. The plate circuit, at full power, is reduced to the load constituted by the diffuser: the order of magnitude of its impedance is between 1,000 Ω and 2,000 Ω. In this operation and according to the technical manuals of the manufacturers, the following resting point can be provided: $V_a = 120$ V, $V_g = -9$ V. The volume control on this instrument is carried out *by the series connection* of a resistance with the diffuser. In a first approach, the operation is simple:

– if the resistance has an infinite value or is open circuit (intensity key released), no current flows through the triode, and therefore, no LF signal flows through the diffuser;

– if the resistance is zero (intensity key pressed), then the entire LF signal available on the plate circuit is applied to the diffuser.

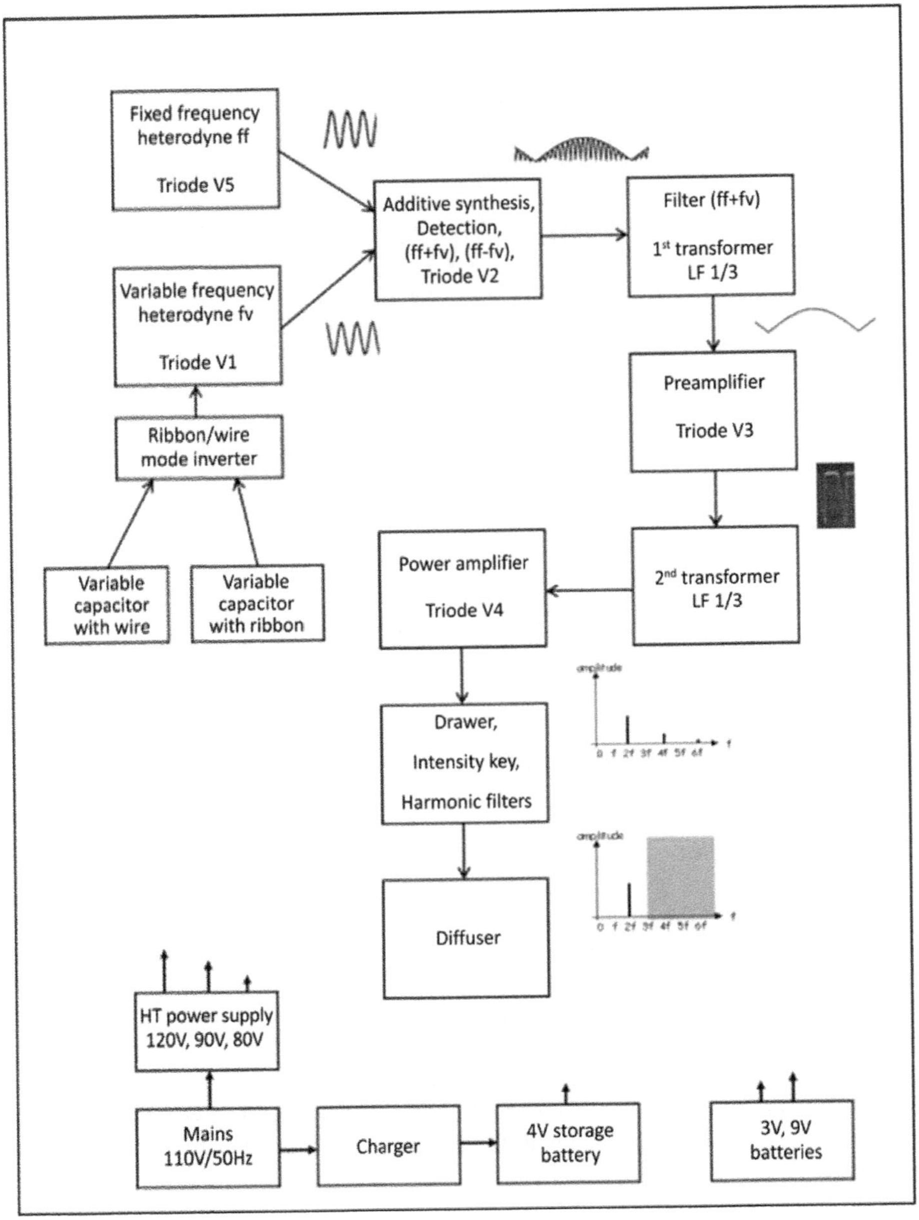

Figure 2.11. *Diagram of ondes no. 15 (the signal represented on a blue background is the measured signal)*

The control of all the resistance values of the intensity key makes it possible to modulate in all its amplitude the intensity of the signal through the diffuser. In addition, an operator can exploit all the subtleties of the playing, in particular for the "attacks".

But a loss of control of the sound volume can occur when the key is released: this is the leakage phenomenon observed by ondists. When the key is in the rest position, the plate circuit of the power triode is normally open circuit, but a very high amplitude LF signal is continuously applied to its grid, as a result of which, through the various parasitic impedances of the stage, we get a weak LF signal that reaches the diffuser.

We present the synoptic diagram summarizing the operation of ondes no. 15 in Figure 2.11.

2.5. The drawer

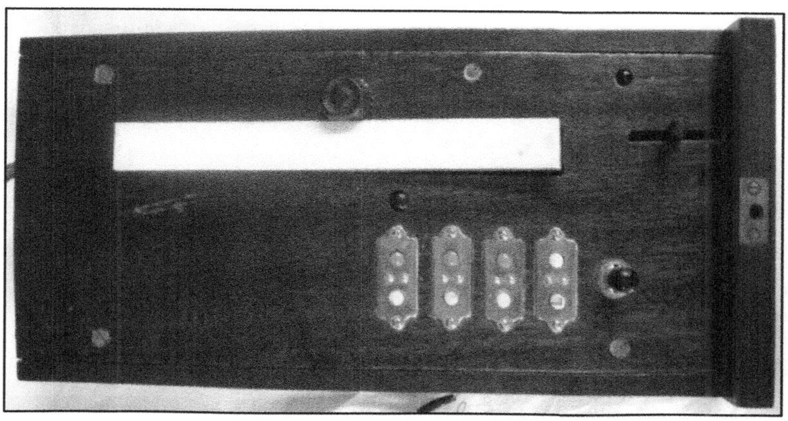

Figure 2.12. *The drawer (Musée de la Musique, Paris)*

This device provides several functions at the level of the instrument: it allows the operator to act on the intensity key, and also to choose different timbres. In addition, it ensures the electrical connections with the diffuser thanks to two two-pin connectors, located on the side of the drawer.

We could play by adopting two different postures:

– In a seated position, the performer played with the index finger of the right hand slipped into the ring and the left hand slightly resting on the drawer. The drawer is then in the open position but secured to the cabinet. The index finger of the left hand controlled the intensity key.

– In a standing position, the performer then holds the wire by means of a finger ring (device designed by Mr. Martenot). This system is described in patent no. 1004406 (Martenot 1947). To raise the note he exerted traction, to lower the note a mechanical system with return spring exerted traction opposite to the previous one in order to rewind the wire. This movement is transmitted to that of the ribbon by a set of pulleys. During the ascending or descending *glissandi*, the wire is thus kept taut.

The drawer is extracted from the cabinet and placed near the instrument. It is placed on a small pedestal table. The index finger of the left hand then controls the intensity key.

We do not have a diagram of the drawer. It is therefore necessary to carry out a survey; this is illustrated on the next page.

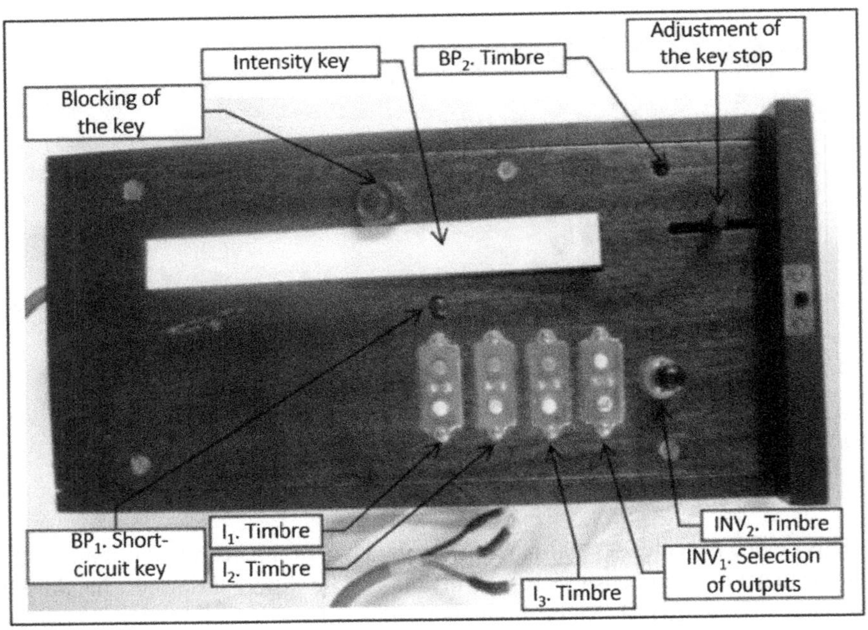

Figure 2.13. *Identification of keys and drawer controls (Musée de la Musique, Paris)*

In this diagram, the switches are all shown in the same position: all the red keys are pressed.

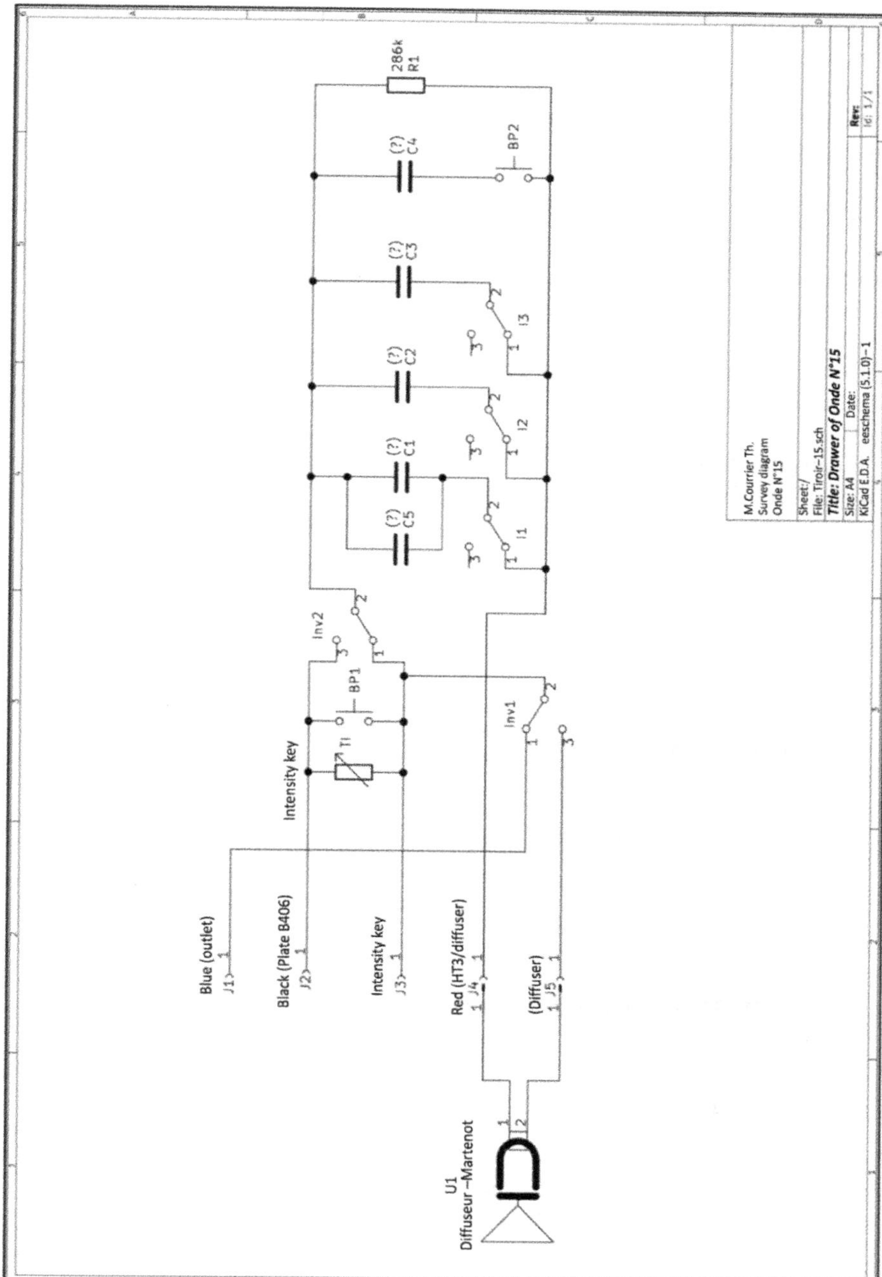

Figure 2.14. *Drawer diagram*

Push button BP_1 short-circuits the intensity key. The INV_2 inverter applies the harmonic filters at the level of the diffuser and in this case the cut-off frequency varies with the gain, or else the INV_2 inverter applies the harmonic filters in parallel to the diffuser + intensity key assembly, and in this case, we get a simple gain control; this is the "g" selection.

All of these effects are achieved by toggle, soft-toggle and high-key switches. The skill of the operator makes it possible to choose them temporarily or permanently. They can, of course, be combined and thus offer the performer a much diversified array of sound resources.

Resistor R_1 provides a discharge circuit for the harmonic filter capacitors and reduces the switching noise.

Finally, the INV_1 inverter directs the output signal either to the drawer connector, or to another connector of the same type installed at the back of the cabinet.

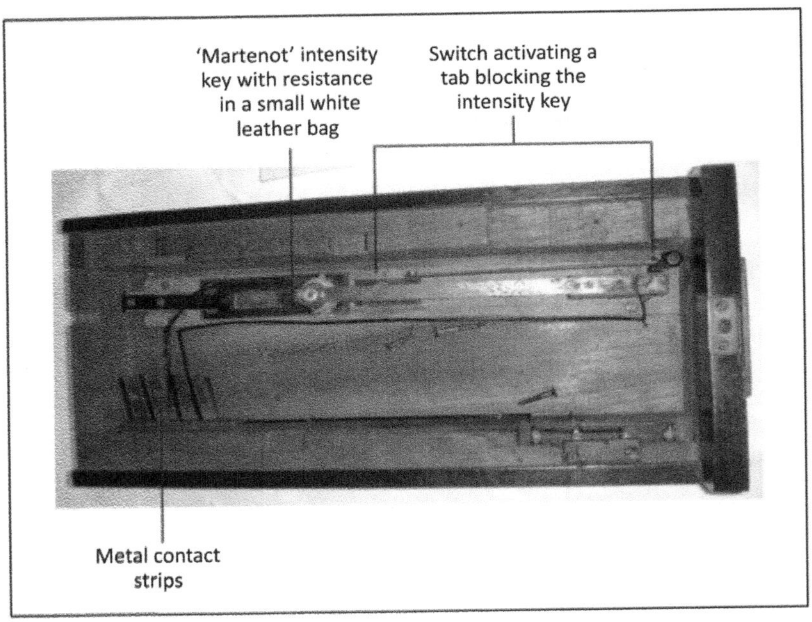

Figure 2.15. *View from the bottom of the drawer: the leather bag of the intensity key (Musée de la Musique, Paris)*

No value could be measured: all these capacitors show significant leaks. But by comparison with the diagram of the drawer of ondes no. 169, we can probably assign the following values to the capacitors: $C_1 + C_5 = 40$ nF, $C_2 = 20$ nF, $C_3 = 10$ nF and C_4 = 250 nF. Further research is needed to verify that these values, when put into use, produce the effects defined by Martenot. On his first instruments, Mr. Martenot identifies these timbre keys by the numbers 1, 2 and 3. He then adds no. 0, then no. 4 and the other keys. The correspondences are not always easy to establish.

2.6. The diffuser

The problem of the diffuser remains for which we have only little information. It is logical to admit that Mr. Martenot used the devices available for his time. We can identify four of them.

– The gooseneck horn.

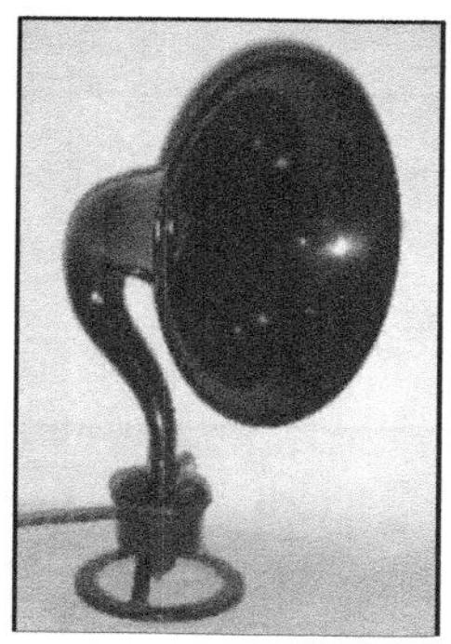

Figure 2.16. *The gooseneck horn*

– The membrane diffuser.

Figure 2.17. *The membrane diffuser*

– The diffuser with rod and cardboard cone.

Figure 2.18. *The cardboard cone diffuser*

– The electrodynamic loudspeaker.

Figure 2.19. *The loudspeaker*

The gooseneck horn works on the principle of the telephone earpiece for which a horn in the shape of a swan's neck has been adapted. By its principle of construction, the bandwidth is very narrow and localized in the treble. This cone has the annoying tendency to deliver a very typical sound: a "duck song". It is likely that Mr. Martenot tried this device and did not remember it.

The membrane diffuser, if it allows original aesthetic achievements, has above all a mediocre performance. Its construction principle is based on a device of the telephone earpiece type for which the metal membrane is connected by a rod to a membrane made of stiffened fabric or stretched paper. The sound power restored remains low; this diffuser is not suitable for the reproduction of musical sounds.

The loudspeaker as we know it today appeared around the years 1927/1930. It was very bulky and heavy, required the use of a matching transformer and possibly an excitation circuit. The probability that Mr. Martenot used this device as soon as it was made available is low. Later, a diffuser model whose external appearance is that of the model 37, will integrate a loudspeaker in duo with a rod and cardboard cone diffuser.

The rod and cardboard cone diffuser is the sound reproduction device which has been essential in particular for radio receivers since 1925. In reality, it is the precursor of the electrodynamic loudspeaker since it uses a permanent magnet and a cone-shaped membrane. The essential difference lies in the transfer of energy between the electrical system and the mechanical system. For the diffuser, the transfer of energy is done through a mobile metal pallet around an axis and connected to the cone by a rod. For the loudspeaker, the energy transfer is direct since it is a magnetic field/coil interaction. The voice coil is attached to the paper membrane.

The rod and cardboard cone diffuser has the best performance of all the devices, excluding the loudspeaker. Even though the bandwidth is narrow, it is higher than that of the horn or membrane diffuser. This is probably the device used by Mr. Martenot from the creation of his instrument.

2.7. Power supplies

The power supplies, which are essential to the functioning of the instrument, must meet design criteria which, if they are not purely musical, can however interfere with the level of the musicality of the instrument: power variation during play, humming noises, various parasites, etc.

The power supplies are divided into four units: the high voltage power supply module, the accumulator battery, the charger and finally, the polarization batteries.

The power supply module installed in the cabinet of the instrument seems to have been made by a specialized workshop. Housed in a black painted case, its disassembly gave us little information. It seems to be a double structure since we observe the presence of two transformers, two type 506 rectifying valves, inductors, filter capacitors and adjustment rheostats. Many elements cannot be identified, either because they have been degraded or they are not accessible. It was not possible to find a diagram.

Figure 2.20. *The high voltage power supply module (Musée de la Musique, Paris)*

The analysis of the overall functioning of the instrument and the reading of the manufacturer's instructions (relating to the electronic tubes used) allow us to circumvent the difficulty and to propose the following probable characteristics:

– one input: 110 V/50 Hz mains;

– a 120 V direct current output to supply the power triode;

– a 90 V DC output to power the preamplifier;

– an 80 V DC output to power the heterodynes.

Two filter capacitors (C_{14} and C_{15}) are installed at the bottom of the cabinet. They ensure the decoupling of high voltages as close as possible to the main plate. On measurement, they show significant leaks, but are not short-circuited.

Remember that to ensure the operation of the instrument, it is also necessary to have an accumulator battery (4 V) for heating the tubes. These batteries are absent from the unit, but the location is clearly identified. Two screw terminals (X_5 and X_7) are placed close to the sliding area of the drawer. They had to be clamped to the corresponding terminals of the storage battery.

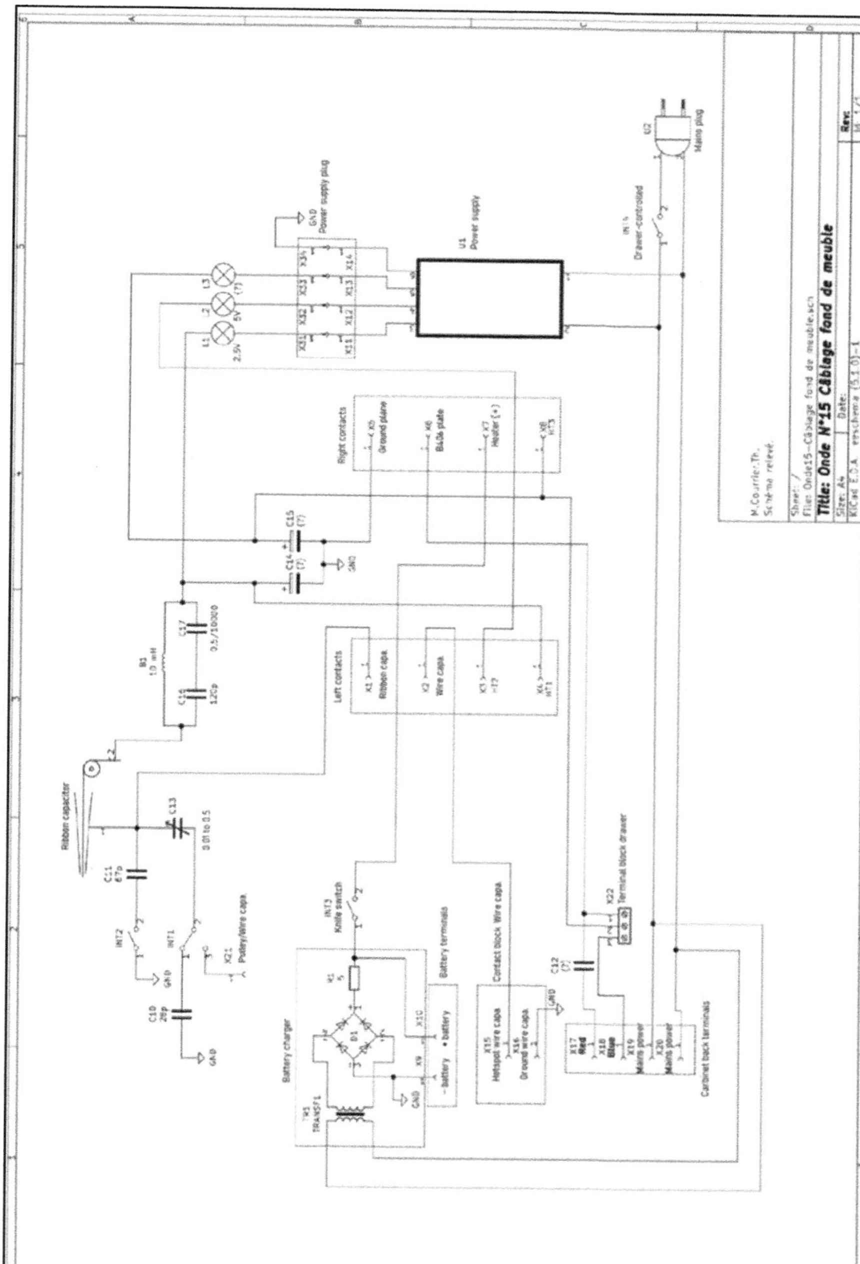

Figure 2.21. *Cabinet bottom wiring diagram*

A battery charging system is installed to the left of the cabinet on the side wall. It consists of a small transformer, connected, on the one hand, to the 110 V/50 Hz network and, on the other hand, to a full-bridge rectifier equipped with metal/oxide or selenium cells. A clever system, imagined by Mr. Martenot, made it possible to put these accumulators into service, using a switch (INT_3), only when the drawer was open. Another switch (INT_4), located to the right of the drawer and also controlled by the drawer, ensured the general powering up of the instrument.

To provide bias (–3 V and –9 V) of the amplifier triodes, a multi-terminal battery is required. It was placed on the main plate: traces of oxidation, due to the leaks that these devices frequently presented, are visible on the chassis. It permanently fed the grids of the amplifying triodes.

The cabling of the cabinet bottom is given in Figure 2.21.

Mr. Martenot had provided a voltage measurement device on his instrument. He therefore placed a two-scale voltmeter on the main plate: 0–180 V/ 0–6 V, as well as two push buttons and an adjustment rheostat. The following measures were thus made possible:

– Both pushbuttons released: measurement of the heating voltage of the heterodyne triodes. An adjustment rheostat, graduated from 0 to 10, made it possible to adjust the heating voltage. Generally, the heterodyne triodes had their filament underpowered (3–3.5 V) in order to improve the stability of the oscillators. This setting value required lowering the voltage applied to the plates to around 80 V.

– Press on BP_1: high voltage measurement (HT_3 = 120 V) across the power triode terminals.

– Press on BP_2: high voltage measurement (HT_1= 80 V) across the terminals of the heterodyne triodes.

The diagram of this measurement system is shown in Figure 2.23.

Also note another curiosity about these power sources: three “mignonnette” or “fish eye” lamps (L_1, L_2 and L_3) are connected in series with each of the high voltage sources. These low voltage lighting lamps were crossed by the current consumed by the triodes. In normal operation, a faint red glow could be seen; in the event of a fault, the overcurrent caused the filament to break, and consequently, the opening and isolation of the corresponding high voltage circuit. These lamps therefore act as a fuse.

Figure 2.22. *The measuring device (Musée de la Musique, Paris)*

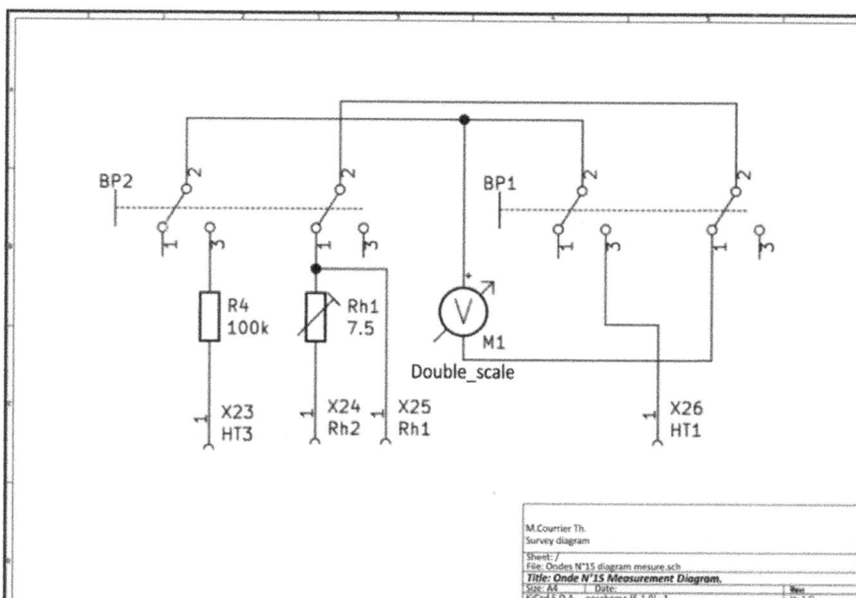

Figure 2.23. *Diagram: measuring device*

Figure 2.24. *"Mignonette" or "fish eye" lamps (Musée de la Musique, Paris)*

2.8. Trials

2.8.1. *Preparations*

All the work carried out on this instrument led us to wonder if it was possible to put this instrument back into service. Indeed, the passive components are few; the use of supports for the triodes allows them to be replaced by equivalent models. An assessment of the general condition of the instrument was essential. We noted and reported in the following table the visual or electrical condition of the components, indicating a possible substitution or not.

These tests are interesting only if we succeed as much as possible in reproducing the sounds as Mr. Martenot heard them. It is therefore necessary to select the elements that have a decisive influence on the authenticity of the ondes musicales.

From a purely instrumental point of view, the key and the ribbon are the two characteristic interfaces of the instrument. The operation of the ribbon will remain authentic, even after the exchange of it.

For the key, the substitution is more delicate but feasible. We made a key with a leather bag and a mixture of graphite powder and mica using the technique described in Chapter 7.

Component	**Type**	**State**	**Necessary**	**Replacement**
Capacitor C_1, 200 pF	Mica	Good	Yes	x
Resistance R_1, 60 kΩ	Agglomerated carbon	Out of tolerance	Yes	x
Triode V_1, A410	Vacuum tube	Not tested	Yes	Yes
Transformer HF_1	Honeycomb	Good	Yes	No
Capacitor C_8 (?)	Variable air	Good	Yes	No
Capacitor C_9 (?)	Adjustable air	Good	Yes	No
Group R_2/C_2 3 MΩ/0.00015	RC cell, carbon/mica	Good	Yes	Yes
Triode V_2, R_{75}	Vacuum tube	Not tested	Yes	Yes
Transformer TLF_1 m = 1/3	Iron/copper	Secondary cut	Yes	Yes
Capacitor C_3 1/1,000	Mica	Good	Yes	x
Capacitor C_4 5/10,000	Mica	Good	Yes	x
Triode V_3, R_{75}	Vacuum tube	Not tested	Yes	Yes
Transformer TLF_2 m = 1/3	Iron/copper	Secondary cut	Yes	Yes
Triode V_4, B406	Vacuum tube	Not tested	Yes	Yes
Capacitor C_5, (?)	Air	Good	Yes	No
Resistance R_3, 60 kΩ	Agglomerated carbon	Out of tolerance	Yes	x
Capacitor C_6, 340 pF	Mica	Good	Yes	x
Capacitor C_{10}, (28 pF)	(?)	(?)	No	No
Component	Type	State	Necessary	Replacement
Capacitor C_{11}, (67 pF)	(?)	(?)	No	No
Capacitor C_{13}, (0.01–0.5)	Adjustable air	Good	No	No
Capacitor C_{14}, (?)	HT Paper	Significant leaks	No	To isolate
Capacitor C_{15}, (?)	HT Paper	Significant leaks	No	To isolate
Capacitor C_{17}, 0.5/10,000	Paper (?)	Low leaks	Yes	Yes

Component	Type	State	Necessary	Replacement
Capacitor C_{16}, 120 pF	Mica	Good	Yes	Yes
Inductance B_1 (?)	Insulated copper	Good	Yes	Yes
Capacitor C_{12}, (?)	(?)	(?)	No	Yes
Transformer TRF_1, 110 V/(?)	Iron/copper	Not tested in good apparent condition	No	No
Rectifier $D_{1\text{-}4}$	Metal/oxide (?)	Not tested in good apparent condition	No	No
Resistance R_1 5 Ω	metal	Good	No	No
Mignonettes L_1, L_2, L_3	Filament lamps	Not tested good apparent states	No	No
Power Supply T_1	Miscellaneous materials	Not tested in doubtful condition	Yes	Yes
Switches INT_1, INT_2, INT_3	Miscellaneous materials	Good apparent states	No	No
Intensity key R (?)	Leather/graphite/mica	Worn out	Yes	Yes
Capacitors drawer C_1(?), C_2(?), C_3(?), C_4(?)	paper	Out of working order	Yes	Yes
Capacitor Drawer C_5 (?)	Mica	Untested, apparent good condition	Yes	Yes
Switches and BP Drawer: INV_1, $INV_{21/22}$, BP_1, I_1, I_2, I_3, BP_2	Miscellaneous materials	Worn but functional	Yes	Yes
Resistor R_4 measurement 100 kΩ	Carbon	Good	No	No
Rheostat RH_1, 7.5 Ω	Metal/porcelain	Good	No	No
Pushbutton BP_1, BP_2	Miscellaneous materials	Good	No	No
Galvanometer	Miscellaneous materials	To be calibrated but functional	No	No
Capacitor ribbon C_{18} (?)	Air/cloth	Ribbon to exchange	Yes	Yes
Diffuser	Various materials	Absent	Yes	Yes

Table 2.1. *Condition report of the components of ondes no. 15*

In terms of electrical operation, it is the detector–preamplifier–amplifier–diffuser assembly that is decisive. Indeed, the two low-frequency transformers have specific characteristics that influence the bandwidth and the dynamics of the system. These performances are directly related to the use of ferromagnetic materials. They will be exchanged with models from the 1920s, strictly equivalent.

The use of battery-type triodes for the detector and amplifiers is also decisive. Especially, since these amplifying triodes operate in class A. We will therefore use the same models, in the new condition.

Capacitors C_5 and C_7[1] are enclosed in a black painted metal casing. These two capacitors are supplied directly under the high voltage HT_1 = 80 V. Neither the general state nor the type of these capacitors can be observed directly. An X-ray is necessary; it was carried out by Mrs. Marie-Anne Loeper-Attia, conservator-restorer in charge of the preventive and curative conservation of the collections of the Musée de la Musique. The shot is of excellent quality. The doubt is removed: they are air-capacitors and in perfect condition. The risk on energizing is eliminated at this level.

Figure 2.25. *Substitution of low-frequency transformers (Musée de la Musique, Paris)*

1 The value of the capacitance C_5 (linked to the inductance of the primary of the transformer HF_2) defines the resonance frequency of the fixed heterodyne. Similarly, the value of C_7 (linked to HF_1) is for the variable heterodyne.

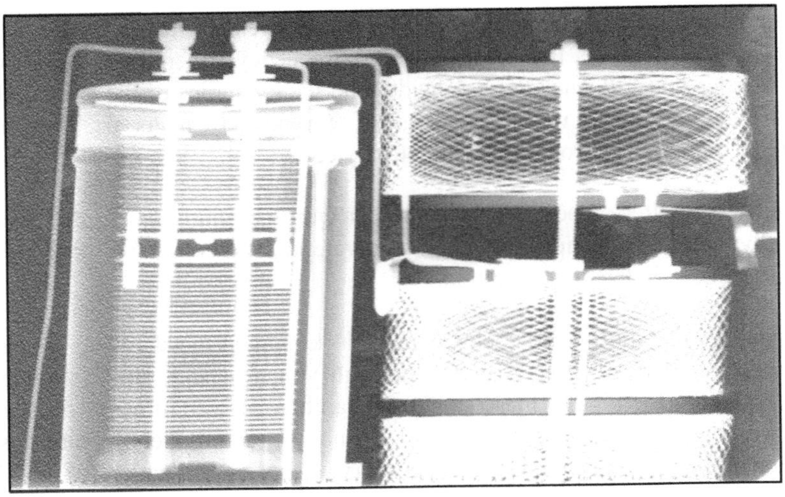

Figure 2.26. *X-ray of the oscillating circuit. On the left, the air condenser. On the right, the honeycomb windings of the transformer (Musée de la Musique, Paris)*

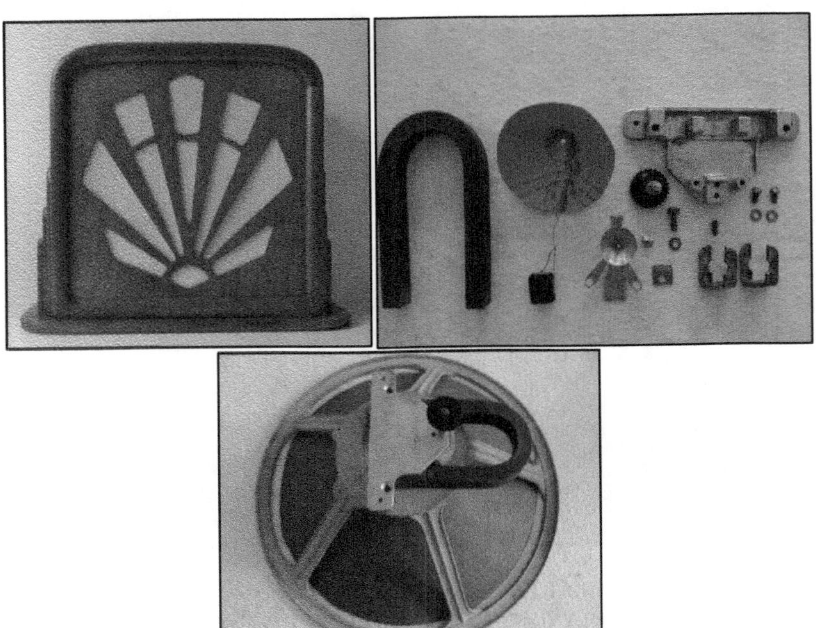

Figure 2.27. *Example of a contemporary diffuser compared to those used by Mr. Martenot. Left: front view; on the right: the different elements that make up the motor. Below: the "salad bowl"*

The diffuser is the last constituent element of the "ondes Martenot" instrument. We do not have at our disposal a representative device associated with the instrument. We have shown that Mr. Martenot used the diffuser model with rod and cardboard cone, these diffusers are very present among collectors of old radios, and we used a model of this type.

2.8.2. *Measurements*

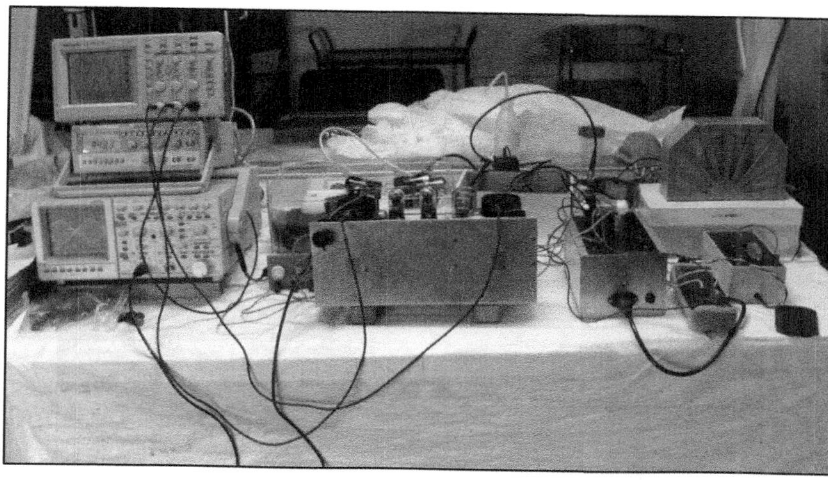

Figure 2.28. *Benchtop measurements*

After replacing the faulty or sensitive components, we carried out the tests on the main plate, outside the cabinet. This was to ensure that both heterodynes could work. Although the fixed-frequency heterodyne had some difficulty starting up, we visualized each of the two signals. These heterodynes deliver very good quality sinusoidal signals.

We very quickly perceive drift problems, in particular frequency shifts, by influence. The major difficulty to overcome is that of the frequency measurements themselves. The capacitance variation to obtain two successive sound frequencies is measured in pF. Under these conditions, the measurement is perturbing. Thus, a frequency measurement (without ribbon) on the variable frequency heterodyne gave the following results:

– measurement by influence using an antenna-type conductor, $f_f = 91.00$ kHz;

– measurement made using a 1/10 probe, $f_f = 90.11$ kHz.

Subsequently, all our measurements will be carried out using a 1/10 probe. To overcome these constraints in terms of measurements, it is necessary to have equipment and measurement conditions that are only accessible in specialized laboratories.

Still on the table, we tested the operation of the detection + amplification chain. First on a pure resistive load, then on a loudspeaker coupled to an impedance transformer. Operation is in accordance with the analysis.

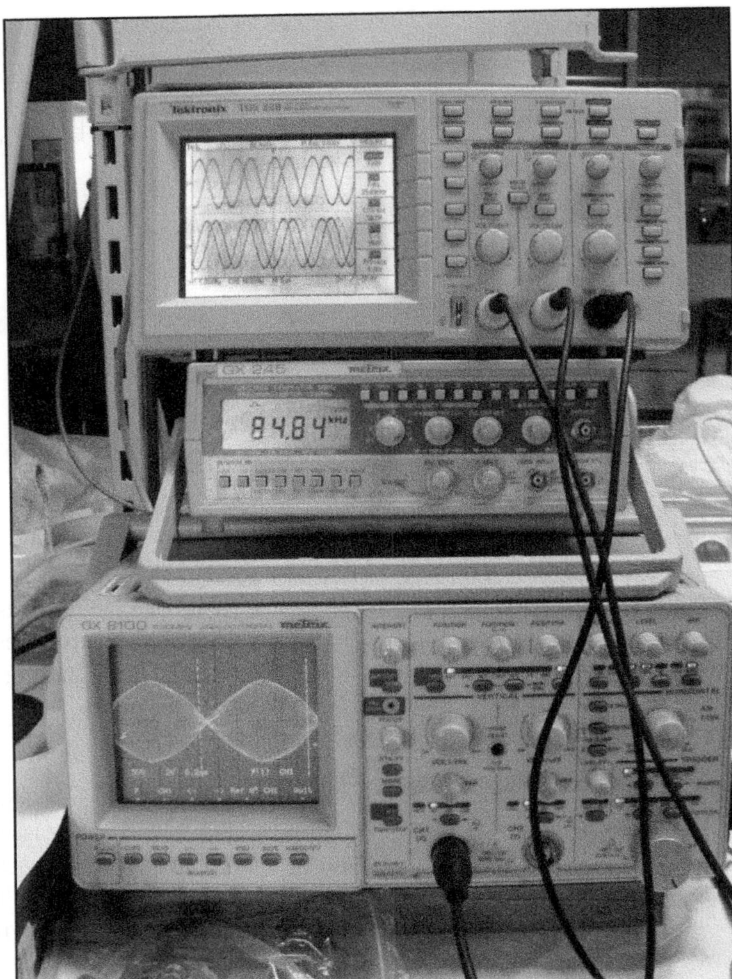

Figure 2.29. *The beat signal at the input of triode V_2*

In a second step, we reintegrated the main plate into the cabinet, and we repeated the tests with the ribbon and a diffuser. The instrument works normally, and *glissandi* are obtained despite the low tension of the existing ribbon.

2.8.3. *Measurement results*

2.8.3.1. *Electronic signals*

We observe a non-negligible variation in the working frequencies of the heterodynes. When these are operating in situ, they drop significantly and the extent of the variation in frequencies measured is 85.50 – 81.41 = 3.66 kHz. It is therefore slightly lower than the audible range Do_0-Si_6, i.e. a range of 3,918 Hz.

Indeed, the instrument is detuned both in centering and in frequency range. The measured frequency of the fixed heterodyne is around 85.51 kHz. It is lowered by the stray capacitances due to the wiring when the ribbon is operated and, of course, by the measurement probe. Frequency shifts are particularly perceptible when the operator's hand approaches near the instrument. It will therefore be necessary during recording tests to operate the instrument with the lid closed.

On each photograph, the signals of the two heterodynes are displayed on the upper oscilloscope, and the LF signal measured at the output of the detector on the lower oscilloscope.

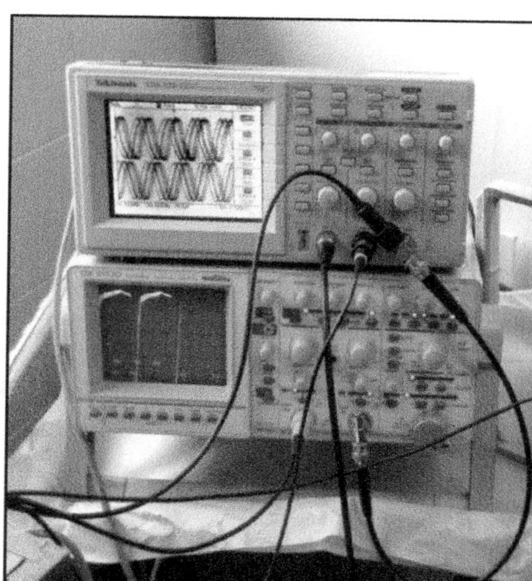

Figure 2.30. *f_v = 85.50 kHz*

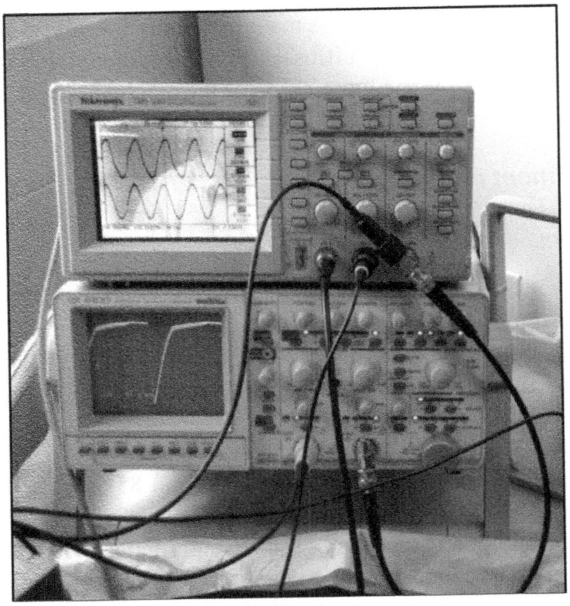

Figure 2.31. f_v = *84.66 kHz*

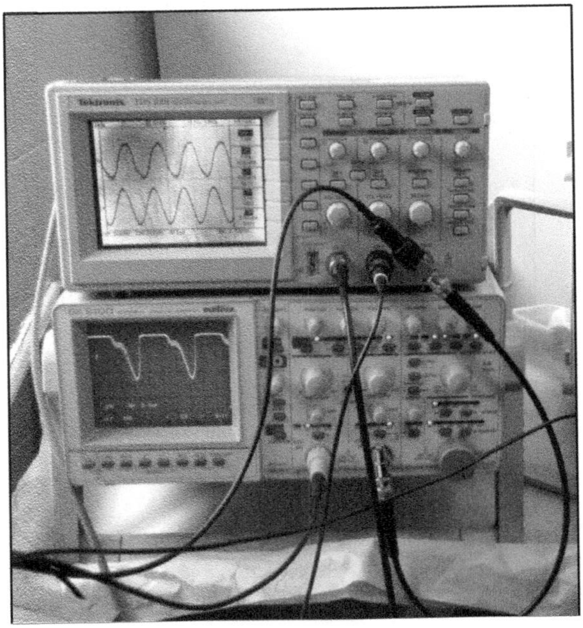

Figure 2.32. f_v = *83.79 kHz*

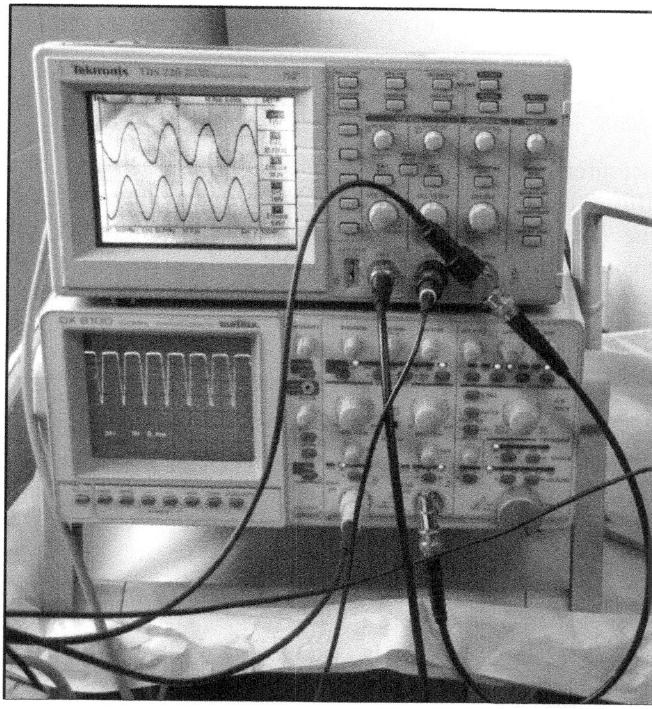

Figure 2.33. f_v = *81.41 kHz*

Over the entire measurement range, we observe, on the one hand, excellent quality of the sinusoidal signals delivered by the heterodynes; on the other hand, a good quality of the LF signal at the output of the detector.

We observe a specific form of the signal which translates a saturation of the LF stages, the saturation originates from the high voltage amplification coefficient for each stage, which in a first approximation has the value $A_v = s * r_i * m = 0.0012 * 7{,}500 * 3 = 27$, this saturation occurs at the level of the triode and also at the level of the LF transformer, this one is crossed by a very high current, the transformation ratio is lowered as well as the width of the bandwidth. Very often in saturation mode, an excessive current value leads to a break in the primary circuit, which is made with an insulated copper conductor of very small cross-section (2/100 mm). In all cases, each triode (LF and power), each LF transformer reacts more or less differently to the saturation regime, it follows the idea that an exchange of one of these components, for maintenance reasons, can modify in large proportions the musicality of the instrument. Similarly, each instrument built according to the same scheme will have its own musicality.

2.8.3.2. *Acoustic characterization of ondes no. 15*

A first measurement is performed to measure the bandwidth of the diffuser alone. The diffuser used is a 1930 Martenot diffuser from the reserves of the Musée de la Musique. This involves injecting a sine sweep (20 Hz–20 kHz) into the diffuser via a controlled source (computer, sound card and amplifier; 7.45 V RMS at the terminals of the diffuser) and measuring the sound emitted via a reference microphone (1 m on the axis, in the Museum's semi-anechoic chamber).

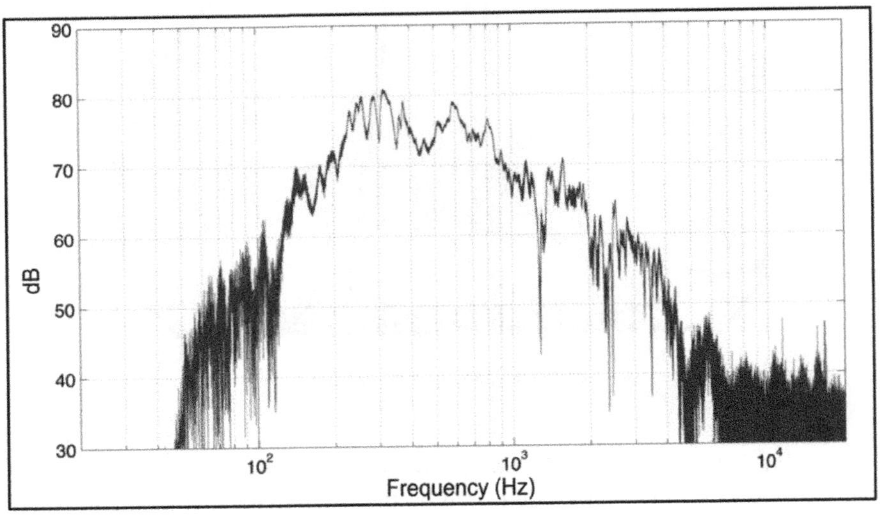

Figure 2.34. *Martenot diffuser response curve*

We obtain a response typical of diffusers with metal rod and cardboard cone: uneven response, and no sound below 100 Hz and above 3 kHz. Therefore, the question arises: how can ondes no. 15 emit a sound over the entire range, i.e. seven octaves (from Do_0 = 32.7 Hz to Si_6 = 3.9 kHz).

We then record the sound emitted by ondes no. 15 connected to the diffuser (via the reference microphone) by performing downward *glissandi* with the ribbon, so as to cover the entire range of the instrument.

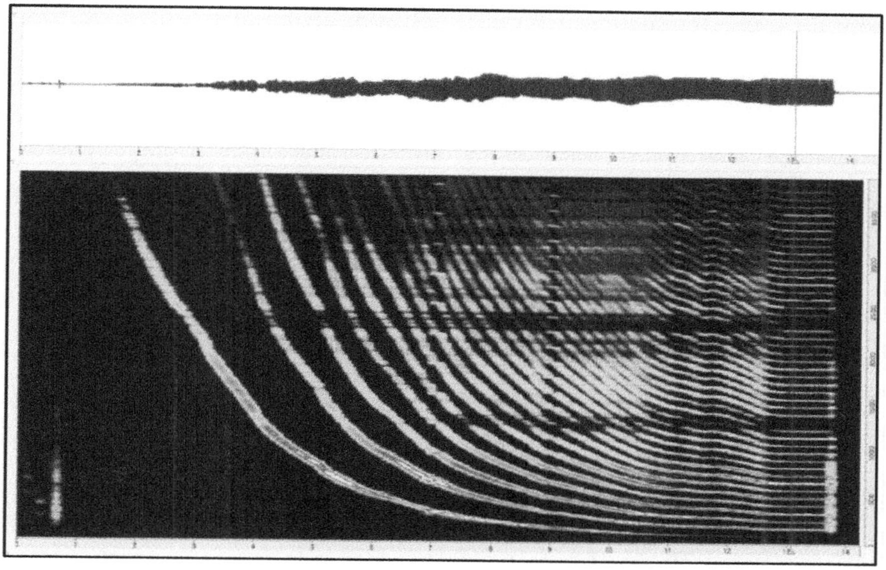

Figure 2.35. *Sonogram of a glissando played on ondes no. 15 (top: temporal signal; main: sonogram)*

Sonographic analysis shows that the ondes Martenot generates a lot of harmonics. Indeed, the signal presents a lot of distortion: this degradation results from the saturation of the low-frequency transformers that equip the amplifier (Boucherot's law).

To illustrate this distortion, we can look at the shape of the time signal at different frequencies.

We observe that the diffuser cuts almost all frequencies below 100 Hz, but *we will be able to hear notes of lower fundamental frequencies, via their upper harmonics*. This effect of reconstructing the musical signal via its harmonics can be find in many low-pitched instruments, for example, the double bass or even the analogic telephone (see, for example, Castellengo (2015) for more information).

By calculation from the previous audio signal, the fundamental frequencies and the associated amount of energy can be obtained. This is the acoustic energy radiated by the diffuser, depending on the note played (Figure 2.37).

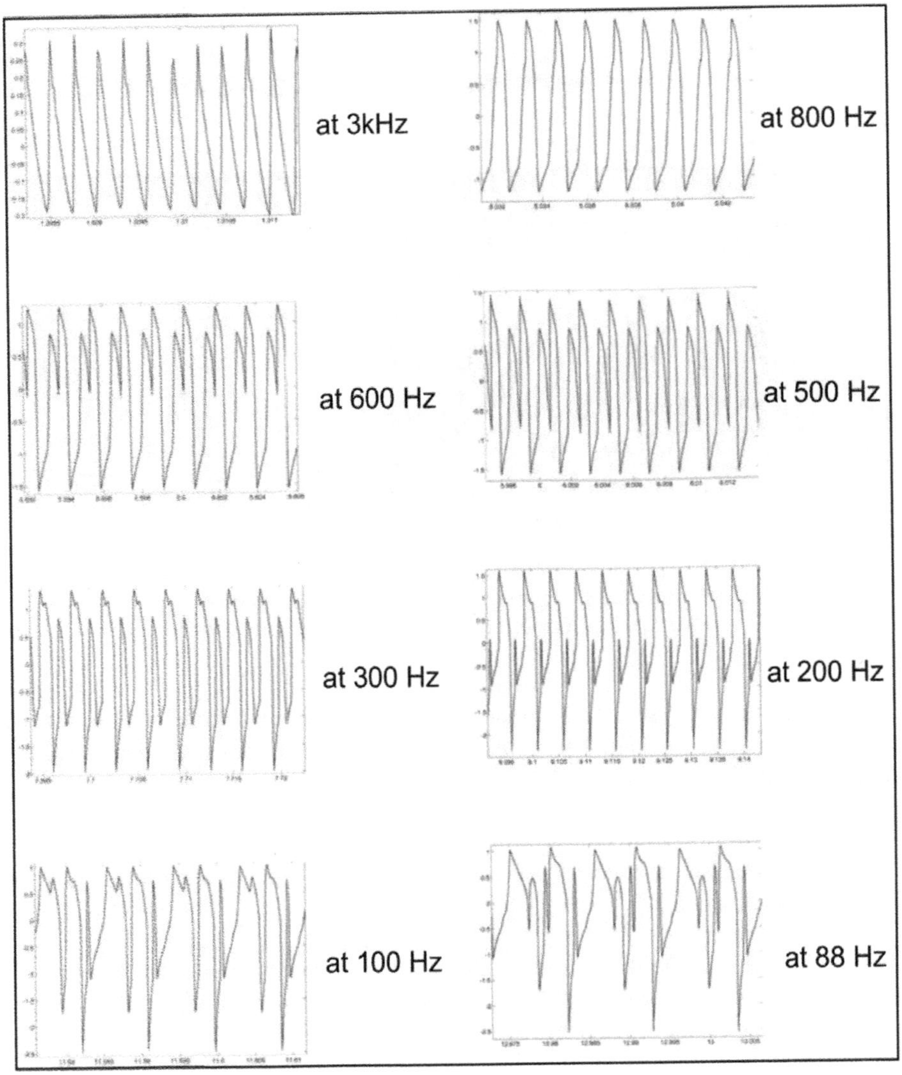

Figure 2.36. *Temporal forms of the* glissando *sound, at different times (therefore at different frequencies)*

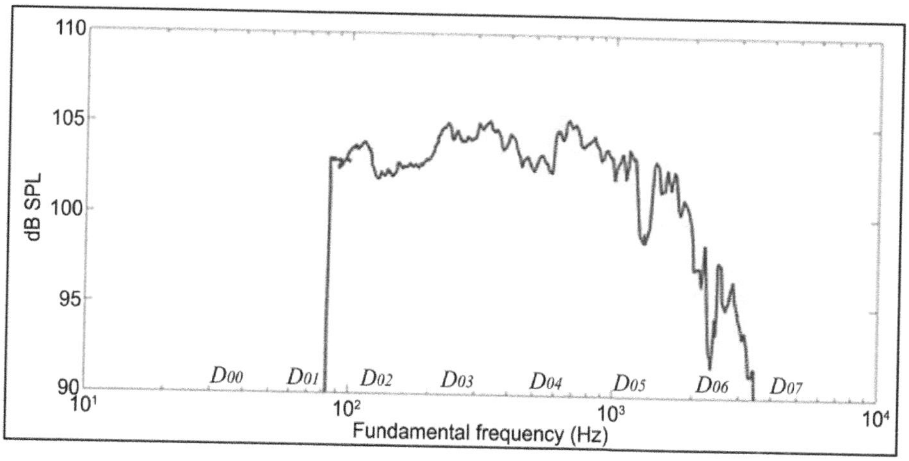

Figure 2.37. *dB SPL according to the notes played (measurements on ondes no. 15)*

It appears that:

a) while the diffuser cuts below 100 Hz, we can hear, for example, the Mi1 (82.4 Hz);

b) for each note, the more the number of harmonics increases, the more likely it is to have harmonic frequencies not falling into the "holes" of the diffuser's response. The result shows that while the response of the diffuser showed dips of 10 dB or even 20 dB, the overall response of the instrument is "smoothed", with differences of only 3–4 dB over almost the entire range.

NOTE.– With the tuning capacitor set to lowest, the lowest note emitted was Fa_0 (43.6 Hz). On the contrary, towards the high frequencies, the capacities of this diffuser (whatever the setting of the tuning capacitor) are limited to around 3 kHz, i.e. Sol_6.

2.8.3.3. *Measurement of the electrical power of ondes no. 15*

To estimate the electrical power of ondes no. 15 in play, the diffuser is replaced by a resistive load (optimized at 4.5 Ω). The measurement diagram is presented in Figure 2.38.

From a *glissando* with the ribbon over the entire range, the acquisition of the electrical signal makes it possible to calculate the fundamental frequency variation and the associated electrical power.

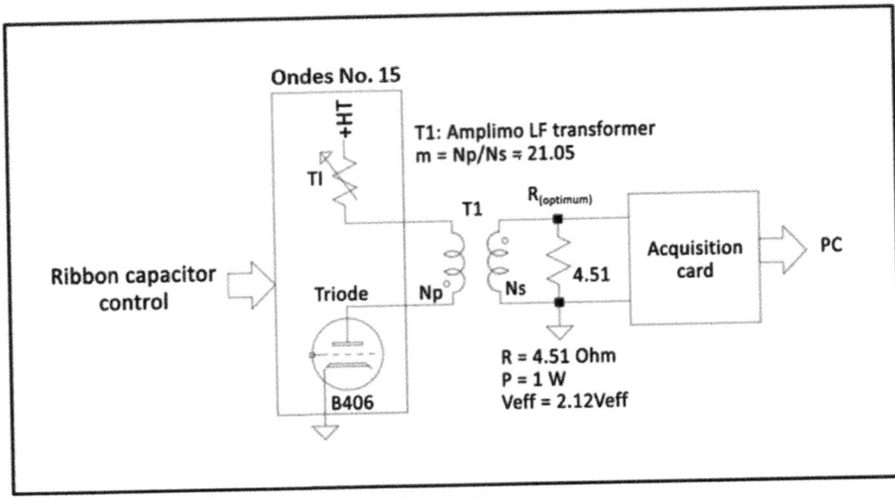

Figure 2.38. *Measurement diagram of the electrical power of ondes no. 15*

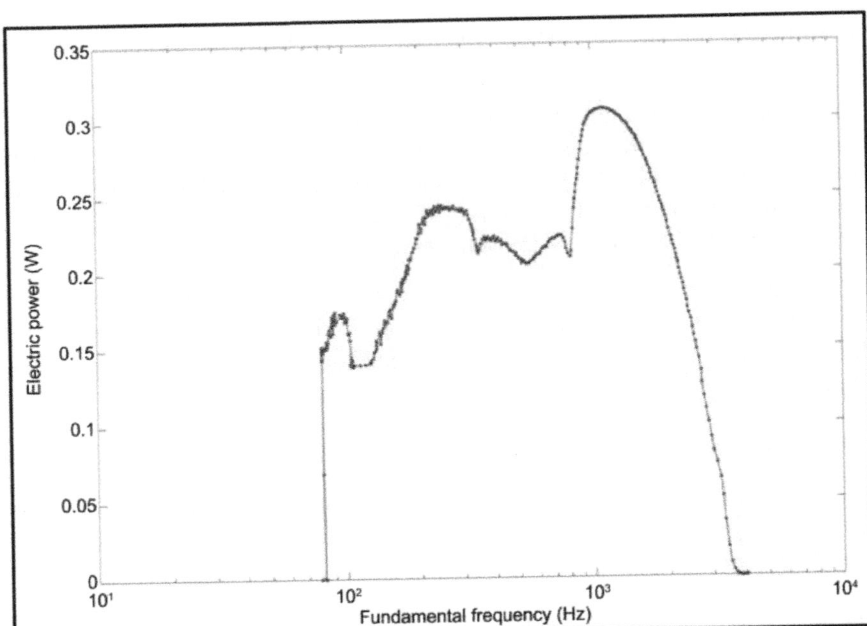

Figure 2.39. *Measurement of the electrical power of ondes no. 15*

3

Ondes No. 169 (1937)

3.1. Presentation of the instrument

Ondes no. 169 can be found at the Museum of Music in Paris (*Musée de la Musique*) under the inventory number: E.982.9.2. It belongs to the group defined by the 5th model, according to the classification proposed by Antoine Cuvelier (Loriod 1987). This model has both ribbon playing and keyboard playing with vibrato. It is probably the concert model presented at the 1937 Exhibition. Another important characteristic is that the keyboard and the ribbon allow us to cover seven octaves: $Do_0 – Si_6$ (32 Hz – 3,951 Hz).

It is within the framework of a condition report carried out in the laboratory of the Museum of Music in Paris that we studied the instrument. From the electrical point of view, it is complete, except for the sound diffusers: no diffuser associated with this instrument appears, it seems, in the inventories of the Museum of Music.

3.2. Organization of the instrument

The general configuration of the instrument conforms to the descriptions made by Mr. Martenot in his various patents. We give in the following figure, the structure of this ondes.

The system operates on the principle of frequency subtraction generated by two heterodynes, one with a fixed frequency (f_f) and the other with a variable frequency (f_v) controlled either by the ribbon or by the keyboard. Neither of these two frequencies is perceptible to the human ear, but the subtraction of one by the other

For the color versions of the figures in this chapter, see www.iste.co.uk/courrier/ondesmartenot.zip.

generates a third audible frequency ($f_{musical}$). Thanks to the variable frequency heterodyne, it is possible to cover a wide spectrum of musical frequencies.

This signal is pre-amplified by an amplifier, which is followed by a power amplifier feeding the diffusers. The particularity of this amplifier is the structure of the sound volume control; in fact, the intensity adjustment is made at the power level using an intensity key, the operation of which we will specify.

The harmonic processing of the signal is carried out at the power level; a set of switches located on the drawer makes it possible to select the filter(s). Finally, the choice of diffuser(s) is obtained by operating four switches grouped together and also located on the drawer. In the synoptic diagram, all of these commands are grouped together in the "drawer".

All these sub-assemblies work with electronic tubes. A device supplies the various necessary voltages to these sub-assemblies from the 220 V network.

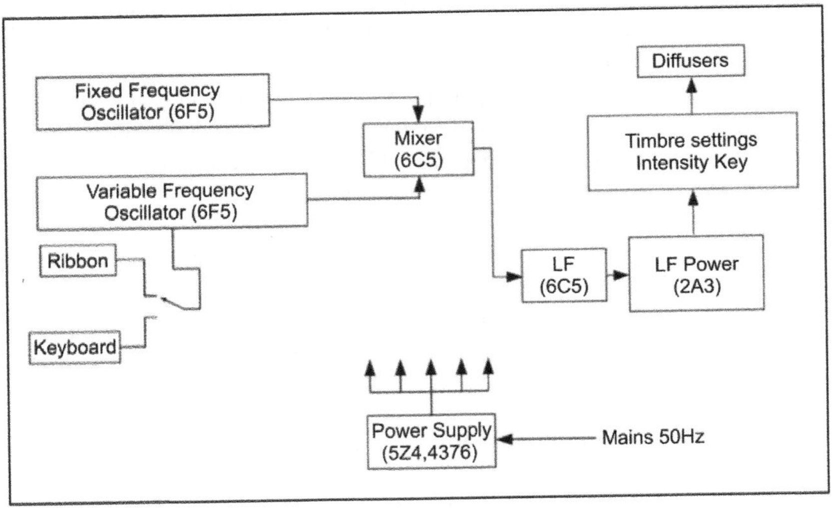

Figure 3.1. *Synoptic diagram of ondes no. 169*

To carry out the functional analysis of the instrument, we will group the various devices into six subsets:

– the "heterodynes + mixer + audio amplifier LF + power amplifier system" is mounted on a frame numbered no. 151;

– ribbon;

– keyboard;

Note that these two subsets are not mechanically independent; in particular, the ribbon guide pulleys are fixed on the same frame as the keyboard (frame numbered no. 169). But the ribbon system has a separate part: blade capacitors, protected by a shield.

– the "drawer" that can be identified by a number appearing on the intensity key "M37";

– the power supply whose chassis can be identified by no. 169;

– the sub-assembly made up of the diffuser(s).

Mr. Martenot's modular design of his instrument greatly facilitates all maintenance, adjustment and calibration operations.

3.3. The heterodyne system + mixer + LF audio amplifier + power amplifier

3.3.1. *General organization*

Figure 3.2. *The block heterodynes + mixer + LF amplifier + power amplifier (Musée de la Musique, Paris)*

We had a ***"generator block"*** diagram from the archives of the "Martenot" collection of the Museum of Music, in file 18 (Martenot n.d.b).

This document is a drawing of a tracing diagram or one of its reproductions. It is perfectly usable, and it made our task easier when analyzing the instrument. There are some differences, and we prefer to focus our analysis on the diagram we have transcribed and presented on the next page.

Moving contacts I_1 and I_2 are positioned for keyboard playing. Components whose values could not be identified are marked with (?).

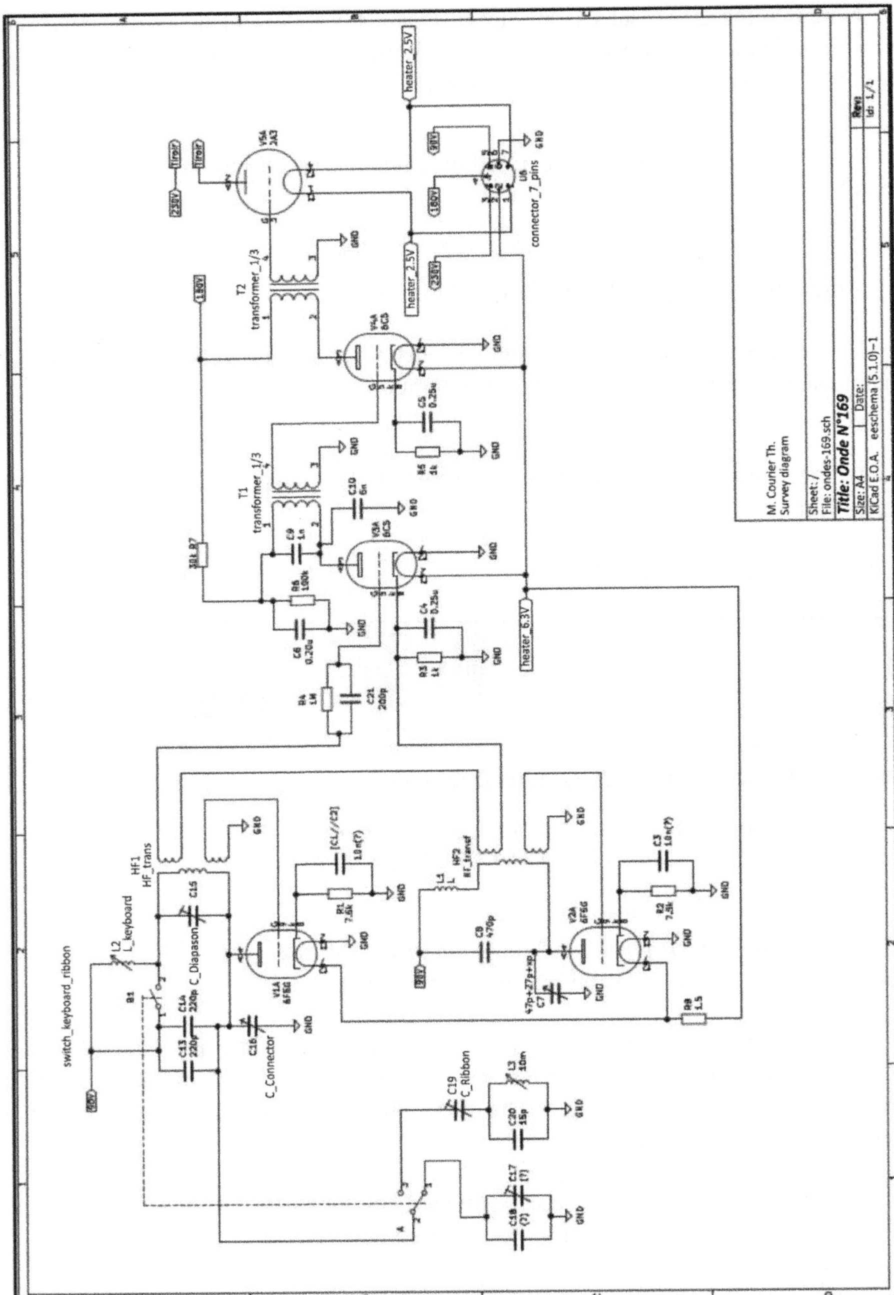

Figure 3.3. *Diagram of the block heterodynes + mixer–pre-amplifier + amplifier*

The V_{1A} and V_{2A} electronic tubes of the 6F5 triode type are the oscillator tubes. The V_{3A} triode of the 6C5 type performs the function of mixer and LF amplifier. The V_{4A} triode of the 6C5 type is also the second LF amplifier which controls the 2A3 triode. This last triode performs the power amplification function and delivers the amplified signal to the diffusers. This very tall electronic tube ($h \approx 130$ mm) cannot be mounted vertically on the frame of the heterodyne sub-assembly + mixer–pre-amplifier + amplifier: the cover of the instrument case could not be closed. It will therefore be mounted horizontally.

This triode set is strictly identical in its structure to the older model that we have been able to observe: ondes no. 15. Mr. Martenot therefore initially updated his instrument using new tubes and components available on the market.

The 6F5 triode with a high μ coefficient in Octal all-metal series was proposed by RCA as early as 1935. The 6C5 type of the same series, but with a lower μ coefficient, was presented the same year by RCA with the necessary recommendations for the use in detectors and amplifiers with LF transformers. Finally, the 2A3 triode, a directly heated 2.5 V power triode, was available with a four-pin US base from RCA in 1933.

3.3.2. *Heterodynes*

Let us now analyze in more detail the operation of oscillators. We propose to return to the block diagram before modification, which is to say for operation only with the ribbon.

Inductor L_1 and keyboard inductance L_2 are removed, as well as capacitors C_{15}, C_{17} and C_{18}. We then find the diagram as it is defined by Mr. Martenot on his patent no. 1.004.406 (Martenot 1947).

At the level of the fixed oscillator (HF_2/V_{2A}), the assembly is classic, it is a triode oscillator and tuned plate circuit. The reaction by the grid is obtained by a first secondary whose number of turns is clearly lower than that of the primary. The polarization is fixed by R_2 decoupled by C_3. Capacitor C_8 and primary inductance of transformer HF_2 set the resonant frequency. This is adjustable thanks to the capacitor C_7, which consists of a parallel connection of two fixed capacitors and one adjustable one.

It is not possible to determine with precision the resonance frequency, and therefore, the adjustment frequency, of this circuit. But in patent no. 1.004.406 (Martenot 1947), we read on page 3: "***for a main oscillating circuit O with a***

capacitance of 500 cm[1] and an inductor of 4 to 5 mH" which allows us to give a framework of possible values of the oscillator frequency:

$$95 \text{ kHz} < f < 107 \text{ kHz}$$

Figure 3.4. *Heterodynes. The masses of cotton immobilize all the coils in relation to the shielding (Musée de la Musique, Paris)*

The HF_2 transformer is made according to the rules of implementation of the time: winding in honeycomb, adjustment of the coupling of the coils by adjustment of the distance, absence of ferrite core allowing the easy adjustment of the value of the inductance and finally, shielding.

The signal is taken from a second secondary winding located on the upper end of the assembly, thus making it possible to limit the reaction of the external circuit on the oscillator.

1 The cm is an old measure of capacitance, 1 pF = 0.9 cm.

At the level of the variable frequency oscillator (HF_1/V_{1A}) controlled by the ribbon, we find the same structure: triode with tuned plate circuit. The reaction by the grid is obtained by a first secondary whose number of turns is much lower than that of the primary. The polarization is fixed by R_1 decoupled by $\{C_2//C_3\}$, the primary inductance of the transformer HF_1 and the capacitors C_{13}, C_{14}, C_{16} and C_{19} set the resonance frequency, C_{16} controlled by the wheel allows the tuning, and C_{19} is the resulting capacitance of the ribbon with respect to a series/parallel set of armatures. In accordance with the patent no. 1.004.406 (Martenot 1947), the dipole $\{C_{20}//L_3\}$ makes it possible to reduce the increase in capacity according to the frequency. By identification on document **PLIII** of the patent already cited, we have L_3 = mark 10 = 10 mH and C_{20} = mark 11 from 15 pF to 35 pF. The HF transformer is strictly identical to that of the fixed heterodyne, and a secondary winding is placed at the high end, which makes it possible to recover the signal while reducing the reaction of the external circuit. In his patent no. 841.866 (Martenot 1939), page 2, line 65 onwards, Mr. Martenot suggests that the frequency of the variable oscillator is lower than that of the fixed heterodyne. If we set the frequency of the fixed oscillator to 95 kHz, for a range of seven octaves, the frequency of the variable oscillator must evolve from 91,048 Hz (Si_6) to 94,967 Hz (Do_0). In ribbon mode, the oscillator structures are simple; it is the ribbon capacitor that fixes the resonant frequency of the variable oscillator, and it is physically outside the enclosure of the oscillator, which is a source of drift. Mr. Martenot raises these problems of drifts in his patent no. 841.866 (Martenot 1939).

In keyboard play mode, the fixed heterodyne consists of the HF_2 transformer, an additional inductance L_1 and the V_{1A} triode wired as an oscillator. This inductance is added in series to the oscillating circuit so as to find a configuration identical to that of the variable oscillator, in order to operate a constant frequency shift on the two oscillators.

The variable oscillator (HF_1/V_{1A}) undergoes an identical transformation, namely: the addition of the keyboard inductance L_2 in series with the primary of the heterodyne transformer HF_1. The diagram is identical to Figure 4 of patent 841.866 (Martenot 1939). S_1, S_2 and S_3 represent the total inductance selected by a key, with inductance represented by L_2 in our diagram. C_1 (patent) is the resulting tuning capacitor $\{C_{13}//C_{14}//C_{16}\}$ of our diagram, capacitor C_4 (patent) is capacitor C_{15} of our diagram, called the "diapason" whose role is explained by Mr. Martenot in his patent. Capacitors C_{17} and C_{18} are also added, which make it possible to compensate for the inevitable frequency shifts when switching from keyboard mode to ribbon mode and vice versa.

The following remarks can be added to this presentation:

– originally, chassis no. 151 is probably that of a previous model, with the single play on the ribbon or the single play on the keyboard;

– the addition of the “keyboard playing” or “ribbon playing” functions results in the addition of the necessary components at the level of the oscillating circuits, the play switch and the socket plate for the keyboard and ribbon connections;

– we have observed several modifications that have taken place over time: exchange and replacement of fixed tuning capacitors, exchange and replacement of adjustable capacitors (re-drilling, conductor with plastic insulation, etc.), as shown in the following photos.

Figure 3.5. *Tuning variable capacitor (Musée de la Musique, Paris)*

Variable oscillator:

– placement of the play mode switch: top right corner;

– replacement of the adjustable capacitors at the top and bottom right (modified drill hole);

– replacement of the tuning capacitors (in red): significantly newer model than the other components;

– rewiring with gray plastic insulated conductors;

– we also observe that some fixed blades of the left capacitor have been removed.

Figure 3.6. *Fixed oscillator: addition of the inductance for playing the keyboard, top left of the photo, replacement of the tuning capacitors (new model in red) (Musée de la Musique, Paris)*

Figure 3.7. *Drilling for the placement of the adjustable capacitor (Musée de la Musique, Paris)*

Figure 3.8. *Re-drilling of the keyboard and ribbon connection terminal board and placement of another adjustable capacitor (Musée de la Musique, Paris)*

These different interventions suggest that it is a workshop prototype, which allowed the development of the '37 model.

Due to the principle used, precision and frequency stability are difficult to obtain. Indeed, let us assume an ideal oscillator with a fixed, strictly constant frequency F_0, equal to 95 kHz. To obtain Do_1 65.4 Hz, we need a frequency of the variable oscillator of 94,934.6 Hz. Imagine a drift of 8 Hz on this frequency, i.e. a precision lower than 1.10^{-4}, the frequency then becomes:

$$94,934.6 - 8 = 94,926.6 \text{ Hz}$$

The note produced has a frequency of 95,000 – 94,926.6 = 73.4 Hz, i.e. the note Re_1.

The adjustment precision must be very high. Indeed, suppose a simple oscillating circuit {L//C} whose inductance value is L = 5 mH (probable value of the system). To produce the note La_{440} with a fixed heterodyne frequency of 95 kHz, the tuning capacitor must be set to the value of 566.572 pF and the frequency of the variable heterodyne will be 94,560 Hz. Suppose a tuning is necessary at 445 Hz: it is therefore necessary to adjust the capacitor to obtain a frequency of 94,555 Hz in order to find the same frequency difference; the capacitance value becomes: 566.632 pF. It is therefore necessary to vary the tuning capacitor by:

$$566.572 - 566.632 = 0.060 \text{ pF}$$

This is an extremely low value. This explains that on his instrument, Mr. Martenot removed fixed blades from the tuning capacitor (see photograph of the variable heterodyne block) and mounted a three-stage motion reducer between the wheel and the control shaft of this variable capacitor.

Special precautions have been taken to stabilize the operation of these heterodynes: the triodes are powered by a stabilized high voltage of 90 V, which is relatively low compared to the current operating voltages, which are of the order of 150 V – 250 V for these electronic tubes; similarly, the heating filaments of the oscillator tubes are slightly underpowered by the insertion of a power series resistance: $R_8 = 1.5\ \Omega$, which brings the heating voltage to $6.3 - 0.6 \times 1.5 = 5.4$ V in place of the standard heater voltage of 6.3 V.

3.3.2.1. *Ribbon*

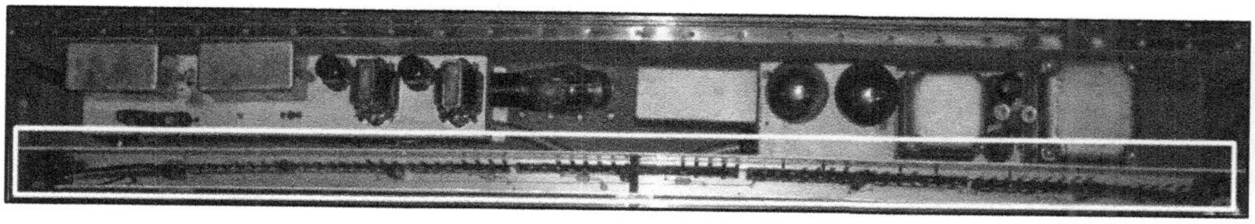

Figure 3.9. *The ribbon system (white box) (Musée de la Musique, Paris)*

The ribbon is one of the interfaces between the musician and the instrument. Mr. Martenot gave the first description in his patent no. 666.807 (Martenot 1929), when a first version, which used a conductive thread, was made. The ribbon is the logical development of this invention. The first capacitors were made from continuous walls with a flared shape. Adjusting the value of these capacitors was therefore problematic. Hence, Mr. Martenot will subsequently make the following improvements:

– the splitting of capacitors by octave or more, and the implementation of series/parallel associations with other capacitors;

– the adjustment of the capacitor value by the use of split blades, which will facilitate the tuning of the instrument;

– the addition of an auxiliary resonant circuit (as for ondes no. 15) in order to reduce the opening of the capacitors with split blades;

– guidance by bottom skids and lateral pins, as well as the installation of a line break in order to reduce the vibration of the ribbon during its movement.

All these improvements are present in ondes no. 169. They are perfectly described by Mr. Martenot in patent no. 1.004.406 (Martenot 1947).

We represent the structure of the ribbon system with some dimensioning elements with a sketch in Figure 3.10.

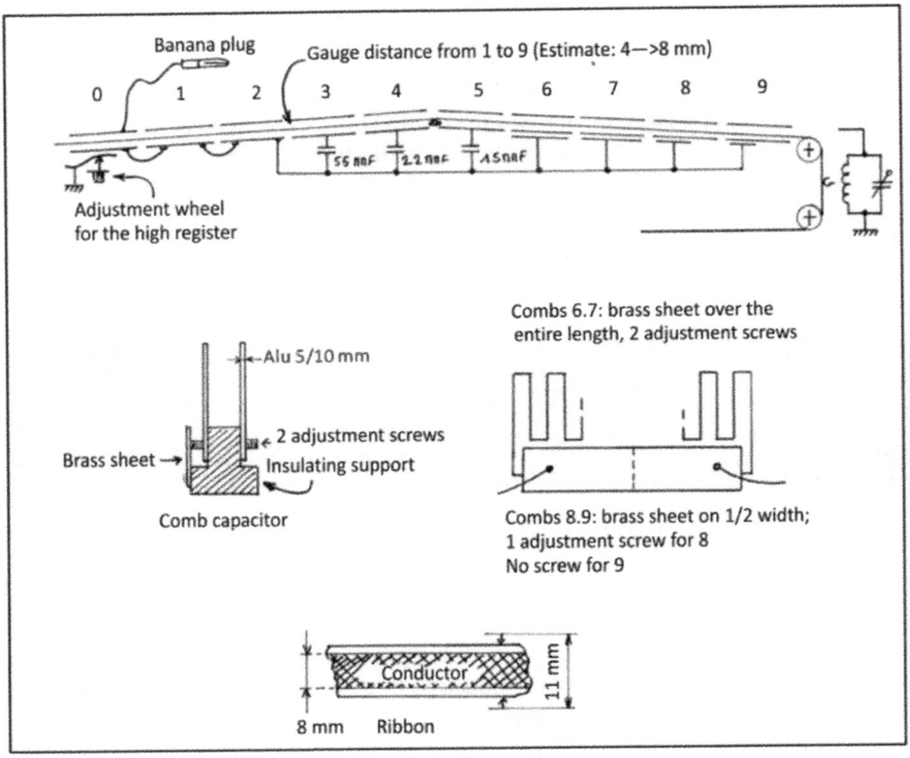

Figure 3.10. *Sketch representing the elements of the ribbon system*

The device has 10 blade capacitors, marked 0–9. This number is therefore greater than the number of octaves. The ribbon with a height of 11 mm enters from the right; the conductive surface of the ribbon has a useful height of approximately 8 mm.

When the performer's ring is on the far left of the index mark, the conductive part of the ribbon reaches the right end of capacitor 9. The frequency generated by this note is therefore the lowest, which means that the frequency generated by the variable heterodyne is the highest, and that by applying the properties of the oscillating circuit, the value of the capacitance must be the lowest.

A single capacitor is not easily realizable because as Mr. Martenot shows in his patent, the capacitor has a very flared shape. On the keyboard: on the right, towards the high musical frequencies, the walls are very close (< 0.5 mm); on the left, towards the low musical frequencies, the gap between the walls is large (> 20 mm).

To reduce the amplitude of the variation in the distances between walls and ribbon, Mr. Martenot added a second resonant circuit (10 mH//25 pF) in series with the resonant circuit, whose resonance frequency is much higher, which has the effect of reducing the increase in the useful value of the capacitor in the high musical frequencies. The complete analysis of the operation of this circuit is carried out in section 5.4.2.

But this was still insufficient, since the adjustment of the ribbon capacitor with continuous walls over the entire displacement of the ribbon was very problematic. He therefore proceeded to split the capacitor into 10 separate modules. The disadvantage was that each module regained the flared shape over a reduced length; in fact to pass from one module to the next, there is a value discontinuity. To remedy this problem, he added a capacitor of suitable value in series with the module, which had the effect of linearizing the shape of the capacitor. But this remained insufficient, the capacitance variations were still too great and the facing surfaces had to be reduced. Mr. Martenot would finally give a comb shape to the fixed walls of the ribbon capacitor. On ondes no. 169, the following arrangements have been made: capacitors 3–9 each have a series capacitor. After capacitor 5, the series capacitor is of very small value; it is then made with a sheet of conductive metal attached to the base of the comb and separated from it by adjusting screws (see Figure 3.10).

Capacitors 0–2 are connected by conductors. When the high register is to be played, the capacitor has its maximum value, and the ribbon is fully engaged up to capacitor 0, which has a wide parallel blade adjustable by a knob. Each capacitor has deformable blades, allowing note-by-note adjustment.

The sliding and guide pads are placed between these comb capacitors.

Finally, so that the performer can spatially locate his index, Mr. Martenot imagined and manufactured a graduated marker by a succession of recesses and studs (round heads of brass points) on a strip probably made of ebonite; a hollow corresponds to a white key, and a bump corresponds to a black key.

Figure 3.11. *The index mark (Musée de la Musique, Paris)*

3.3.2.2. *The keyboard*

Figure 3.12. *The keyboard (Musée de la Musique, Paris)*

In keyboard mode, inductance L_2 is put into service: it is split into as many inductances as there are notes to play. The operating principle is described by Mr. Martenot in his patents (Martenot 1929, 1939, 1947). This wiring diagram configuration allows monophonic playing. If several keys are pressed at the same time, it is the lowest note that is audible.

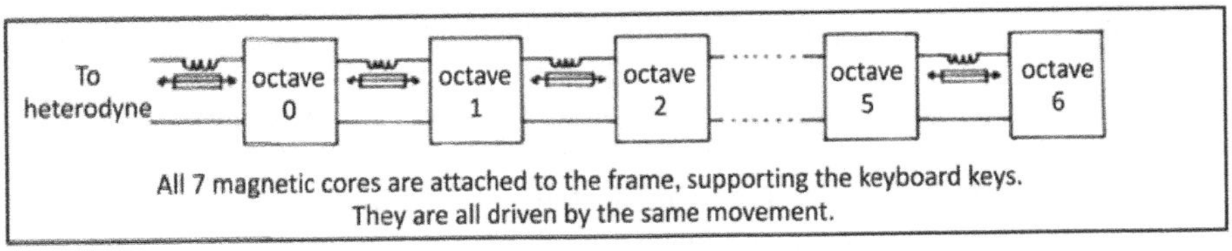

Figure 3.13. *Diagram of the keyboard of ondes no. 169*

The decisive particularity is the possibility of playing a note with a vibrato, thanks to a mobile keyboard. We will not return to the mechanical aspect of the suspension of the keyboard, which is perfectly described by Mr. Martenot himself. On the contrary, we propose an electrical study of this keyboard control.

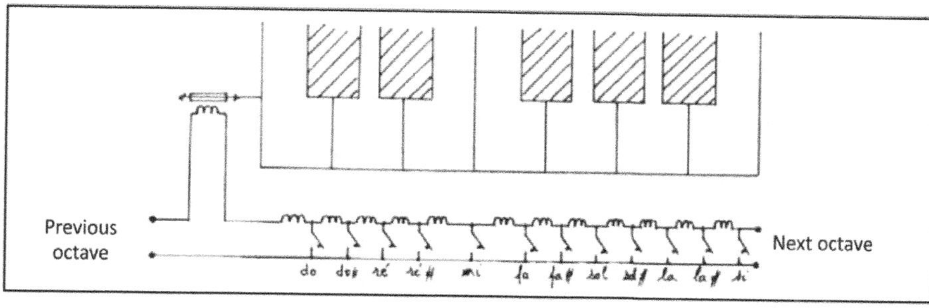

Figure 3.14. *The keyboard: structure of an octave*

The device is made by octave. For an octave, we have one inductance per note, but for the note Do, the inductance is split into two parts: a part wound on a support through which a magnetic core can move, the movement of which is linked to that of the keyboard, and another part which is fixed. It is this inductance that allows the play of vibrato. So, there are seven split inductors. This makes it possible to obtain a vibrato whose depth depends only on the play of the performer, independently of the frequency of the note played, the perception of the vibrato being more or less the same in the interval of an octave.

Figure 3.15. *The vibrato inductor (Musée de la Musique, Paris)*

The magnetic core of the vibrato inductor is probably made with a transformer sheet cutout.

Let us give an example of calculating the values of the inductances for the octave containing the La_{440} and under the same conditions as the ribbon (F_0 = 95 kHz, C_0 = 350 pF).

For Do_3 (261.63 Hz), f_v = 94,738.37 Hz, which gives us the value of the inductance of the resonant circuit L_{DO} = 8.0634 mH and for the note immediately above $Do\#_3$ (277.18 Hz), f_v = 94,722.82 Hz → $L_{DO\#}$ = 8.0661 mH. The corresponding increase ΔL = 2.7 µH, it is therefore the value of the coil that must be wound and inserted between the Do_3 and $Do\#_3$. This value is very small; it is obtained by winding a few turns of insulated copper wire around a very small mandrel Ø = 4 mm, but it is necessary to manufacture 12 x 7 = 84 of different values for a seven-octave keyboard!

On the 37 model, with the exception of the vibrato ones, the coils are all made without a magnetic core, the adjustment of the value being done by modifying the number of turns. When the keyboard is made, we can only set an overall tuning. Subsequently, Mr. Martenot will use magnetic core coils, which will greatly facilitate the tuning of the notes played on the keyboard.

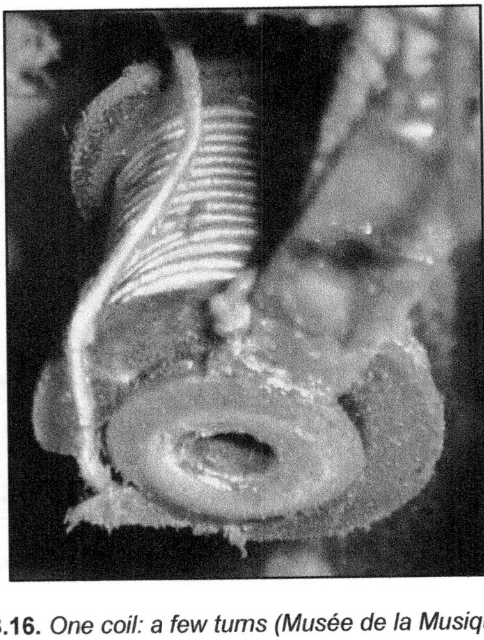

Figure 3.16. *One coil: a few turns (Musée de la Musique, Paris)*

Figure 3.17. *The inductors are placed under the keys of the keyboard (Musée de la Musique. Paris)*

3.3.3. *The mixer–amplifier*

The first type of 6C5 triode is wired as a detector through the grid with the dipole $\{R_4//C_{21}\}$ in series with it. The signals of the two heterodynes are taken from the secondary windings of each HF transformer placed in series. This triode also performs the pre-amplifier function; it is charged in its plate circuit by the primary of an LF transformer with a ratio of 1:3. It is followed by a second type of 6C5 triode mounted as an amplifier and also charged in its plate circuit by a LF transformer with a ratio of 1:3. Each triode has its operating point fixed by the cathode resistance R_3 or R_5 (1 kΩ) in parallel with the decoupling capacitor C_4 or C_5 (0.277 µF). The second triode is supplied directly under a stabilized voltage of 180 V. The first is supplied via the divider bridge R_7 (30 kΩ) and R_6 (100 kΩ) decoupled by C_6 (0.2 µF), a divider bridge which, considering the current drawn by this triode (3.3 mA), fixes the voltage at around 62 V.

This assembly with two triodes and LF transformers has reduced performance in terms of bandwidth. If it was widely used at the beginning of wireless radio technology, it was quickly abandoned in favor of the amplifier with pentode tubes and resistance/capacitor connections available from 1931.

Figure 3.18. *The mixer–pre-amplifier (Musée de la Musique, Paris)*

Capacitor C_9 connected in parallel to the primary of the transformer of the first triode has the role of limiting the bandwidth at high frequencies. Indeed, the mixer stage transmits the frequencies f_f, f_v, f_f+f_v and f_f-f_v; however, only the last frequency should be kept: $f_{musical} = f_f-f_v$. A second capacitor C_{10} (6 nF) was added and apparently determined by experimentation (connection to the ground). It will also be noted that these triodes are insufficiently decoupled since the frequency from which it is considered that the bias voltage across R_k is constant is given by $f \geq 2/RC = 2/(1{,}000 \text{ x } 0.277.10^{-6}) \geq 7{,}220$ Hz. At that time, capacitors with values greater than 1 µF were bulky, expensive and unreliable components.

Finally, note that the cascading of two amplifier stages ($g \approx 20$) constitutes a high-gain amplifier ($\approx$ 400). The amplitudes of the voltages delivered by the oscillators are greater than 1 Vcc, as a result the amplitude of the output voltage at the terminals of the secondary of the second transformer is very large and the signal is greatly distorted by saturation.

3.3.4. *The amplifier*

3.3.4.1. *The amplifier stage*

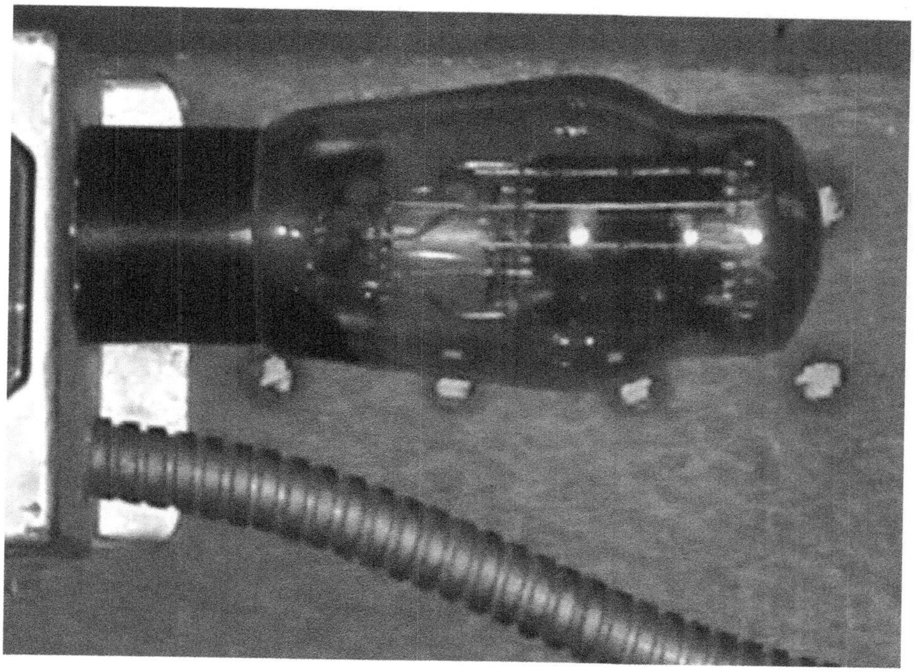

Figure 3.19. *The amplifier (Musée de la Musique, Paris)*

This amplifier stage delivers the signal to the diffusers. It is fitted with an RCA 2A3 power triode, supplied at 230 V and whose operating point is fixed by the cathode resistance R_1: 1,370 Ω, itself decoupled by a 10 μF capacitor C_1 (see the diagram in the part dealing with power). This decoupling is insufficient, in fact for this last LF stage, it is essential to decouple the cathode circuit suitably, and otherwise, there will be a variable grid bias. Indeed, the time constant of this circuit R_1C_1 is τ_{RC} = 1,370 x 10E-6 = 13.7 ms and for the lowest frequency to be transmitted, the time constant is $\tau_{max} = 1/f_{min} = 1/32 = 31.25$ ms.

These two components are wired to the power supply chassis. Although the isolation voltage of this capacitor is low (~ 40V), it has a large footprint; the installation of the triode on the power supply chassis allows it to be easily accommodated. This triode is charged by the diffusers in series with the intensity key resistance, and in parallel with the various capacitors constituting the harmonic filters. These components are grouped together in the drawer.

A first remark relates to the high value of the cathode resistance R_1: 1,370 Ω, which is the measured value. However, in the power supply diagram in the archives (Martenot n.d.b), this value is 750 Ω, which is more in line with the operating point recommended by the manufacturer of the 2A3 triode. Is this an aging of the resistance or was it a choice by Mr. Martenot?

3.3.4.2. *Sound volume control*

The volume control is carried out on this instrument using a resistance of a particular type, developed by Mr. Martenot, and placed in series with the diffuser in the plate circuit: the intensity key (see section 3.4.4).

3.3.4.2.1. The operating point

Further analysis shows that the operation is not so simple: by varying the resistance of the plate circuit, the operating point of the amplifier changes.

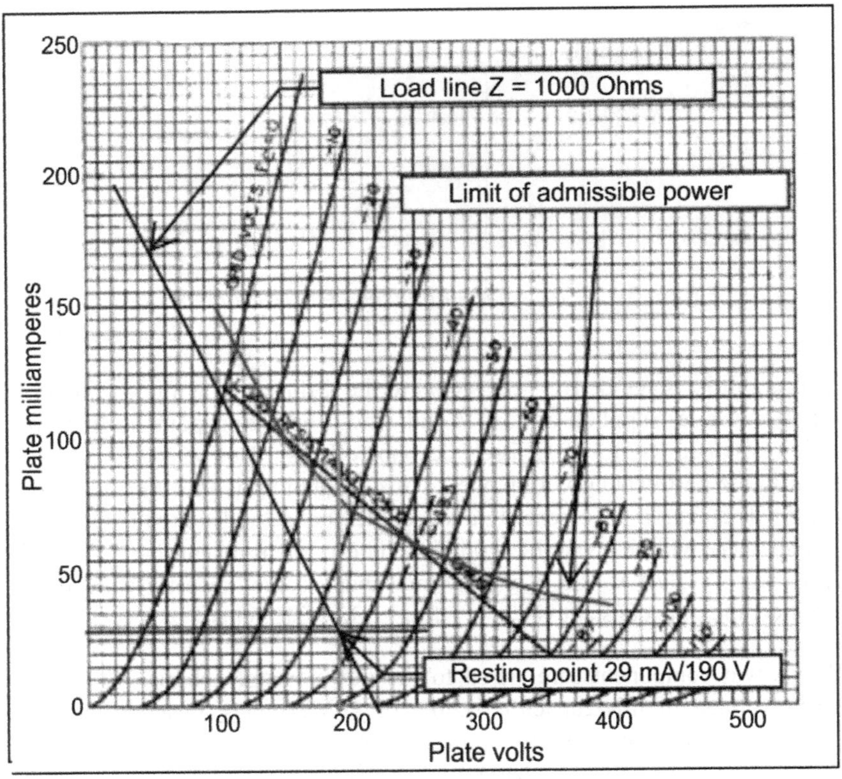

Figure 3.20. *Characteristic $I_a = f(V_a)$ of triode 2A3. RCA 1933*

The characteristic used is that of the plan: the anode current I_a (mA) as a function of the anode voltage V_a (V). We assume that the load comes down to negligible dc resistance diffuser impedance and a zero key resistance.

We have set the operating point for a cathode resistance $R_k = 1{,}370\ \Omega$.

In the absence of a signal, the triode is biased with a grid voltage of the order of –38 V. It is crossed by a no-load anode current equal to approximately 29 mA, and the anode voltage is fixed around 190 V. Under these conditions, the supply voltage necessary for the operation of the assembly is:

$$V_{alim} = V_{ak} + R_k I_a = 190 + 1{,}370 \times 0.029 = 230\ \text{V}.$$

This operating point is well below the hyperbola (curve in red) of the power admissible by the tube $P_{max} = 15$ W. The load line (in black) corresponds to all the points used when signal is applied to the grid, but only for the case where $R_t = 0$. It will be very complicated to establish a relationship between pressing the key or of its resistance and the sound level obtained in the diffuser. In effect:

– the continuous operating point of the triode is determined by the equation:

$$V_{power} = V_{ak} + (R_k + R_t) * I_a;$$

– each value of the R_t resistance of the intensity key corresponds to a different continuous operating point of the triode;

– the V_{power} supply voltage is not constant and varies from 300 V at no load to 220 V at full load.

3.3.4.2.2. Experimental approach

We will circumvent the difficulty by carrying out the assembly. We will use a 2A3 triode manufactured by Sovtek available on the market today. It will be charged by an LF transformer with a ratio m = 0.057. The secondary of this one will be closed on a resistance of 3.3 Ω/5 W; the load seen by the triode is therefore $Z_d = 1{,}000\ \Omega$. The key resistance is obtained by putting discrete 500 Ω resistances in series. In order to limit the error on the final measurement, each node is measured with an ohmmeter of sufficient precision.

This triode is driven by the 6C5 two-triode pre-amplifier in accordance with the diagram of ondes no. 169. Everything will be powered by power sources identical to those produced by Mr. Martenot, in order to take account of fluctuations in supply voltages depending on the load.

The first triode (6C5) will be powered by an LF generator whose frequency of the sinusoidal signal will be fixed at 440 Hz, and the amplitude adjusted so that the distortion observable on the oscilloscope is negligible. This value is 1.55 V_{eff} at the load terminals. We have chosen as the output quantity the effective power available at the terminals of this same load, because it is representative of the level of the perceived sound signal.

The following table shows the measurements. This is, of course, an indication without any claim as to the calculation of errors.

R_t (Ω)	V_{power} (V)	V_k (V)	V_d (V)	P_d (W)	I_k (mA)
0	239	41.5	1.550	0.728	30.3
493	243	38.9	1.280	0.496	28.4
994	245	36.7	1.120	0.380	26.8
1,474	248	34.8	1.000	0.303	25.4
1,976	250	33.2	0.903	0.247	24.2
2,476	253	31.7	0.818	0.203	23.1
2,979	254	30.0	0.746	0.169	21.9
3,480	255	29.0	0.685	0.142	21.2
3,982	256	28.0	0.633	0.121	20.4
4,484	258	27.0	0.589	0.105	19.7
4,986	259	26.0	0.549	0.091	19.0
5,487	260	25.0	0.515	0.080	18.2
5,987	260	24.0	0.470	0.067	17.5
10,000	268	19.0	0.302	0.028	13.9
20,000	277	12.7	0.146	0.006	9.3

Table 3.1. *Measurements for gain adjustment*

We find the operating point of the triode for zero key resistance. The slight offset in the bias voltage is due to the fact that in the previous modeling, the DC resistance of the LF transformer was assumed to be zero.

We can better visualize the relationship between key resistance and available power on the following graph.

A nonlinear curve is obtained but with good continuity, with most of the adjustment being carried out for a resistance between 6,000 Ω and 0. It can be seen that the triode is not used at its maximum power $P_{signal\ max} < 3.5$ W.

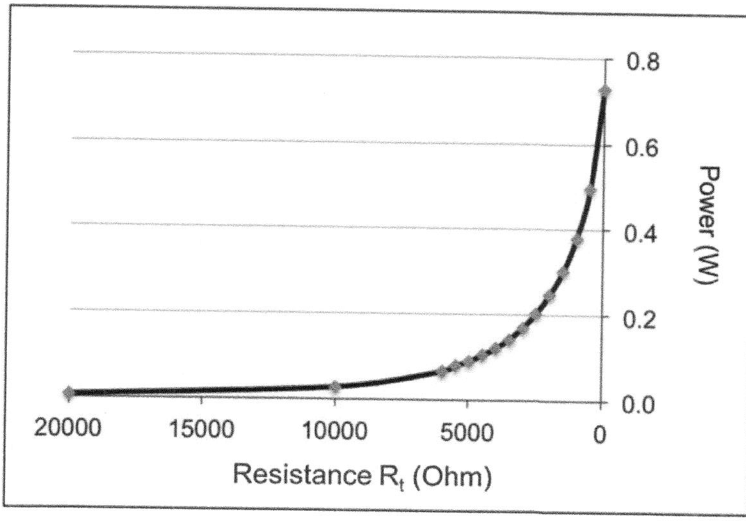

Figure 3.21. *Measurement of the power supplied at the output of the triode as a function of the key resistance*

3.3.4.2.3. The leakage phenomenon

For very high values of resistance R_t, or in open circuit, Mr. Martenot, in his workshop notebook 21 (Martenot n.d.c) (*Note, instrument modification* 1937), observes a malfunction that he calls a "***leakage***": when the key is released, a sound is perceptible in the diffuser. In fact, the amplitude of the LF signal applied at the input is very high because, as we have seen previously, the last LF stage operates in saturation. Although the triode has its plate circuit open, the signal circulates through the various interfering impedances of the stage and generates an audible signal of low intensity in the diffuser: this is the phenomenon of "***leakage***".

To remedy this malfunction, Mr. Martenot suggests, in the workshop notebook and on the power supply diagram (Martenot n.d.b, n.d.c), to negatively polarize the triode plate. To do this, he proposes a modification of the power supply: he cuts the (-) conductor of the power supply from ground and connects it through a 750 Ω resistance, decoupled by a 0.1 μF capacitor, to the ground (this modification can be seen in pencil in Figure 3.22). It thus has a negative DC voltage (≈ -16 V), which it applies to the plate of the triode 2A3 via a 150 kΩ resistance. When the key is released, the triode should be blocked (see Figure 3.23). This modification was not applied to ondes no. 169.

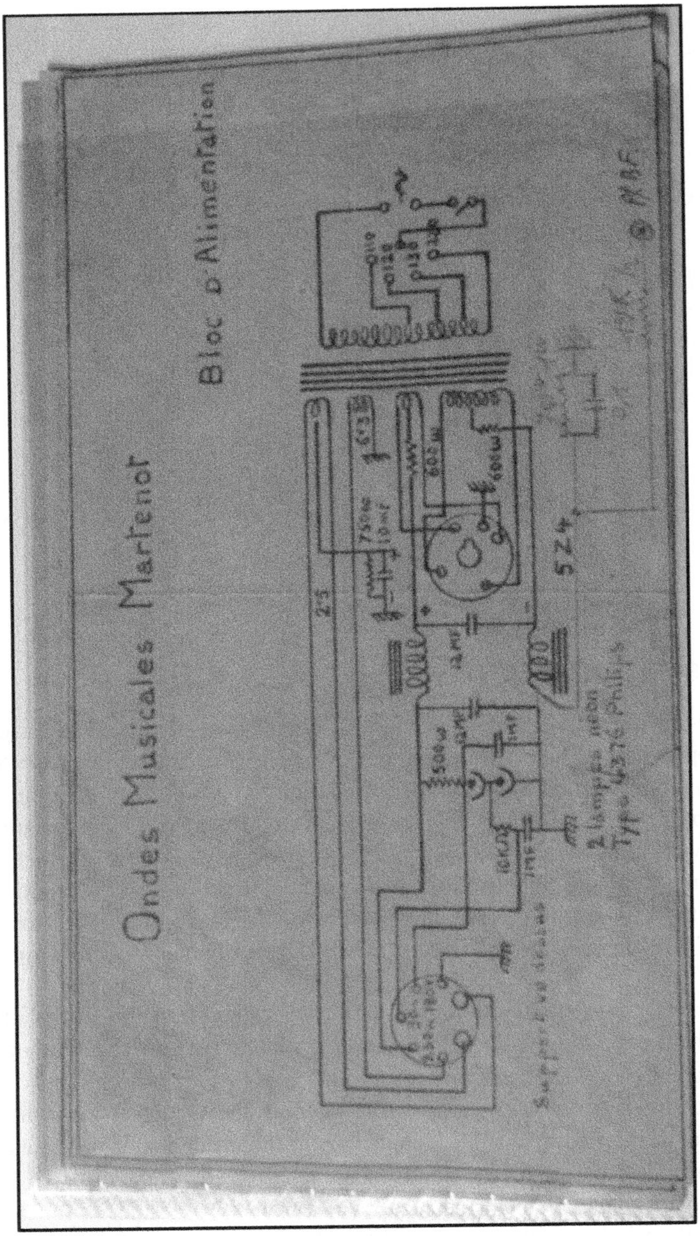

Figure 3.22. *Schematic diagram of the power supply unit with modification (Musée de la Musique, Paris)*

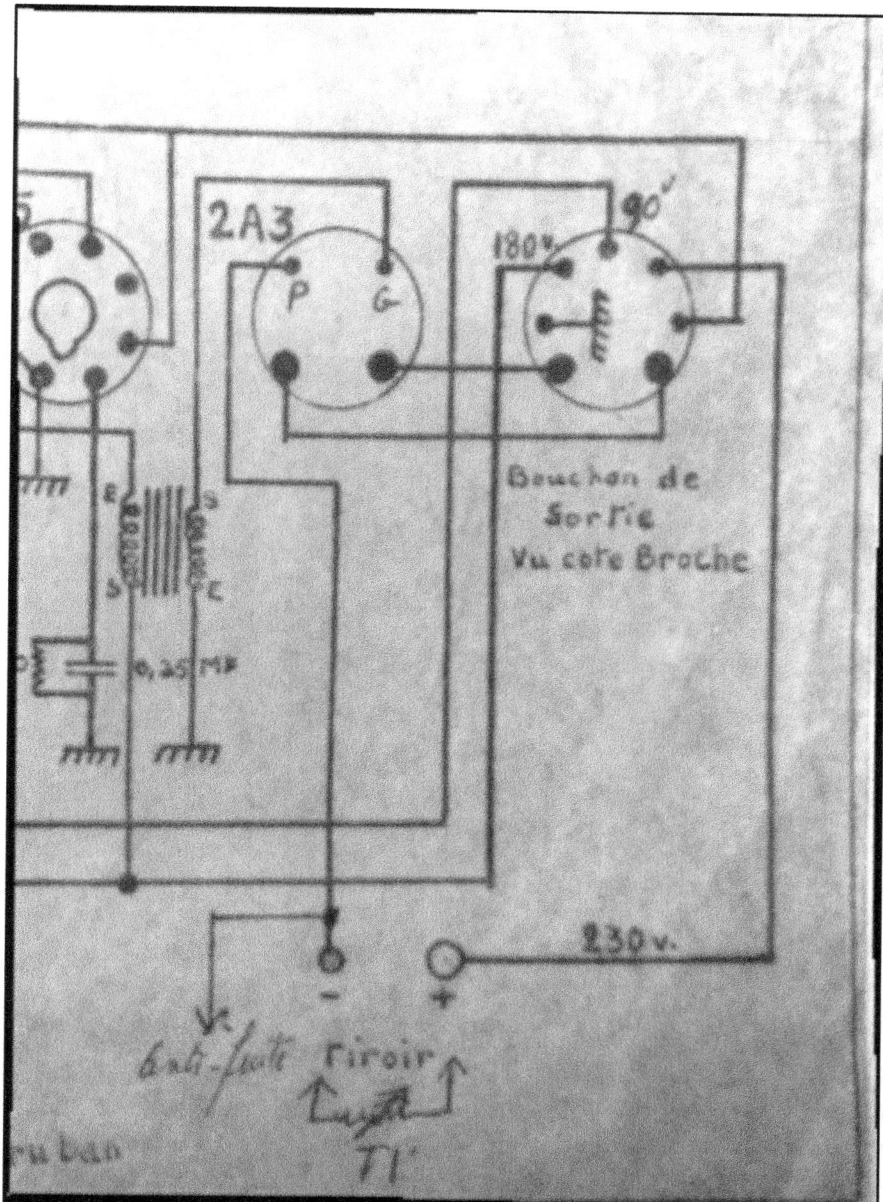

Figure 3.23. *Modification of the diagram for the insertion of a negative polarization of plate (Musée de la Musique, Paris)*

3.4. The drawer

3.4.1. *Description*

Figure 3.24. *The drawer of ondes no. 169 (Musée de la Musique, Paris)*

This device performs several functions at the level of the instrument: it allows the operator to act on the intensity key, to select the "ribbon playing" mode or the "keyboard playing" mode, and also to choose different timbres. In addition, it ensures the electrical connections with the diffuser(s) thanks to a five-pin connector located under the drawer.

We have three drawer diagrams (Martenot n.d.b). One of them is likely to be suitable, but the connections with the chassis are made with three conductors. However, we only observe two on ondes no. 169. There are commonalities; it is therefore necessary to carry out a survey.

3.4.2. *Circuit diagram*

The following photograph shows a view of the inside of the device.

Figure 3.25. *Bottom view of the drawer (Musée de la Musique, Paris)*

NOTE.– The “ribbon play” or “keyboard play” selection lever is located at the top right in the photo. It is a mechanical cable control that transmits the movement of the lever to the switch located on the frame of the heterodyne.

We located and identified the components by assigning connection markers (α, β, γ, λ, ε) and determined the equipotentials.

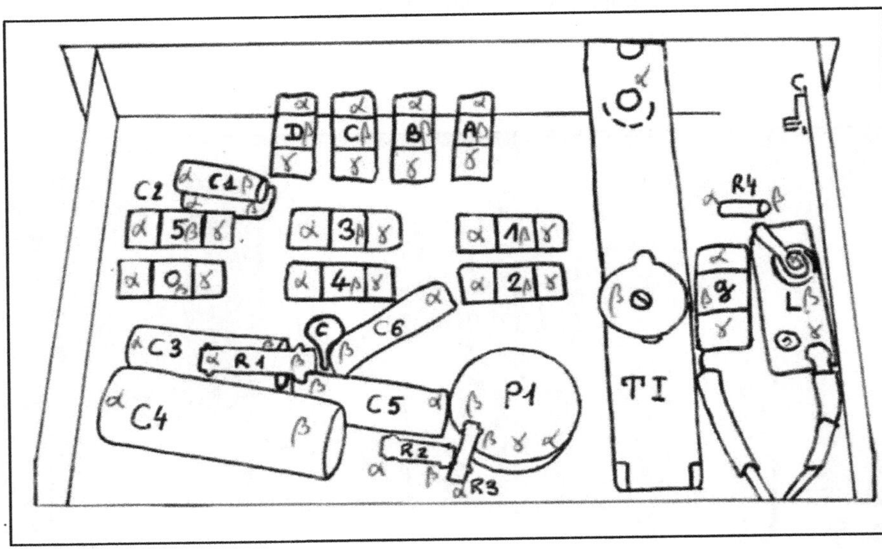

Figure 3.26. *Identification of components*

The following table lists the equipotentials.

By cross-checking with the available documents, the following values can be assigned to capacitors C_1 and C_2: 1 nF.

The diagram in Figure 3.27 shows the pins for the diffuser connector located under the drawer.

In this diagram, the switches are all shown in the same position (which corresponds, on the instrument, to all red keys pressed).

Component	Value		Equipotential	Comment
C1	(?)	2 br	Gamma connector to beta int D	Green wire
C2	(?)	2 br	Beta connector to alpha int C	Red wire
C3	0.04 mF	2 br	Epsilon connector to alpha int B	White wire
C4	0.25 mF	2 br	Lambda connector to alpha int A	Beige wire
C5	0.02 mF	2 br	Alpha connector to alpha int D	Brown wire
C6	0.01 mF	2 br	Alpha int D to beta int C	
Five-pin connector	Brown, red, green, white, beige	5 studs	Alpha int C to beta int B	
Tube support		2 br	Alpha int B to beta int A	
Intensity key		2 br	Beta int D to “C”	
Int A	nf	2 br	Alpha int A to gamma int 5	
Int B	nf	2 br	Int gamma 5 to alpha R2	
Int C	nf	2 br	Beta int 5 to alpha C2	
Int D	nf	2 br	Beta int 5 to alpha C1	
Int 0	nf	2 br	Beta int 5 to alpha int “g”	
Int 1	nf	2 br	Int gamma 5 to beta C2	
Int 2	nf	2 br	Alpha int 4 to beta C1	
Int 3	nf	2 br	Beta int 0 to equip “R”	
Int 4	no	2 br	Int gamma 3 to “R” equip	
Int 5	nf	2 br	Gamma int 1 to equip “R”	
Int “g”	Inverter	3 br	Int gamma 2 to “R” equip	
R1	100 kΩ	2 br	Beta int “g” to equip “R”	
R2	7.5 kΩ	2 br	Beta int 3 to alpha C6	

Component	Value		Equipotential	Comment
R3	13 kΩ (?)	2 br	Beta int 1 to beta C3	
R4	≅ 500 kΩ (?)	2 br	Beta int 2 to alpha C5	
Potentiometer	(P1) 0.050 MΩ	3 br	Beta int 4 to gamma int "g"	Connection "y"
Conductor plate	Yellow color		Beta int 0 to alpha R1	
+V wire	Yellow color		Alpha int 0 to alpha C4	
Common point "C"	Welding	8 bindings	Alpha C3 to "C"	
Equipotential "R"	Brown sheathed conductor		Beta R1 to "C"	
			Beta C5 to "C"	
			Beta C4 to "C"	
			Beta C6 to "C"	
			Gamma P1 to "C"	
			Beta P1 to beta R3	
			Alpha R3 to beta R2	
			Alpha P1 not connected	
			Gamma tube support to "C"	
			Gamma tube support to plate	
			Beta "TI" to gamma int "g"	
			Gamma int "g" to beta tube support	
			Gamma int "g" to +V	
			Alpha "TI" to alpha int "g"	
			Alpha tube support to beta R4	
			Alpha R4 to drawer bottom	

Table 3.2. *Marking of the equipotentials of the drawer*

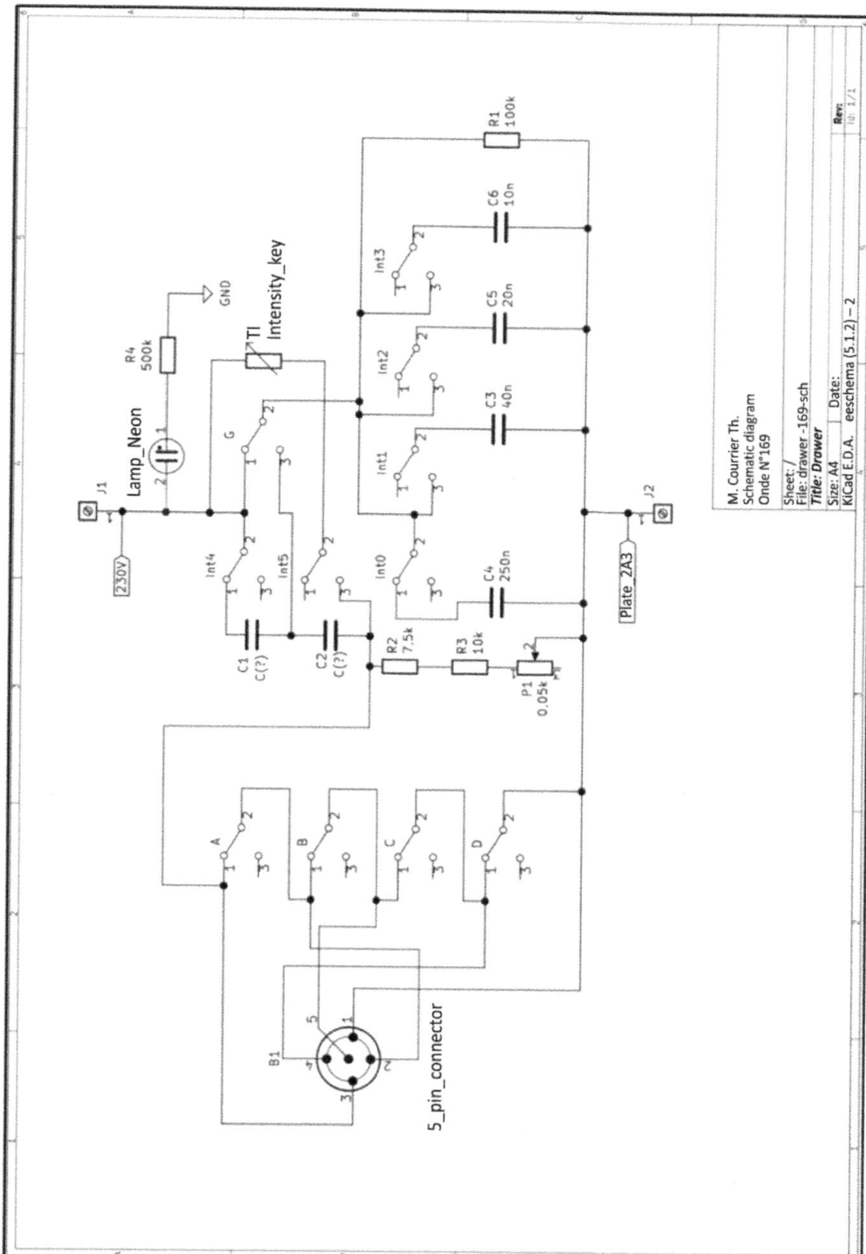

Figure 3.27. *Diagram of the drawer of ondes no. 169*

3.4.3. *Functional study*

It is observed that the selection of timbres is obtained by the commissioning of capacitors (C_3, C_4, C_5 and C_6) that limit the bandwidth of the system, by a high cut-off frequency. This is the choice made by Mr. Martenot, a choice that he confirms in a lecture given at Gravesano (Martenot 1954), during which he specifies that "***the dosage of harmonics by frequency bands seemed to [him] quite sufficient***".

All the switches do not have a clearly identified role, but it is probably a prototype that has undergone successive modifications. The diagram noted is a variant. We noted in the workshop documentation of Mr. Martenot three distinct diagrams associated with this drawer.

It should be noted that the connections of capacitor C_1 on our diagram are not identical to those proposed by Mr. Martenot for the capacitor marked a (diagram with the mention: ***corrected on 9 September 49***). However, two of the three diagrams in file no. 18 (Martenot n.d.b), which also contain errors, have the same connections for C_1. Moreover, taking into account the incompatibilities existing between these three diagrams and, on the contrary, by noting that the drawer of ondes no. 169 has been modified, nothing allows us to conclude that the wiring of the drawer itself is correct. It is therefore necessary to continue research in order to propose the right diagram.

However, switch "G" applies the harmonic filters either upstream of the intensity key or downstream of the intensity key. This is the "petit gambé" command.

Effects are achieved by toggle, soft-switch and high-key switches. The skill of the operator allows them to be chosen temporarily or permanently (except for C_2). They can, of course, be combined and thus offer the performer a much diversified array of sound resources.

Resistance R_1 provides a discharge circuit for the harmonic filter capacitors and reduces switching noise.

The components in series R_2, R_3 and P_1, as they stand, do not have a well-defined role:

– A comparison with one of the diagrams in the file (Martenot n.d.b), ***the drawer***, shows us that the three components are operational if a third wire is present.

– Note that the sum of the values of the resistances R_2, R_3 (10 kΩ, 7.5 kΩ) is close to the 20 kΩ value of the resistance which appears in the original diagram by Mr. Martenot.

– We observed the presence of an intermediate electrical connection device between the drawer and the chassis, which validates the existence of this third conductor. This device with three 4 mm diameter sockets can be used to operate the return diffuser. With three conductors, according to the following diagram, the diffuser is operational, and the sound level can be adjusted through it by P_1.

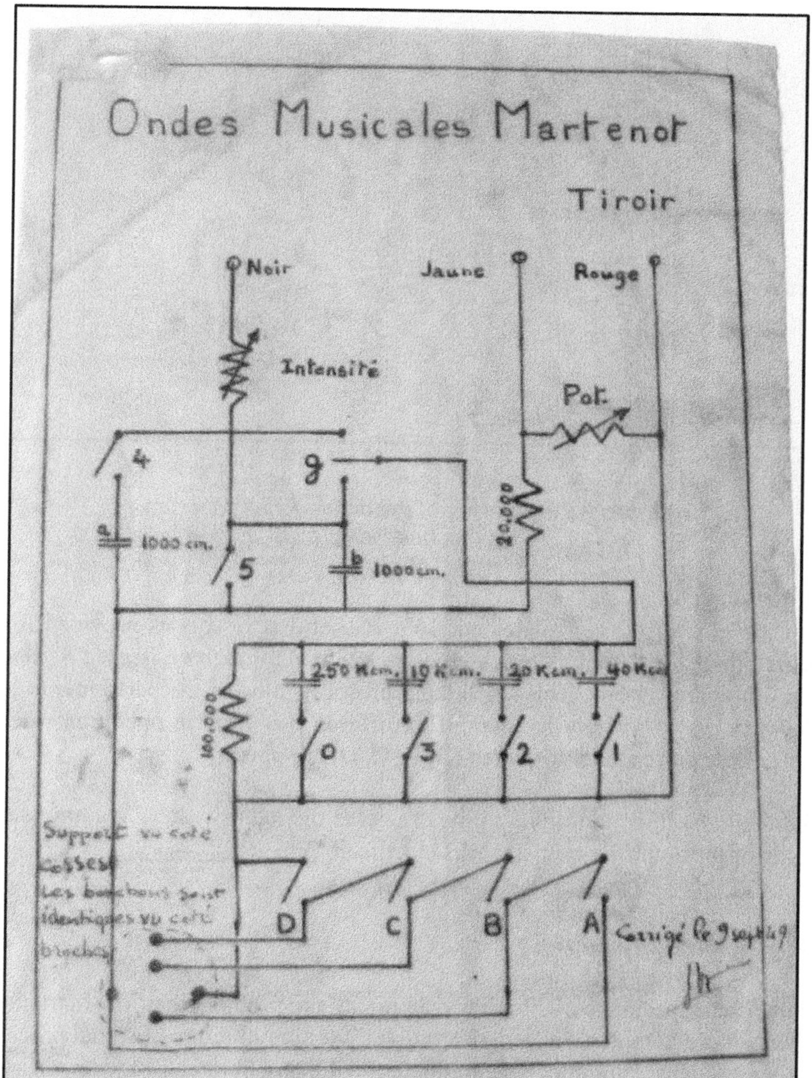

Figure 3.28. *Diagram justifying the role of the potentiometer (P_1) of ondes no. 169 (Musée de la Musique, Paris)*

In this hypothesis, the red conductor was connected to the plate of the triode 2A3, the black conductor was connected to +230 V, and the return diffuser was connected between the yellow conductor and the red conductor, in parallel with the potentiometer.

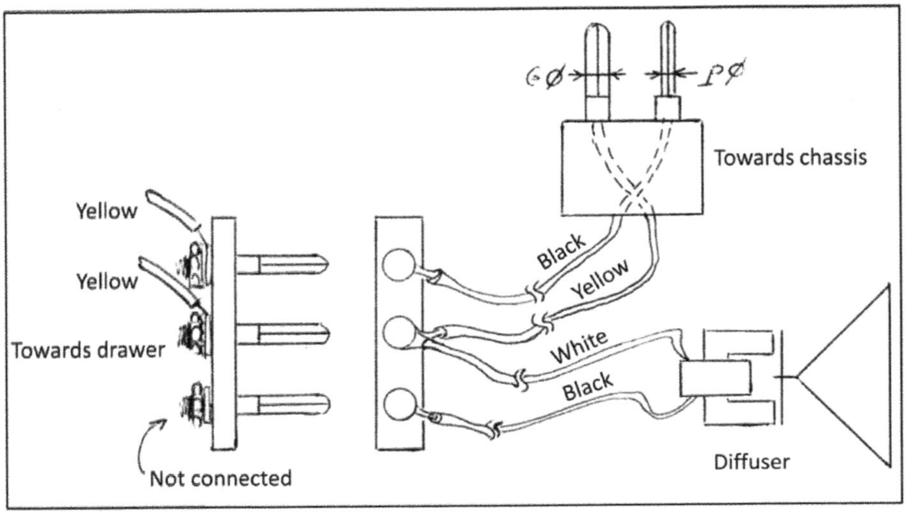

Figure 3.29. *Drawing of the drawer/chassis/return diffuser connections, as is*

But let us return to our ondes no. 169. A useful accessory is provided for this model: a small neon lamp V_1, powered by +230 V through resistance R_4 and the ground of the drawer box. This small lamp probably allowed the performer to locate the drawer keys in a dimly lit room. The diffuser(s) is (are) in operation when the contacts of the corresponding switch (A, B, C, D) are open.

But the most important thing in the drawer is the intensity key. We have seen in the study of the amplifier how it works electrically.

3.4.4. *The intensity key*

Under the appearance of simplicity, in reality the problem that Mr. Martenot has to solve is more complex. Indeed, the "normal" setting in terms of signal processing is to use a so-called "volume" potentiometer on the input circuit of the amplifier. For systems using electronic tubes, the usual value of this component is between 100 kΩ and 1 MΩ. When the cursor is placed towards the cold point, the amplitude of the collected signal is low or null. When the cursor is placed in such a way that the resistance, which opposes the path of the signal, is minimum, the signal has its maximum amplitude. In the 1920s to 1930s, potentiometers were made from a winding of resistive alloy wires, of small diameter on a torus-shaped support in steatite, mica, or even in "Presspahn". A brush in soft alloy (bronze) or in hard graphite, in contact with this winding ensured the sampling of the signal. The movement was therefore circular and the resistance values low, because to obtain high resistance values, a very small wire section had to be used and breaks became very frequent. These devices were called rheostats.

Mr. Martenot only had low value rheostats with circular motion, high inertia and high friction, which were not very compatible with the constraints of use specific to his instrument. Later, rheostats with rectilinear movement or potentiometers with carbon tracks were developed, which had the same disadvantages in terms of fragility and inertia.

To ensure low inertia operation and low movement wear, Mr. Martenot suggests using a variable resistance made from a mixture of graphite and mica powders contained in a leather "pouch".

The realization is described in its patent no. 703.923 (Martenot 1931). If the principle of operation is simple, its manufacture and its adjustment are much more delicate. Indeed, the values of the resistance result at the same time from:

– initial proportions of the components of the mixture;

– the compression of the mixture (increase in surfaces and contact pressures of the graphite grains);

– the variation of the cross-section of the shape of the resistance (straight cylinder in rest position, sphere crushed in maximum pressure);

– the reduction of the distance between the electrodes.

To ensure the reversibility of movements, the elasticity of the leather that constitutes the envelope is decisive. The first resistance we observed is that of ondes no. 15. To obtain the performance necessary for the proper functioning of the instrument, Mr. Martenot manufactured a resistance of small volume, with very fine leather and a long key lever. Thus, the low amplitude of the leather bag is compensated by the amplifying effect of the key lever. A return spring ensures the return of the resistance to its initial state. But the resistance variation amplitude is low, around 5 kΩ to 10 kΩ. Mr. Martenot obtained a low inertia manual control, practically without friction and allowing all the ways of "striking" the key without risking the destruction of the resistance.

In the rest position, the resistance is maximal or in open circuit, Mr. Martenot places this "variable resistance" in series with the diffuser, in the plate circuit of the amplifier tube.

Figure 3.30. *View of the intensity key, ondes no. 169 (Musée de la Musique, Paris)*

We found in file no. 21 (Martenot n.d.c), ***Resistance***, notes relating to the dosages of graphite and mica for model 37.

In this same workshop notebook (Martenot n.d.c), ***§Miscellaneous remarks***, Mr. Martenot points out the problems of noise generated by a badly prepared key and the aging of the key by agglomeration of powders on the internal contacts. For more details on the bag, the dosage of the powder, etc., we refer the reader to Chapter 7.

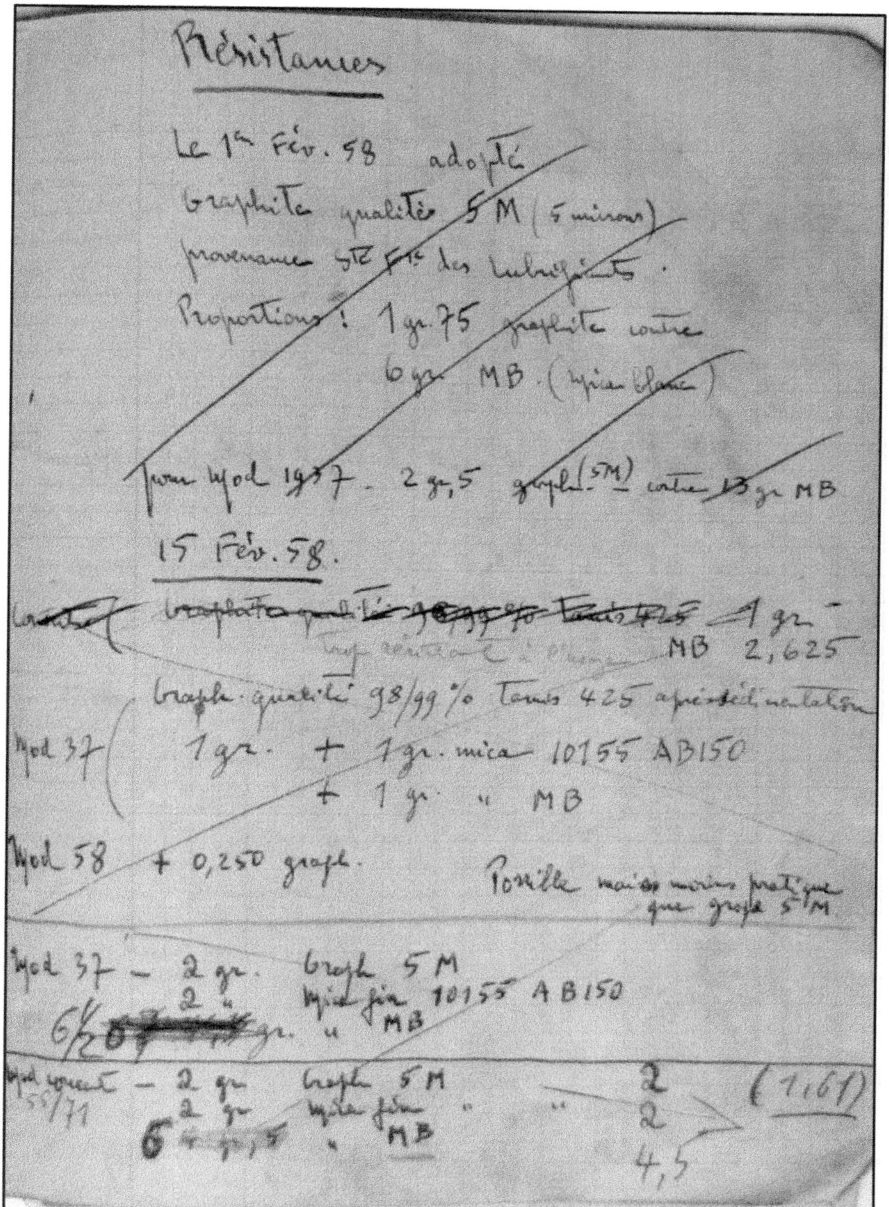

Résistances

Le 1er Fév. 58 adopté
Graphite qualité 5M (5 microns)
provenance Sté Fse des lubrifiants.
Proportions : 1 gr. 75 graphite contre
6 gr. MB. (Mica blanc)

pour Mod 1937. 2 gr,5 graph (5M) contre 13 gr MB

15 Fév. 58.

1 gr.
MB 2,625

Graph. qualité 98/99 % Tamis 425 après décantation
Mod 37 1 gr. + 1 gr. mica 10155 AB150
+ 1 gr. " MB
Mod 58 + 0,250 graph.
Possible mais moins pratique que graph 5M.

Mod 37 – 2 gr. Graph 5M
2 " Mica fin 10155 AB150
6/20 gr. " MB

Mod courant 55/71 – 2 gr Graph 5M 2
2 gr Mica fin " " 2
6 " MB 4,5 (1,61)

Figure 3.31. *The different graphite/mica proportions to achieve the intensity key (Musée de la Musique, Paris)*

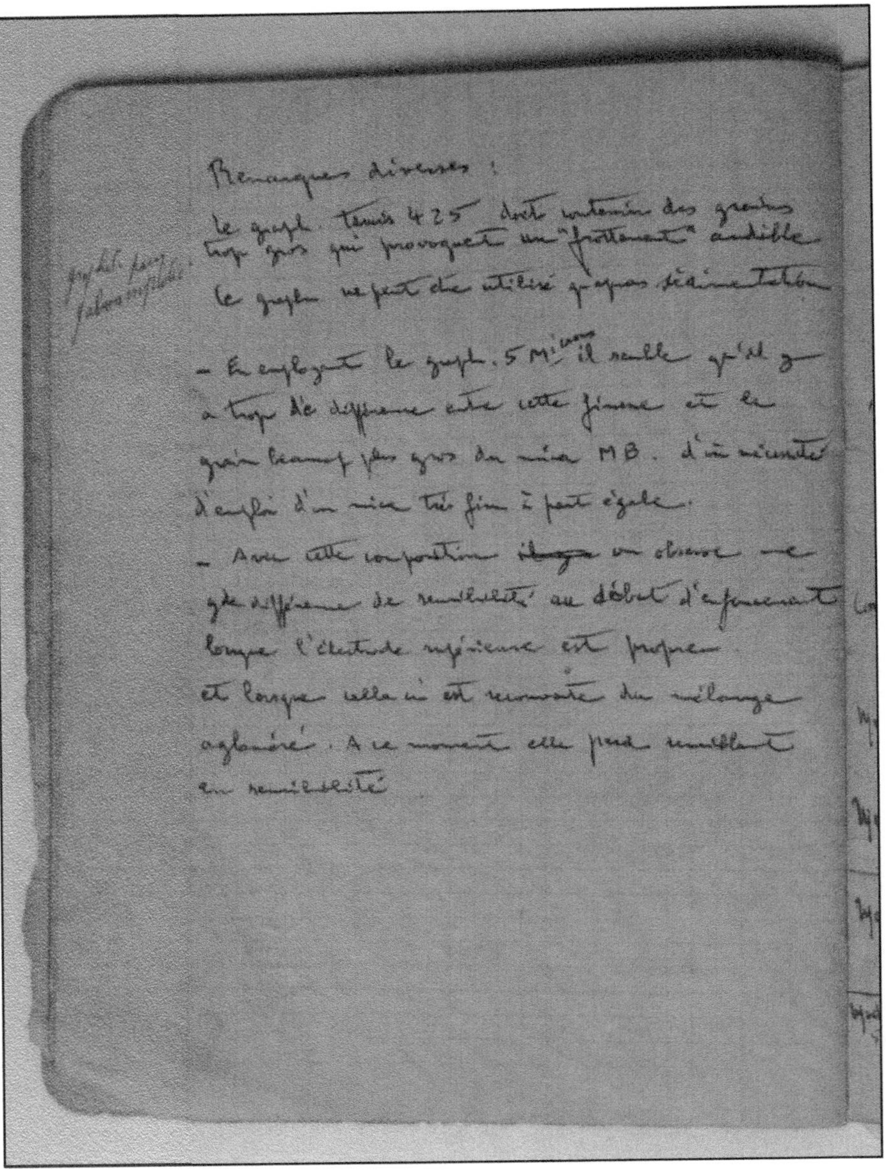

Figure 3.32. *Optimization of the granulometry and the dosage of the powders (Musée de la Musique, Paris)*

3.5. The diffuser

Figure 3.33. *Diffuser No. 982-9-2 (Musée de la Musique, Paris)*

The diffusers are the last constituent elements of the “ondes Martenot” instrument. They too have undergone numerous improvements, linked to the evolution of technologies of sound reproduction devices, and also to the very creativity of Mr. Martenot.

Each type of instrument is associated with one or more types of diffusers.

We observed during a private sale in February 2021 of an ondes no. 159, model 37 that the diffuser consisted of a wooden box supporting in its lower part a cardboard cone diffuser, and in its upper part a small speaker.

We are therefore led to ask the question of whether the 37 model, and more precisely, the concert model, was associated with a single diffuser. In this case, the drawer did not look the same as the one installed in no. 169. Remember that Mr. Martenot presented the Palme in 1950 (Loriod 1987), and that the Gong had been developed a few years earlier (Martenot 1947). The drawer of this ondes was probably modified from this date.

Figure 3.34. *Diffuser for ondes model '37 (private collection)*

The drawer of the instrument that we studied has a five-pin connector allowing the control of four diffusers with the corresponding switches. Implementing multiple diffusers divides the available power, and impedance matching is required to achieve an even distribution of the available sound power. For this model, Mr. Martenot chose serialization of the diffusers, which means that they all have the same current flowing through them.

We also saw a diagram of an amplified Palme in file no. 18 (Martenot n.d.b). This shows us that Mr. Martenot carried out numerous experiments to determine the mode of operation of the diffusers, which were ultimately all passive.

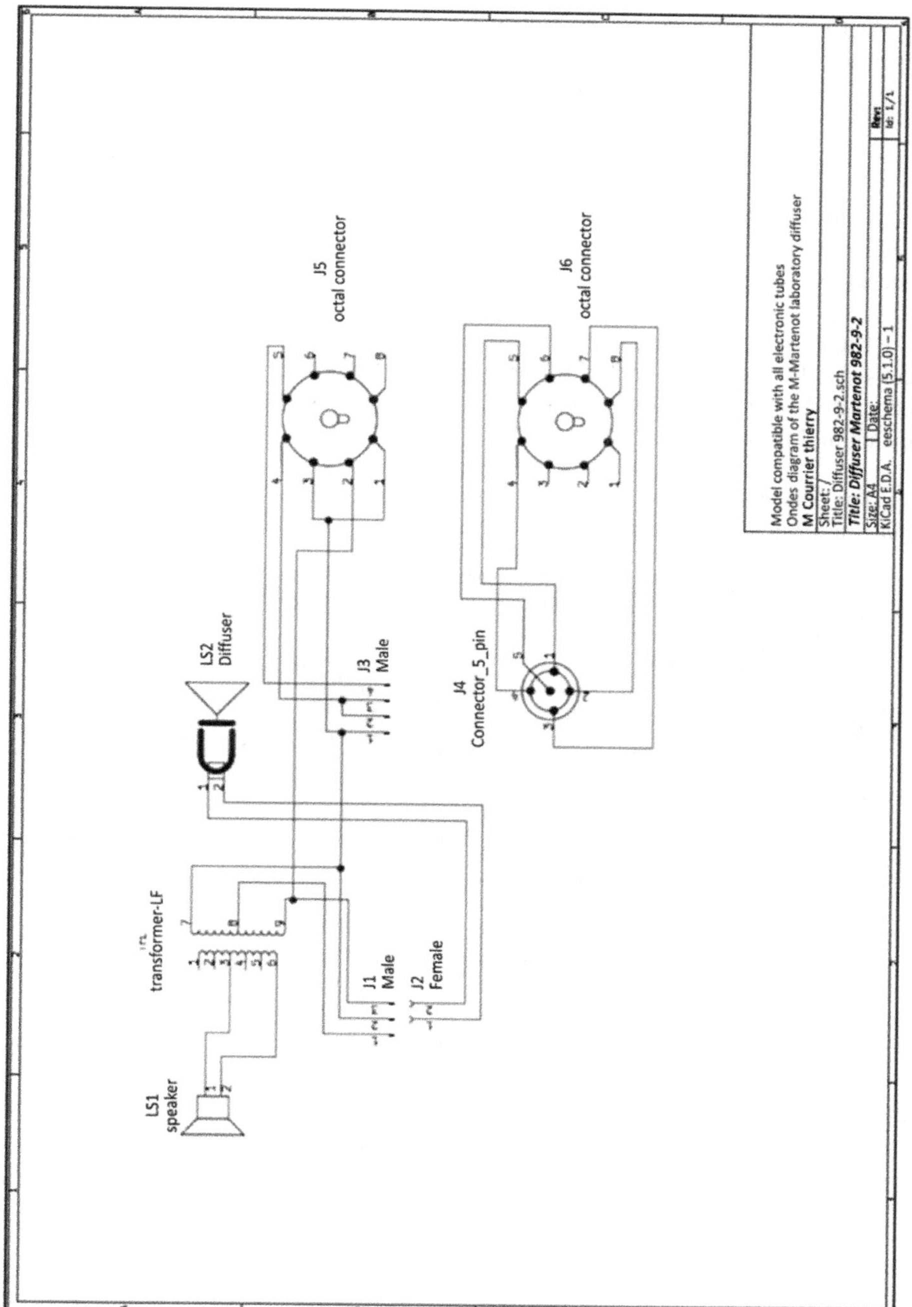

Figure 3.35. *Diffuser wiring diagram 982-9-2*

During the analysis of the operation of the amplifier, we determined two operating conditions:

– the optimal load impedance of the diffuser(s) is between 1 kΩ and 2.5 kΩ;

– the power delivered is of the order of 1 W.

With dissipative diffusers and a relatively moderate available power, it will be necessary to produce large surface diffusers. Also, if we use a loudspeaker, it will be necessary to use an impedance matching device.

These conditions are met in diffuser No. 982-9-2 (see Figure 3.33). Indeed, it consists of a cardboard cone diffuser with a large diameter Ø = 430 mm, which is connected to a permanent magnet loudspeaker Ø = 200 mm and an impedance matching LF transformer.

The electrical diagram shows many possibilities in terms of impedance matching, but the connection itself is not simple.

Figures 3.36 and 3.37. *Rear face and connectors (Musée de la Musique, Paris)*

This diffuser was probably modified after the '37 model, or has undergone modifications in order to adapt to all the new instruments that Mr. Martenot will build thereafter. It has an "Octal" type connector, while the drawer has a five-pin connector. A five-pin/Octal adapter plug was found in its place inside this diffuser. However, the "Octal" connector is the one that Mr. Martenot would use later on in ondes model '53.

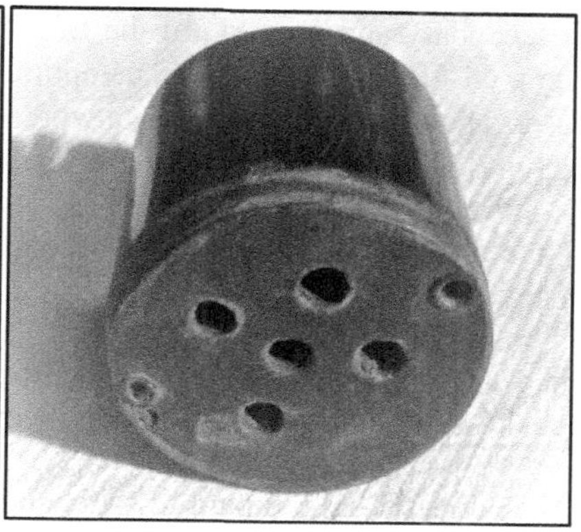

Figures 3.38 and 3.39. *Five-pin/octal adapter plug (Musée de la Musique, Paris)*

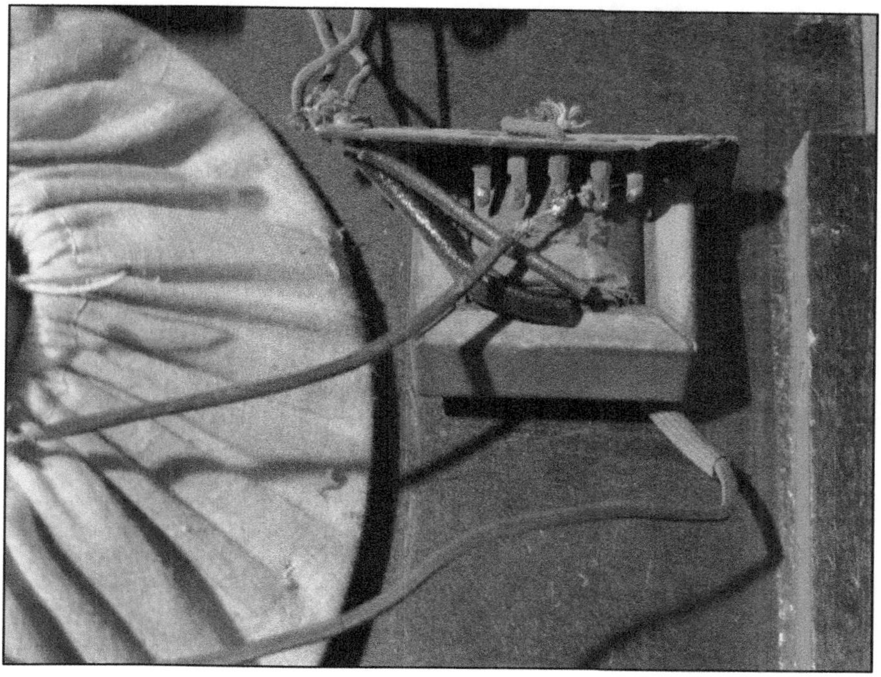

Figure 3.40. *The impedance matching transformer (Musée de la Musique, Paris)*

As it is, the cardboard cone diffuser is electrically in parallel with the primary of the LF transformer of the loudspeaker, and the multiplicity of connection terminals, both secondary and primary, of the LF transformer allows the necessary adaptation of the total load impedance versus optimum impedance. This set of two associated diffusers shows that it is probably a workshop device; it allows the connection of the three types of ondes models with electronic tubes.

Note also, in Figure 3.40, the secondary impedance matching (HP side) made with a piece of insulated wire.

It was not possible to obtain information relating to the values of these impedances, with no numerical mention appearing here or there on the device.

This diffuser is compatible with ondes model '37, its particularities residing above all in the very large diameter of the membrane of the diffuser and in the large size of its motor.

Figure 3.41. *Diffuser membrane (Musée de la Musique, Paris)*

Finally, to complete this study relating to the diffuser, we recall the presence in the very box of the instrument of a small return diffuser for the performer. This diffuser is not electrically connected to the drawer (a connection is missing).

Figure 3.42. *Diffuser motor (Musée de la Musique, Paris)*

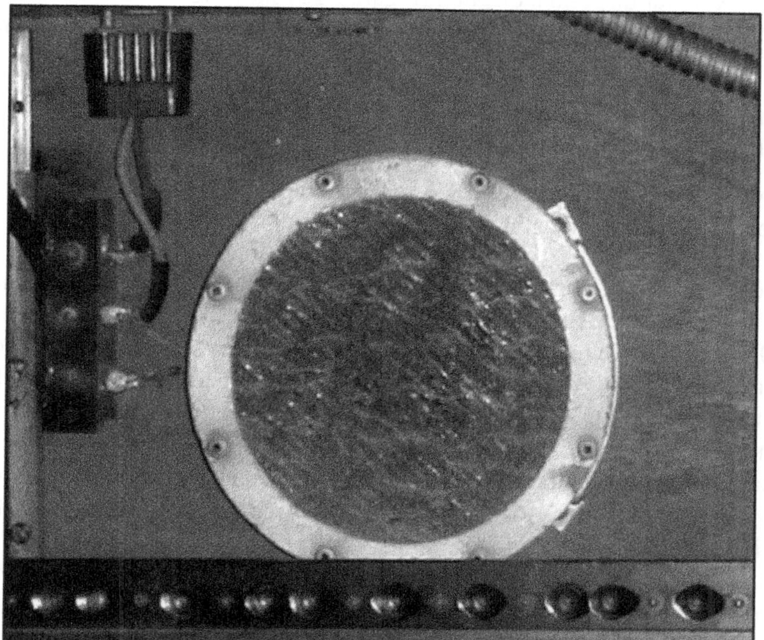

Figure 3.43. *The return diffuser (Musée de la Musique, Paris)*

3.6. The power supply module

The power supply module, essential to the functioning of the instrument, must meet design criteria which, if they are not purely musical, can interfere with the musicality of the instrument: power variation during play, humming noises, various interfering signals, etc.

Figure 3.44. *The power supply (Musée de la Musique, Paris)*

We have at our disposal a diagram (Figure 3.22) showing the modification of the plate polarization for the anti-leakage. We therefore compared this diagram with our survey, and we found two differences:

– the modification for operation with the "anti-leak" circuit has not been made on this model;

– the bias resistance of tube 2A3 has a value of 1,350 Ω and not the value indicated on the original diagram 750 Ω.

In terms of design and construction, this scheme is classic and consistent with the technologies of the time. After a network transformer, which probably delivers 2 x 300 V, a full wave rectification is carried out using the 5Z4 tube, via two 600 Ω power resistances. This tube, indirectly heated, was probably chosen for its height h = 81 mm instead of the 5Y3G (h = 117 mm), which was much more common and therefore less expensive. These two straightening tubes (Octal/1935 series) are electrically interchangeable. The presence of a recess under the enclosure cover and the lowering of the support observable on the power supply chassis show that an adaptation was necessary (see Figure 3.46).

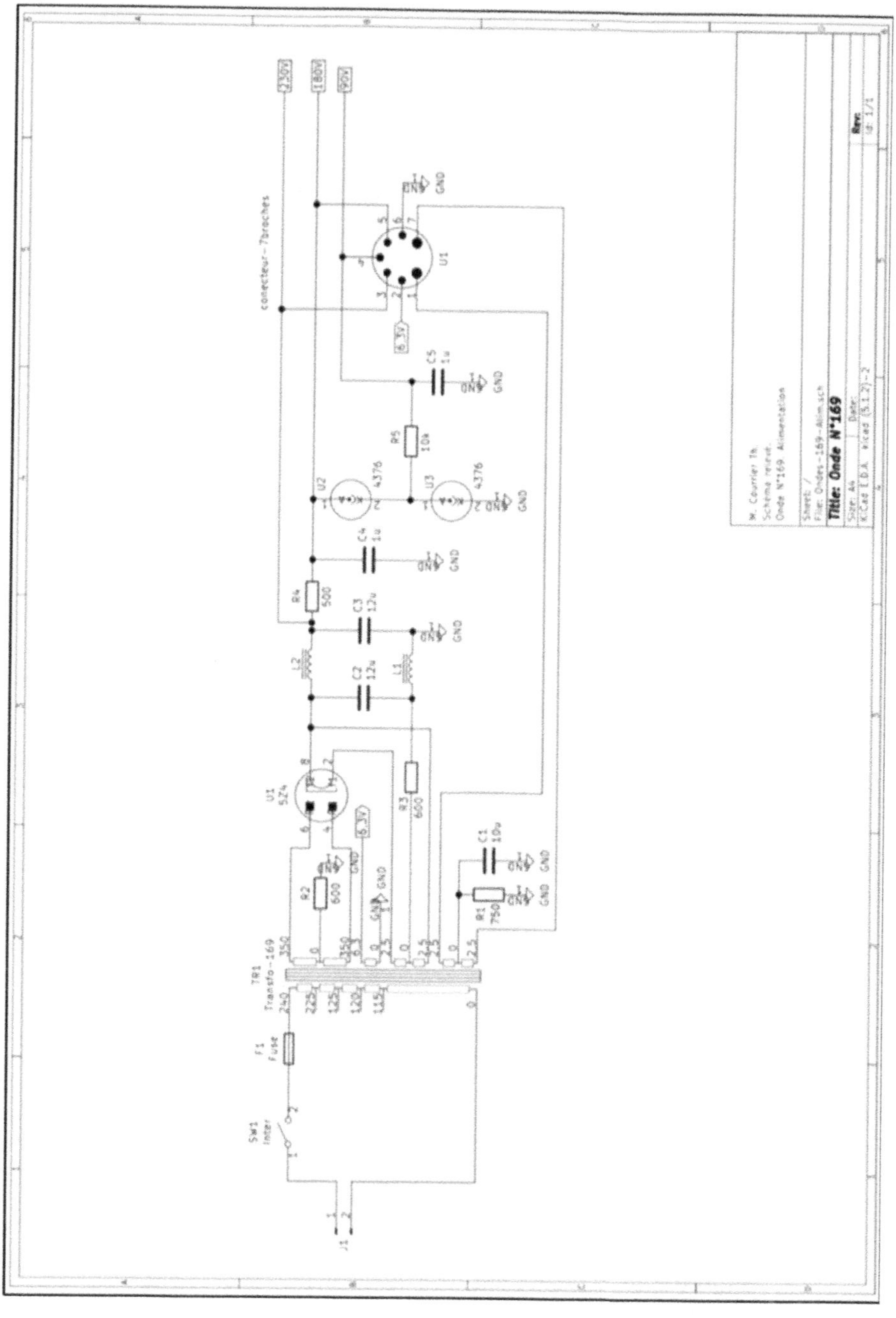

Figure 3.45. *Diagram of the power supply of ondes no. 169*

Figure 3.46. *A recess has been made in the cover of the cabinet. For a 5Y3G? (Musée de la Musique, Paris)*

A symmetrical filter made up of two coils (from 5 H to 10 H on the same magnetic core), preceded and followed by 12 μF capacitors, provides effective filtering. We then obtain the high voltage necessary for the operation of the triode 2A3 that is of the order of 230 V under load.

To obtain the two other voltages 90 V and 180 V necessary for the operation of the heterodynes and the pre-amplifier, stabilization is obtained by placing in series two neon gas regulator tubes of the type 4376 (90 V) with a screw base. This tube, manufactured by Philips from 1936, was very quickly replaced by the manufacturer by models with European four-pin or Octal bases, and of smaller size. Capacitors placed at the output of each of these sources ensure operation as a voltage generator.

The 6.3 V and 2.5 V heating voltages are taken directly from the supply transformer. This is connected to the network by a two-pin socket via an on/off switch, accessible from the underside of the instrument cabinet.

To complete this description, recall that the elements R_k = 1,370 Ω and C_k = 10 μF, of the bias circuit of the triode 2A3 are mounted on the frame of this power supply.

A cord + seven-pin connector device allows the connection of the heterodyne block + pre-amplifier + amplifier to the sources.

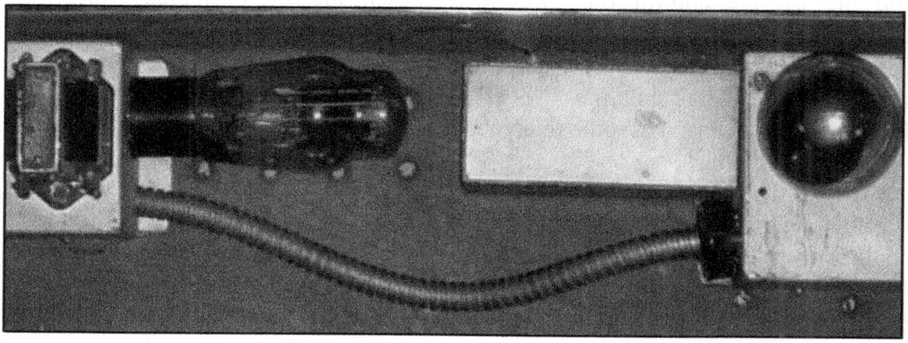

Figure 3.47. *Connecting cord, source/amplifier (Musée de la Musique, Paris)*

3.7. Model

3.7.1. *Presentation*

Ondes no. 169 and ondes no. 15 are built on the same diagram, only the components change generation. Battery type triodes are replaced by indirectly heated triodes. Using the 2A3 triode significantly increases output power. Some differences remain at the levels of the controls of timbres, which are more numerous on the model no. 169 and, of course, the addition of a mobile keyboard. We have produced a model by adapting to the latest technical constraints imposed by the triodes currently available. Only playing with the ribbon will be exploited in this model that we will call: model 15-169.

3.7.2. *Realization*

3.7.2.1. *The main module*

It brings together the heterodynes, the mixer and the LF amplification chain on the same chassis. The diagram is given in Figure 3.48.

For reasons of feasibility, we have adopted Hartley-type oscillators for heterodynes; in this way, it is easy to make the HF windings from commercial inductors. We kept the physical separation of the triodes by excluding the ECC81/82/83 types. For the oscillators, we have adopted the EC92 model whose electrical characteristics are close to those of the ECC81 triode, and for the detector and the LF pre-amplification, we have chosen the EC90 model, which has characteristics identical to those of the ECC82. These are “miniature” series triodes with seven contact pins. There is no power triode in this series, so we have retained the 6S4 model in the nine-pin “Noval” series. The 1/3 ratio LF transformers are from the manufacturer AES, the reference is P-T 156.

The two HF transformers are identical, the value of the self-induction coefficient is 5 mH, and the connection terminal with the cathode is located at 1/3 of the total number of turns from the ground terminal. We used the ferrite core of a commercial 4.7 mH (10 mm * 14 mm) inductance. Indeed, the ferrites manufactured today are much more stable than those manufactured in the 1950s. We therefore rewound it to obtain the value of 5 mH and to take out the plug at 1/3. We have also added the secondary winding to a piece of PVC tube. The operating frequency is 100 kHz.

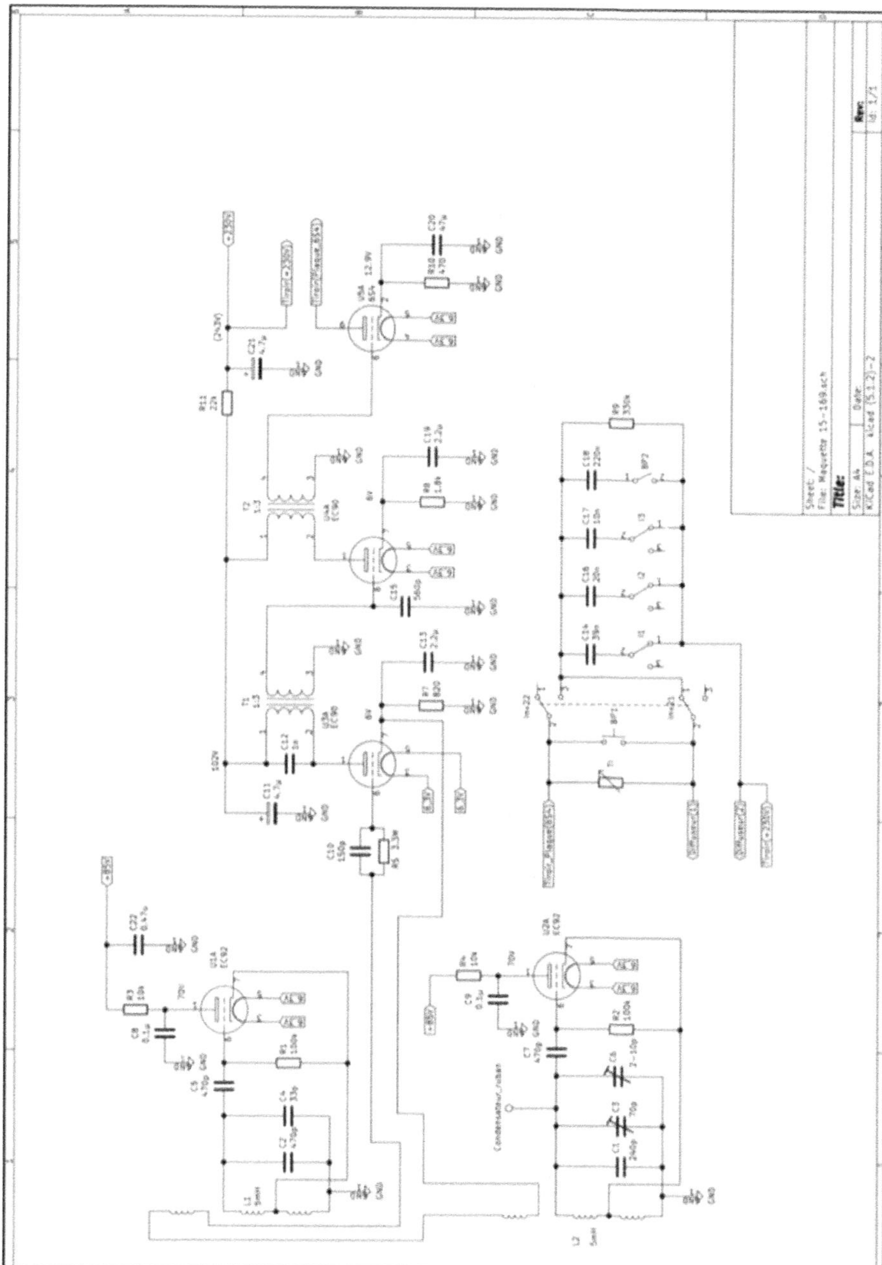

Figure 3.48. *Diagram of model 15-169*

Figure 3.49. *Winding construction*

Figure 3.50. *Winding and its shielding*

3.7.2.2. *The power supply module*

Of classic construction, this power supply built with a small 2 x 205 V / 30 VA transformer provides all the voltages necessary for the operation of the model. The diagram is given in the following.

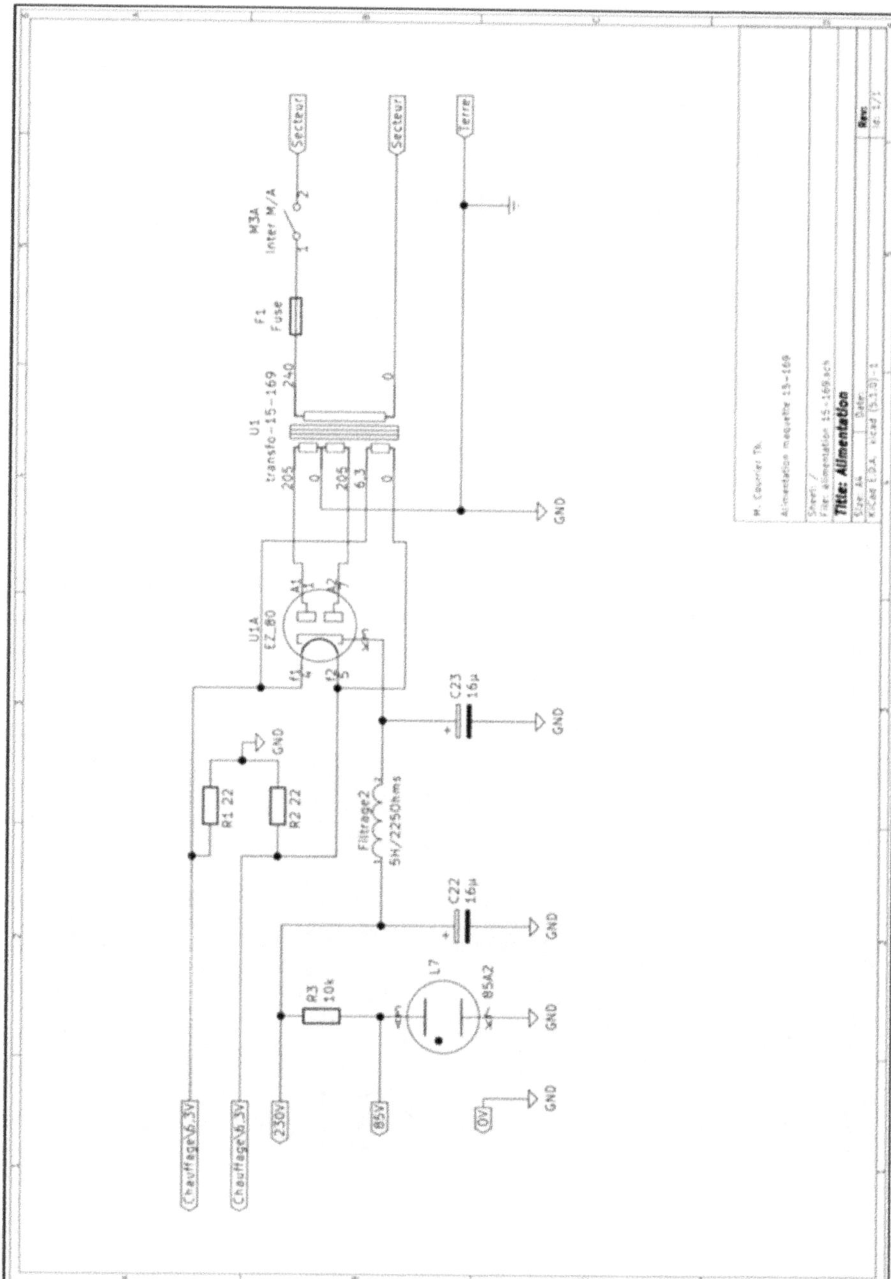

Figure 3.51. *Power supply diagram for model 15-169*

Figure 3.52. *Model 15-169: both chassis*

3.7.2.3. *The drawer*

We made a model from the drawer of ondes no. 15. The diagram is given in section 3.4.2. The intensity key is built according to the ondes model no. 208; all the information relating to its manufacture is given in Chapter 7. The diffuser is an 8 Ω loudspeaker associated with impedance adapter transformer 2,000 Ω/8 Ω/4 W.

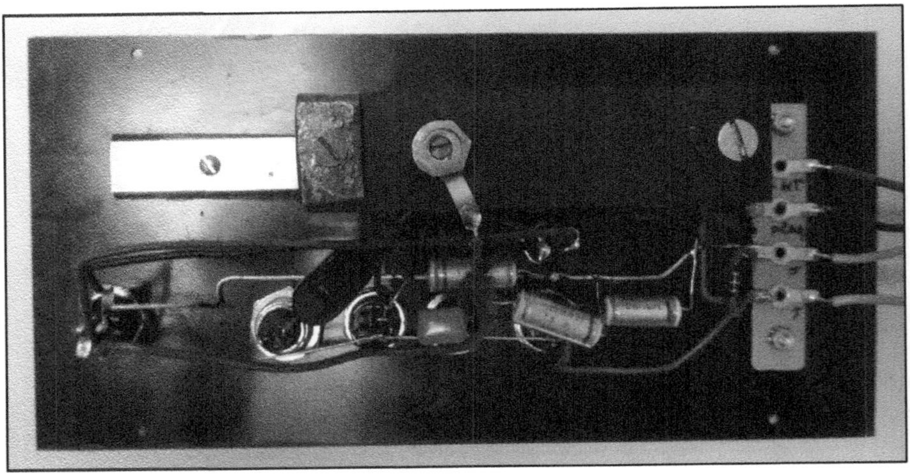

Figure 3.53. *The drawer, view of the wiring*

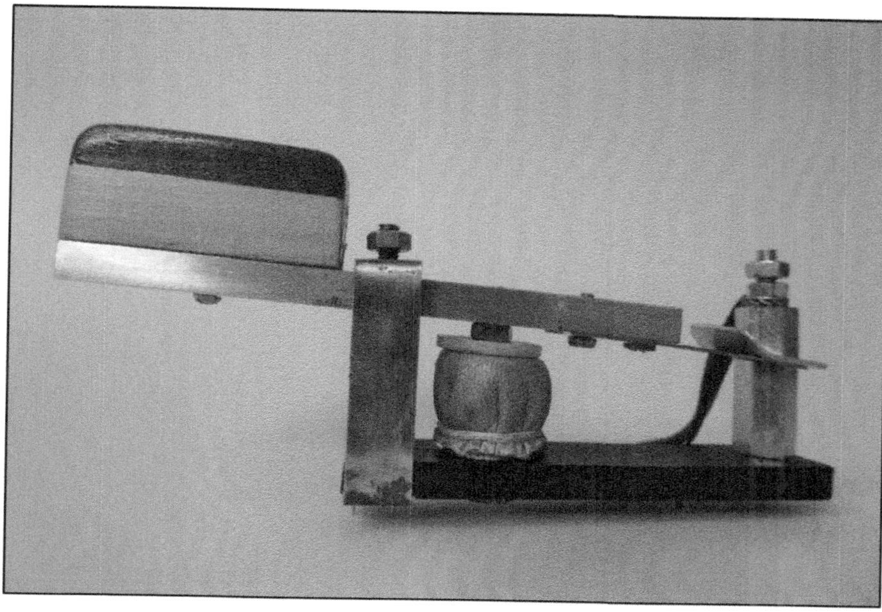

Figure 3.54. *The intensity key*

3.7.2.4. *The ribbon capacitor*

It is built from the ribbon capacitor from model no. 169: 10 pairs of aluminum comb 1 mm thick, 92 mm long and 40 mm high, the blades are 4 mm wide except for the high register where the walls are whole but split. The blades are mounted on an insulating support which defines the spacing between the two walls of the same comb. From bass to treble the spacings in mm are: 8, 8, 7, 7, 6, 6, 5, 5, 4 and 4. The whole is arranged in a 2-mm thick U-shaped aluminum shield 47 mm wide and 52 mm high with a cover.

The ribbon is made from the gold thread braid used in trimmings; its height is 8 mm. The useful displacement length of the index finger on the dummy keyboard is 0.98 m. The resonant correction circuit for the high register consists of an adjustable inductor (10 mH – 30 mH) in an 8 mm * 8 mm box from the manufacturer Toko, and a fixed 15 pF mica capacitor. Two types of rings were made, the first conforming to the original of ondes no. 15 in transparent plastic material ("Rhodoïd"), and the second identical to that used for ondes no. 208.

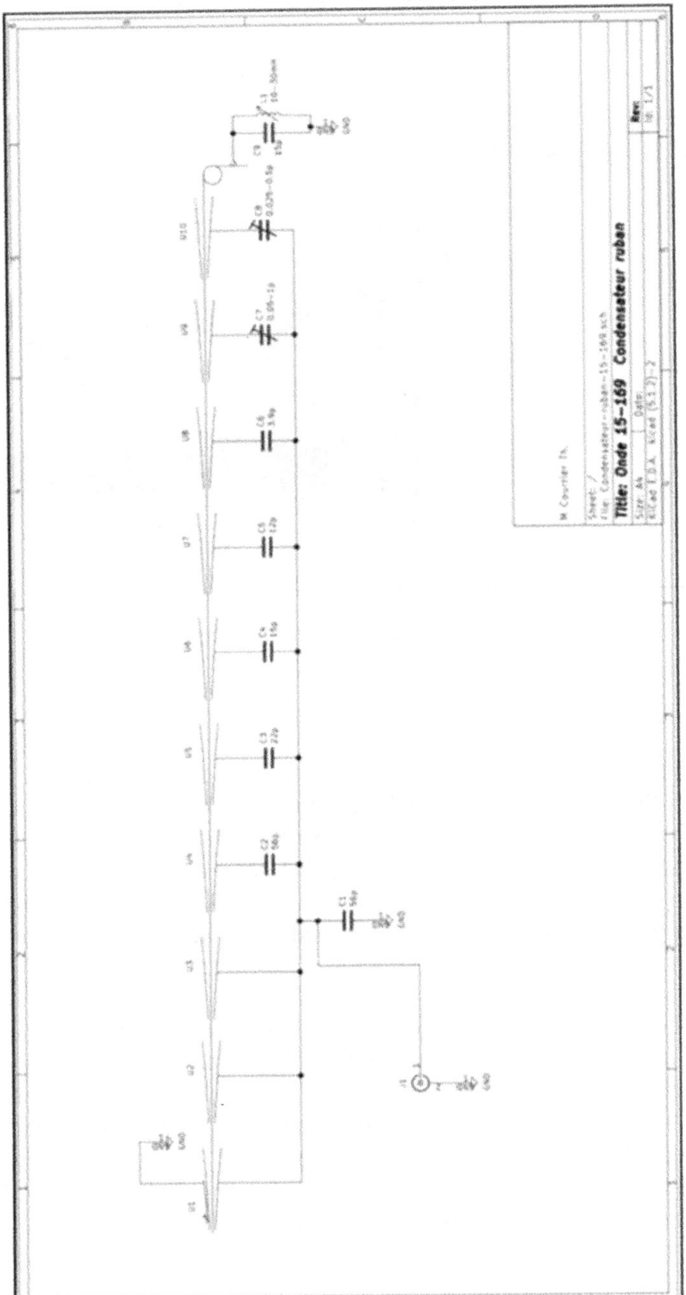

Figure 3.55. *Diagram of the ribbon capacitor*

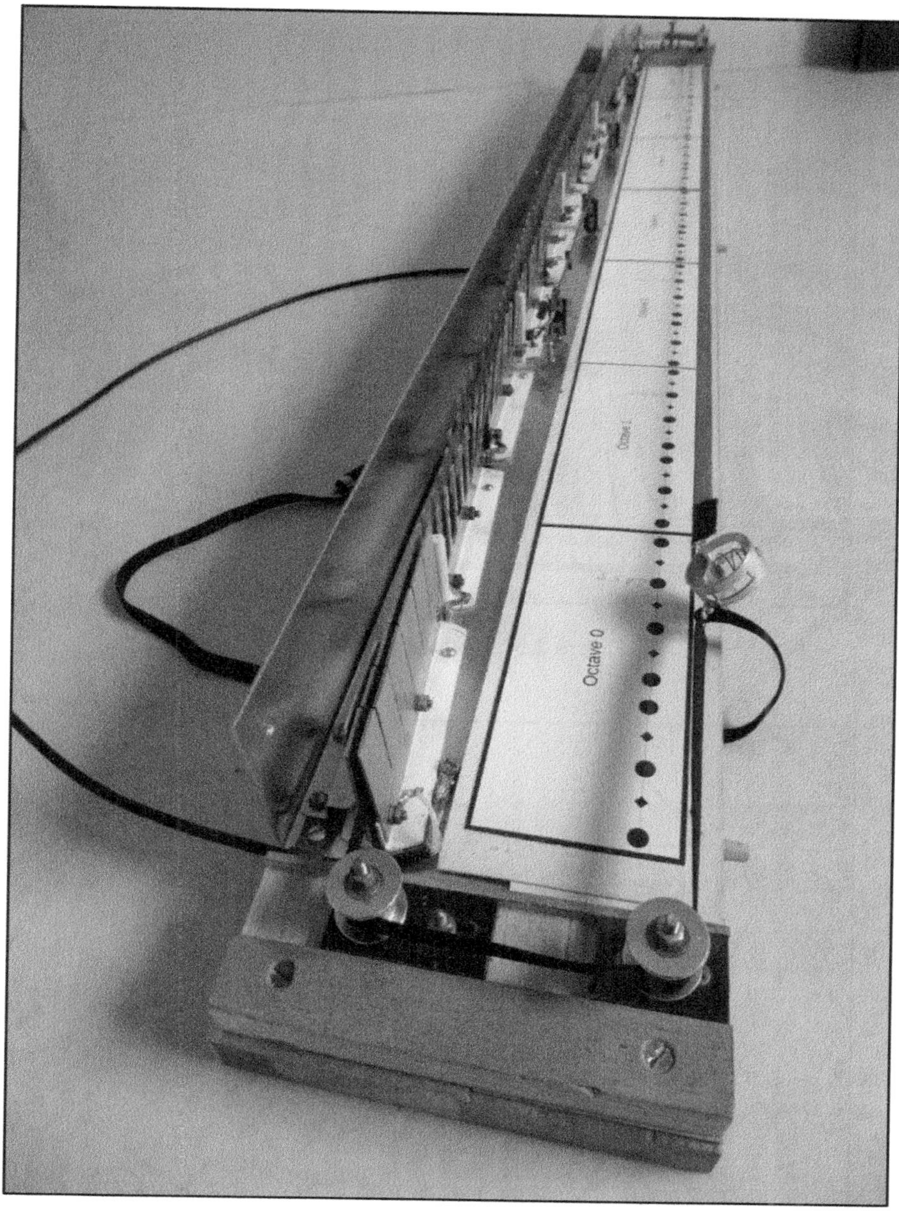

Figure 3.56. *The ribbon capacitor, the front face of the shielding has been removed*

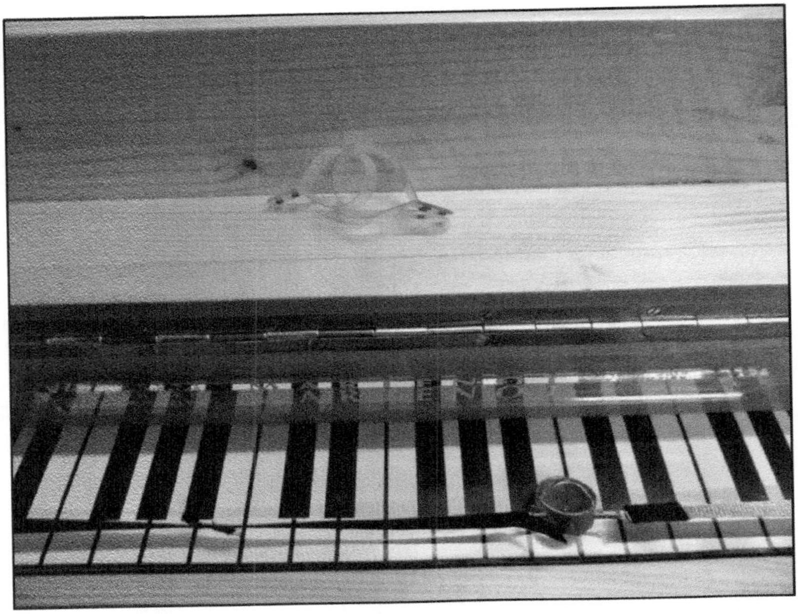

Figure 3.57. *Both types of rings*

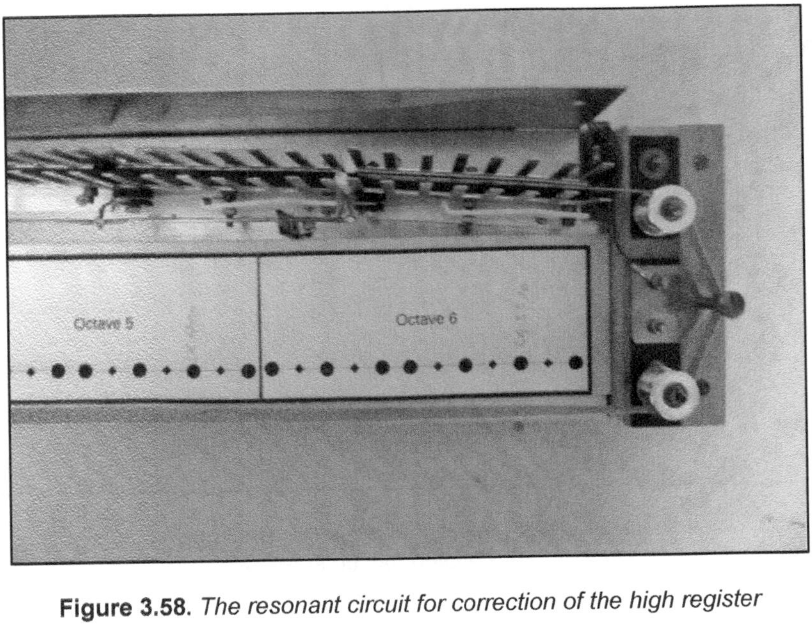

Figure 3.58. *The resonant circuit for correction of the high register*

3.7.2.5. *The case*

We have chosen the simplest possible realization; it is therefore a "box" of h = 20 cm * p = 25.5 cm * l = 112 cm, with a piano-type lid closure. Fixed underneath are 4 feet, 2 cm thick. For playing, the instrument is placed on a table.

3.7.2.6. *The diffuser*

The output of the power LF triode is free of connection. As in the case of ondes no. 169, we have kept this arrangement in order to be able to connect as desired: a classic old radio diffuser whose characteristic impedance is between 1,000 Ω and 2,500 Ω, or a low impedance loudspeaker (8 Ω) associated with its LF output transformer (2,000 Ω/8 Ω). We carried out the tests on each of the devices.

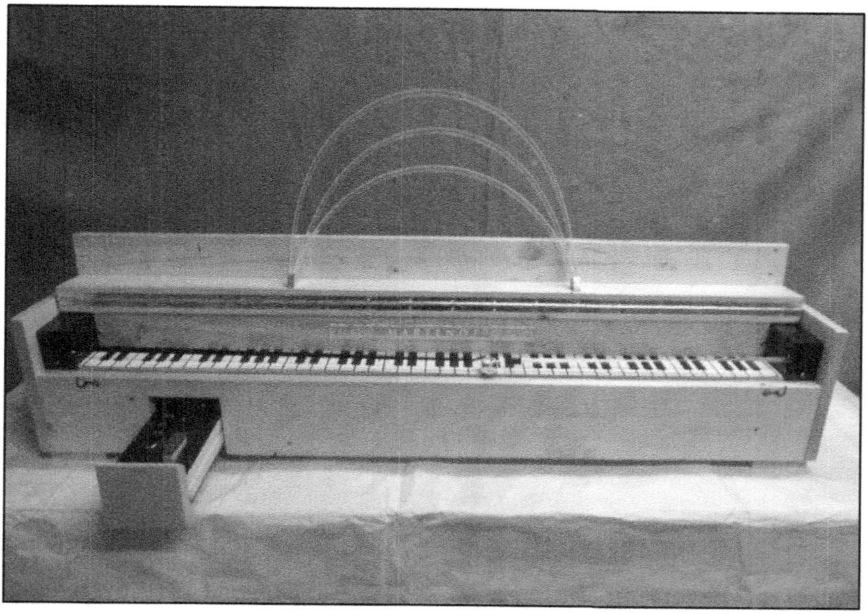

Figure 3.59. *The instrument in playing position*

3.7.3. *Tests and measurements*

The operation of this instrument is entirely consistent with our analyses. The adjustment of the ribbon capacitor is extremely delicate, it is difficult to obtain a correct agreement on the seven octaves, and it is the low register that is the most difficult to adjust. The following oscillograms summarize the measurements performed on this instrument. The signal in yellow is to be analyzed, and the signal in blue is generated by the fixed frequency oscillator.

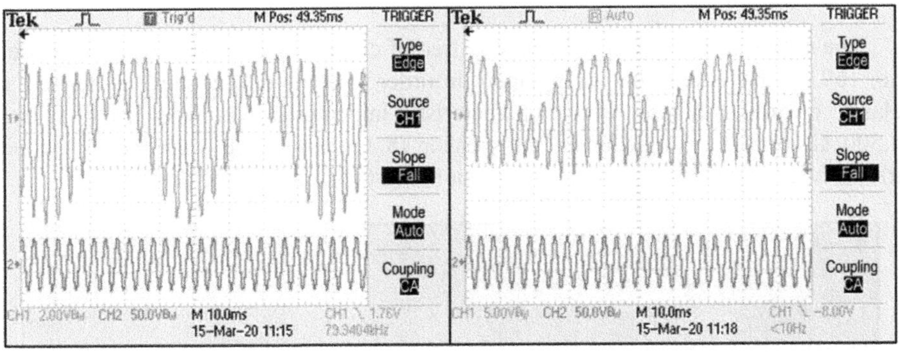

Figure 3.60. *Measurements on model 15-169: (left) detector output and (right) plate output 1st triode LF*

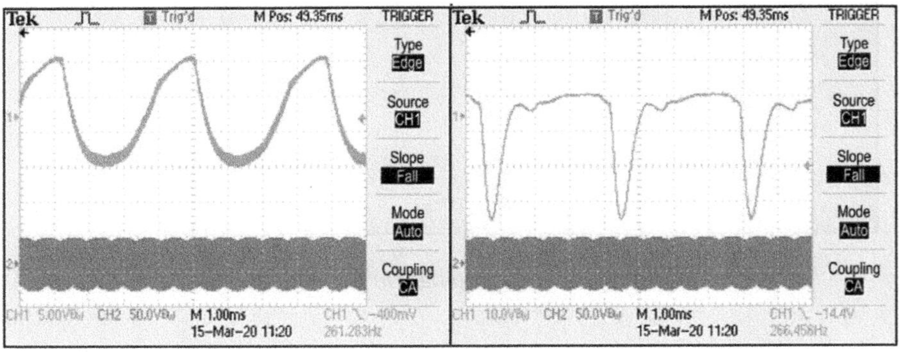

Figure 3.61. *Measurements on model 15-169: (left) 2nd LF triode input and (right) power triode input*

And to end this series of measurements.

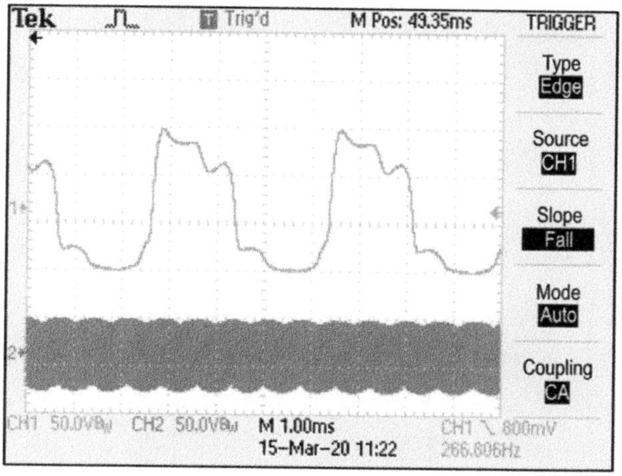

Figure 3.62. *Signal available at the primary of the LF transformer*

4

Model '47

4.1. Presentation

This model is a prototype made by Mr. Martenot after the 39–45 conflict. To date, we have no trace of a manufacture. On the contrary, we have three diagrams, two appear on the workshop book 21 (Martenot n.d.c) and the third which is a drawing of a plan showing the chassis seen from below and its wiring. On this last document, a mention of "1947" appears at the top right hand corner.

In the absence of an existing prototype, the study of operation is based on analyses of diagrams, tests on a model, and also when possible on simulations using the LT Spice software from Linear Technology. This prototype implements all the innovations that we will find on the '53 model.

4.2. Synoptic diagram

The synoptic diagram is given in Figure 4.1.

4.3. Operation analysis

4.3.1. *Heterodynes*

The diagram is given in Figure 4.2.

For the color versions of the figures in this chapter, see www.iste.co.uk/courrier/ondesmartenot.zip.

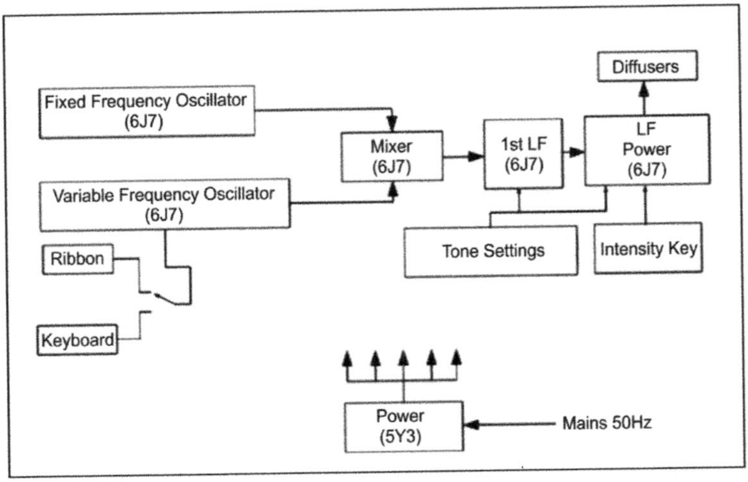

Figure 4.1. *Synoptic diagram of Ondes '47*

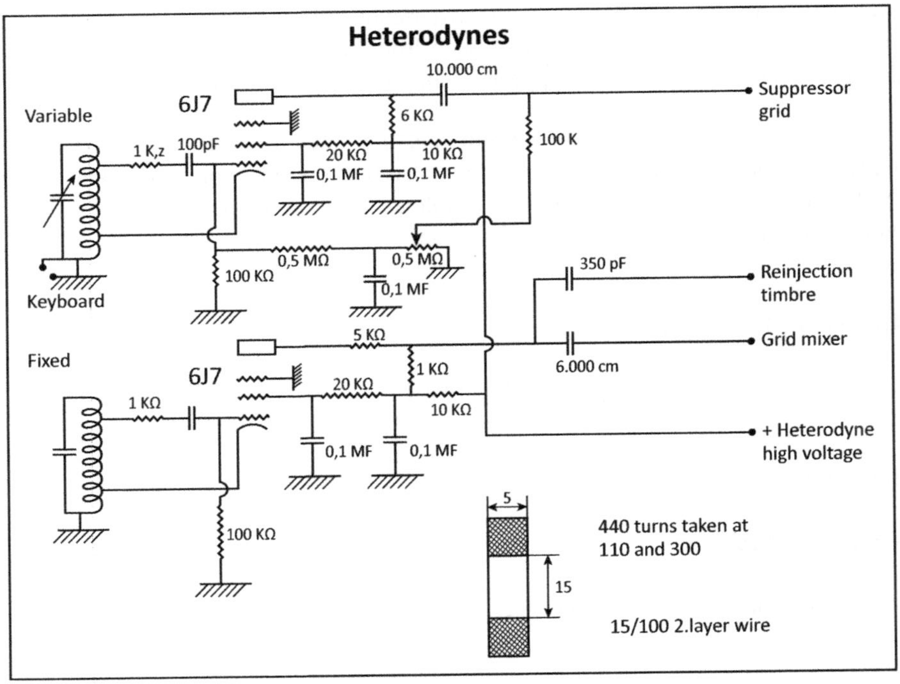

Figure 4.2. *Diagram of the oscillators and sizing of the inductor. Redrawn from workshop notebook 21 (Martenot n.d.c) (Musée de la Musique, Paris)*

The fixed oscillator is built around the Octal 6J7 pentode; it is wired as an E.C.O. oscillator, renowned for its great stability. The oscillating circuit consists of a 2.8 mH four-pole inductance, in parallel with a 910 pF capacitor; the resonance frequency is therefore close to 99 kHz. The plate load is made by putting the 5 kΩ and 1 kΩ resistors in series; at the nodal point, the output signal is taken by a 6 nF capacitor and injected on the G_1 grid of the mixer tube. This signal is not sinusoidal, but rather has the form of a rectangular signal with a low duty cycle, it is rich in harmonics.

We will see later the role of the 350 pF capacitor. The screen grid is powered via the 20 kΩ resistor which is decoupled by the 0.1 μF capacitor.

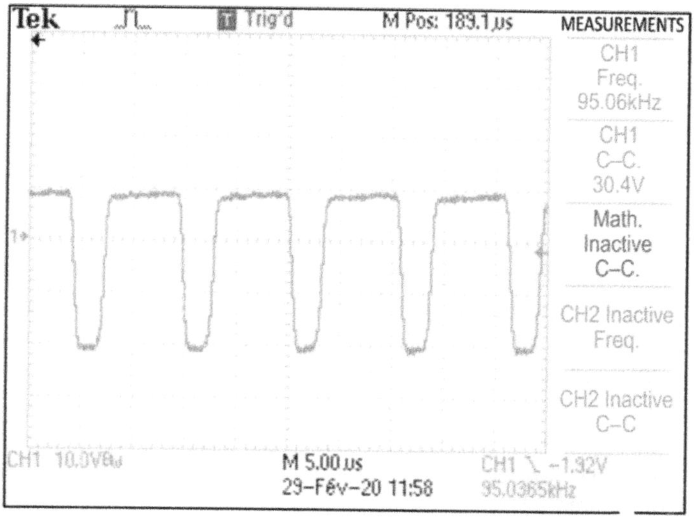

Figure 4.3. *Output signal of the fixed oscillator (measurement carried out on a model of the author)*

A particular arrangement attracts attention; the winding that constitutes the coil of the resonant circuit in an E.C.O. type oscillator assembly has three connection terminals: grid, cathode and ground. The ratio of the number of turns between the mass and the grid and the number of turns between the ground and the cathode is of the order of 1/3, in order to ensure the conditions necessary for the production of oscillations.

However, in the configuration adopted by Mr. Martenot for the winding, the grid connection point is extended again by a fraction of the coil connected to the capacitor of the resonant circuit, and the coil therefore has four terminals.

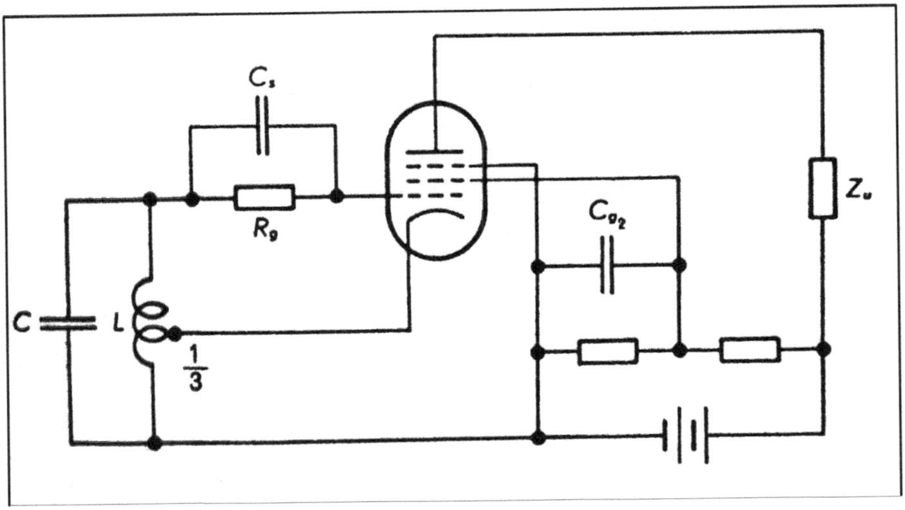

Figure 4.4. *E.C.O. oscillator (Electronics and Radio Electricity, G.Thalmann. Dunod, 1964)*

One finds in the Anglo-Saxon literature on the stability of oscillators, an article published in 1939 by Lampkin (1939), which shows that in the case of a Hartley-type oscillator using a triode tube, the displacement of the connection of grid towards the cathode connection point decreases the load of the oscillating circuit by the oscillator tube, which allows for a more stable operation of the latter. This technique also applies in the case of an E.C.O. type oscillator (which uses a pentode tube). Thus, the frequency of the oscillator is relatively independent of variations in the characteristics of the tube.

In this same article, the technical choices which make it possible to obtain a very stable oscillator are indicated, most of these technical recommendations are respected in the achievements of Mr. Martenot, in particular on the ondes '53 model. We quote some of them as follows:

– wiring of HF circuits and oscillators with conductors of large sections or twisted in pairs, the mechanical rigidity thus obtained reduces frequency variations by mechanical vibrations;

– use of coils without magnetic cores and placed in generously sized shields. Indeed, if the use of ferrites provides adjustment flexibility and reduces the sizes of the coils, the first ferrites manufactured in the 1940s are very sensitive to temperature variations; the inductors will therefore be air core coils that are suitable to magnetically isolate external fields by shielding;

– the oscillator tubes and the elements of the resonant circuits are geographically separated; this arrangement is intended to reduce the thermal effects generated by the heating of the electronic tubes on the components of the oscillating circuits.

The inductance construction parameters are given at the bottom of the photographic document given in Figure 4.5 (Martenot n.d.c).

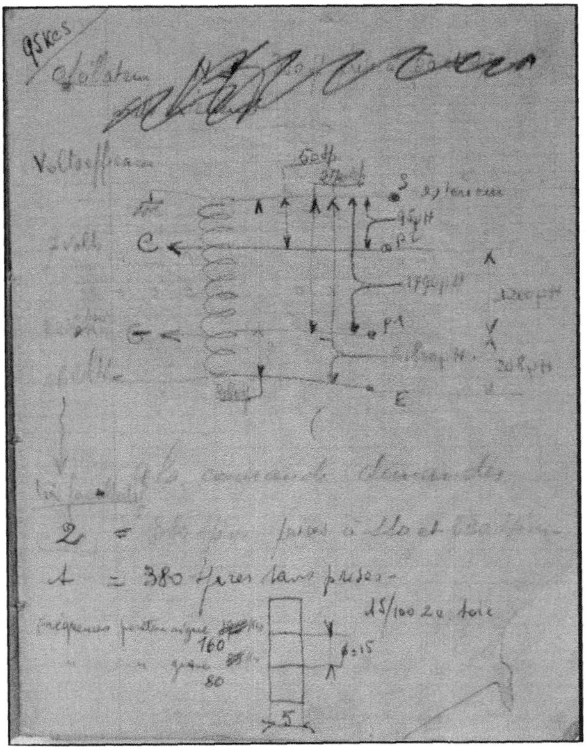

Figure 4.5. *Construction parameters of HF inductors. Workshop notebook 21 (Martenot n.d.c) (Musée de la Musique, Paris)*

The variable frequency oscillator is built in the same way as the previous one. We do not repeat the operation analysis in detail except for this: a reinjection circuit was created by Mr. Martenot; it reinjects a fraction of the signal taken from the grid circuit onto the plate of the oscillator tube. A 500 kΩ potentiometer is used to dose this signal fraction. A technical justification of this initiative is difficult to propose because the oscillator is already a looped system; this inefficient circuit will be applied to ondes no. 208 and then abandoned. Finally, the output signal taken by the 10 nF capacitor is injected on the grid G_3 of the mixer tube.

In accordance with the operating principle of Martenot instruments, the frequency of this oscillator is controlled either by a ribbon-type variable capacitor or by a set of coils in series activated by the keys of the keyboard.

4.3.2. *The mixer*

This is the first time that Mr. Martenot has implemented electronic signal mixing using a pentode tube. The diagram is given in Figure 4.6.

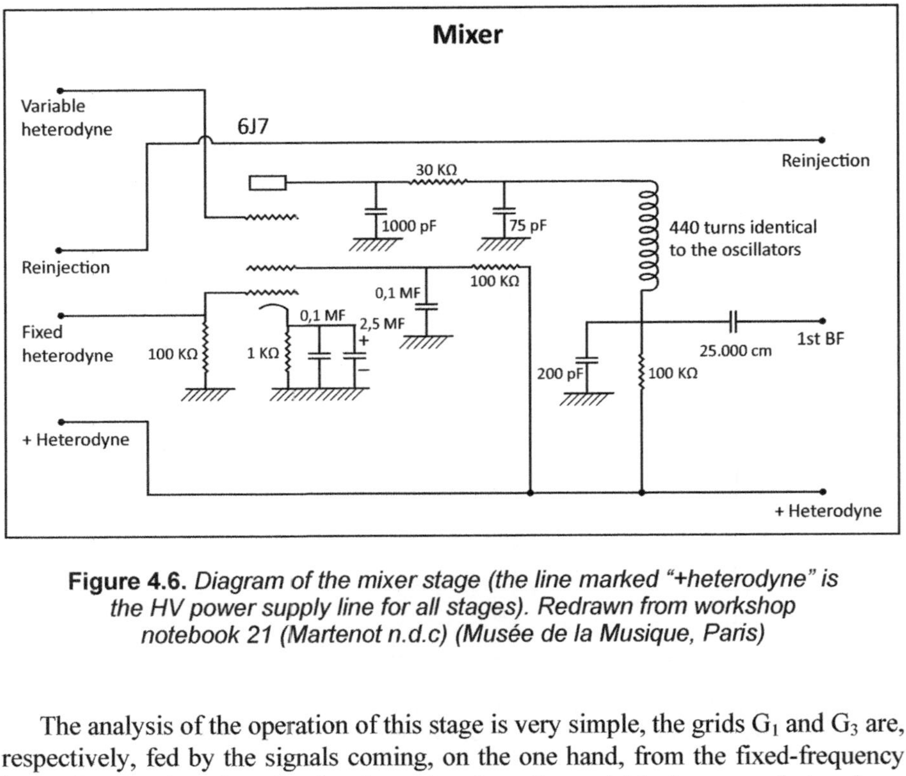

Figure 4.6. *Diagram of the mixer stage (the line marked "+heterodyne" is the HV power supply line for all stages). Redrawn from workshop notebook 21 (Martenot n.d.c) (Musée de la Musique, Paris)*

The analysis of the operation of this stage is very simple, the grids G_1 and G_3 are, respectively, fed by the signals coming, on the one hand, from the fixed-frequency heterodyne, and on the other hand, coming from the variable-frequency heterodyne. The cathode polarization is provided by the 1 kΩ resistor in parallel with the 0.1 μF and 25 μF capacitors. The screen grid is supplied with high voltage by the 100 kΩ resistor decoupled by the 0.1 μF capacitor. Plate charging is achieved by connecting the 30 kΩ resistor, a 2.8 mH inductor and the 100 kΩ resistor in series. The inductor is an HF coil. The capacitors 1 nF, 75 pF and 200 pF provide HF decoupling and harmonic filtering of high frequencies. The LF signal resulting from the product of the high-frequency signals is injected into the next stage by the 25 nF capacitor.

It is therefore a new way for Mr. Martenot to produce a musical signal, indeed until the '37 model the synthesis was additive.

For an HV of 300 V, a sinusoidal signal of 94.670 kHz / 5.7 Vdc applied to G_1 and a sinusoidal signal (in blue) of 94.270 kHz / 31 Vdc applied to G_3, we obtain the LF signal (in yellow) of frequency equal to 400 Hz and 46 V peak-to-peak amplitude.

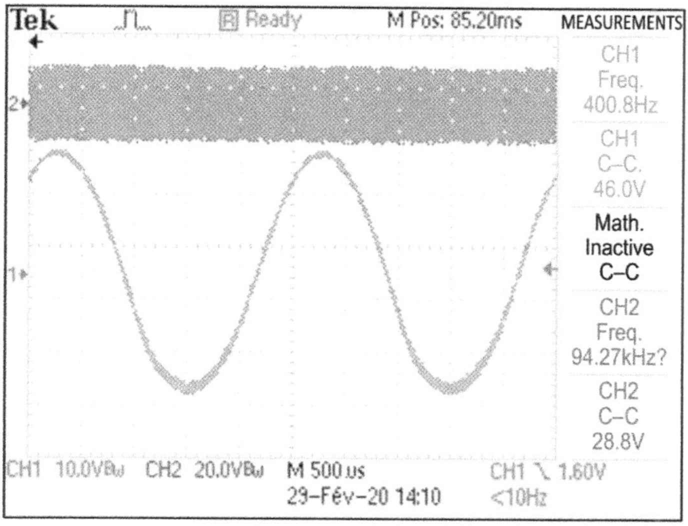

Figure 4.7. *Mixer stage output signal (measured on a model by the author)*

4.3.3. *First LF*

This stage, introduced for the first time by Mr. Martenot in his prototypes, is not an amplifier stage, but a signal shaping circuit, it precedes the final amplifier stage or power stage. A diagram is given on the next page.

We observe the presence of many switches: T4, T5, T6 and T7, they intervene in three different ways: screen grid polarization (T4 and T6), cathode polarization (T5) and screen decoupling (T7). The analysis of the operation of this assembly is complex, because, on the one hand, cathode polarization is not in conformity insofar as the decoupling capacitor does not have a sufficient value to ensure a constant potential (it seems that its value is rather chosen as in the case of a frequency filter), and on the other hand, the polarization of the screen grid is variable according to the presence or not of another decoupling capacitor, in the absence of this one the screen grid potential evolves in the same way as the anode potential.

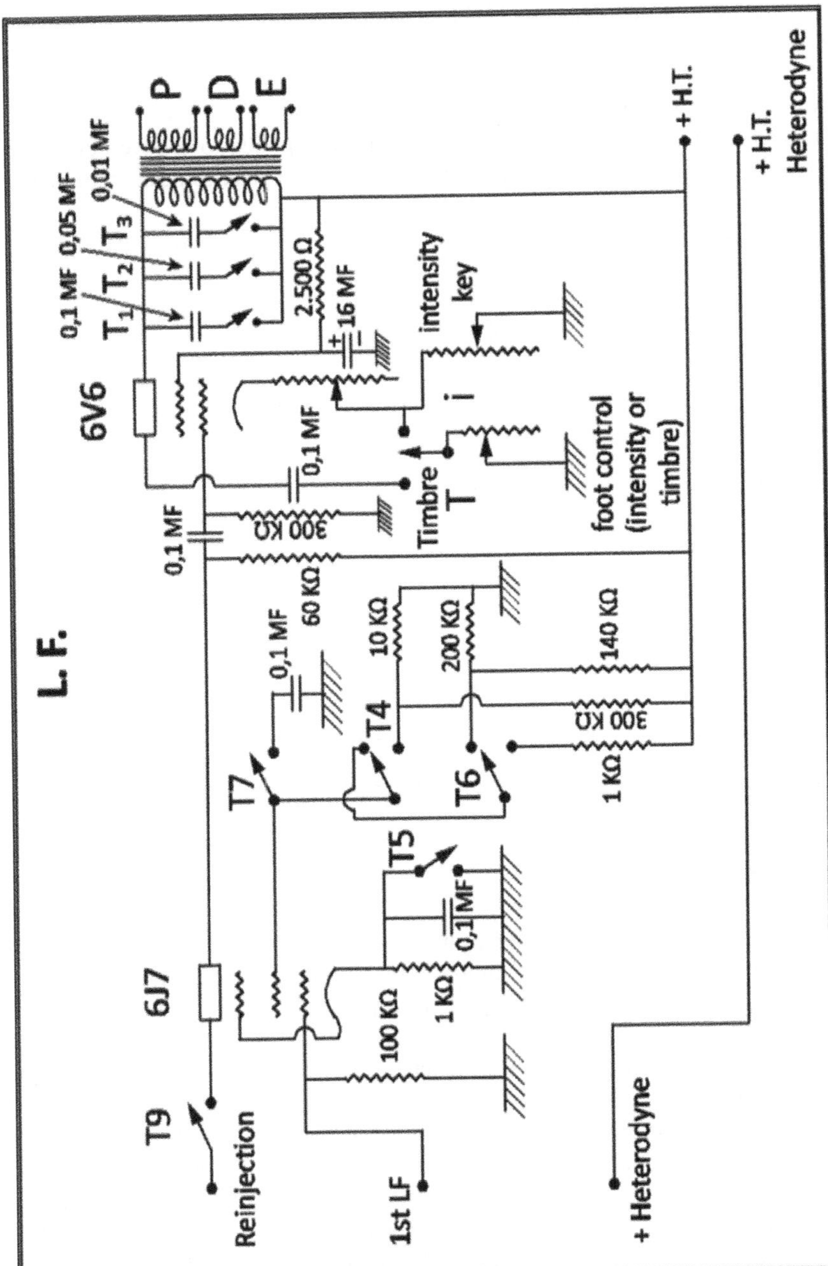

Figure 4.8. *Diagram of the first LF and the power stage. Redrawn from workshop notebook 21 (Martenot n.d.c) (Musée de la Musique, Paris)*

We preferred to list the effects of the different selections of timbres on an 800 Hz signal (in yellow on the oscillograms below) applied to the input grid of the 6J7 pentode. In all cases, the assembly supply voltage is 300 V, and the sinusoidal input voltage is 46 Vdc. The diagram in Figure 4.9 shows the labeling of the positions of the timbre switches.

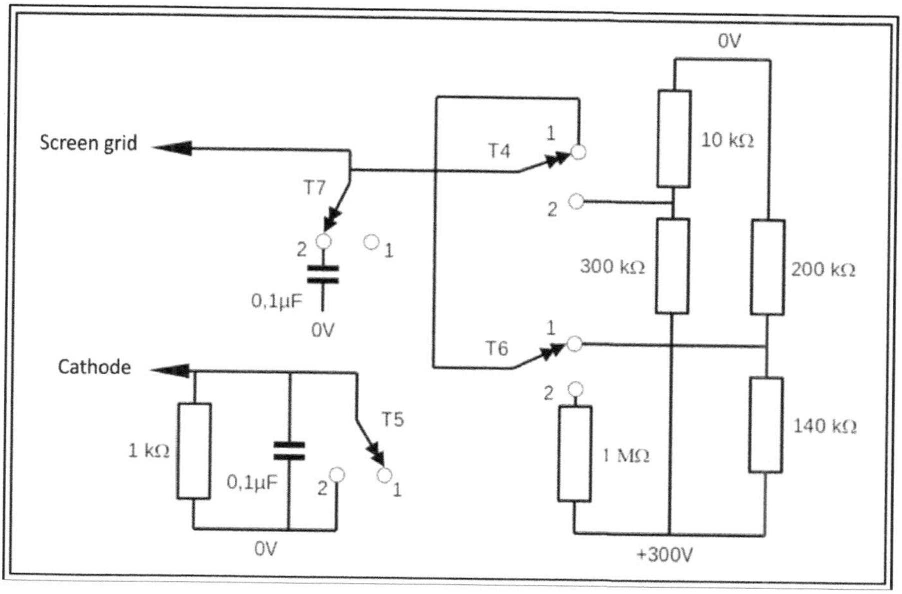

Figure 4.9. *Identification of switches T4, T5, T6 and T7*

Table 4.1 indicates the identification of the positions of the switches with the oscillograms.

We gather in the following table all the oscillograms. The output signal of the stage (in blue) is measured on the plate of the 6J7 pentode. The input signal is an 800 Hz sinusoid.

From all of these measurements, it appears very clearly that the determination of these circuits is more experimental than technological. Some solutions are redundant; others lead to a significant reduction in the amplitude of the output signal. Finally, Mr. Martenot will retain only two solutions for ondes no. 208, those which correspond to oscillograms no. 3 or no. 5 for which the pentode operates in a saturated mode, the clipping of the input signal is asymmetrical. The first solution (no. 3) consists of operating the pentode with a screen grid series resistor, which is not decoupled, the second solution (no. 5) consists of supplying the screen grid via a decoupled resistive voltage divider. The two solutions adopted by Mr. Martenot for ondes no. 208 generate highly asymmetrical signals which are therefore rich in harmonics. The resistance values will be adapted to the pentode used (6AU6).

Note that the output signals have large amplitude variations (from 138 Vdc to 296 Vdc).

Oscillogram	Screen grid polarization	Screen grid decoupling	Cathode polarization	T4	T6	T7	T5
1	Resistance of 1 MΩ	Yes	Yes	1	2	2	1
2	Resistance of 1 MΩ	Yes	No	1	2	2	2
3	Resistance of 1 MΩ	No	Yes	1	2	1	1
4	Resistance of 1 MΩ	No	No	1	2	1	2
5	Voltage divider 140 kΩ / 200 kΩ	Yes	Yes	1	1	2	1
6	Voltage divider 140 kΩ / 200 kΩ	Yes	No	1	1	2	2
7	Voltage divider 140 kΩ / 200 kΩ	No	Yes	1	1	1	1
8	Voltage divider 140 kΩ / 200 kΩ	No	No	1	1	1	2
9	Voltage divider 300 kΩ / 10 kΩ	Yes	Yes	2	1	2	1
10	Voltage divider 300 kΩ / 10 kΩ	Yes	No	2	1	2	2
11	Voltage divider 300 kΩ / 10 kΩ	No	Yes	2	1	1	1
12	Voltage divider 300 kΩ / 10 kΩ	No	No	2	1	1	2

Table 4.1. *Choice of timbres, identification of possible combinations*

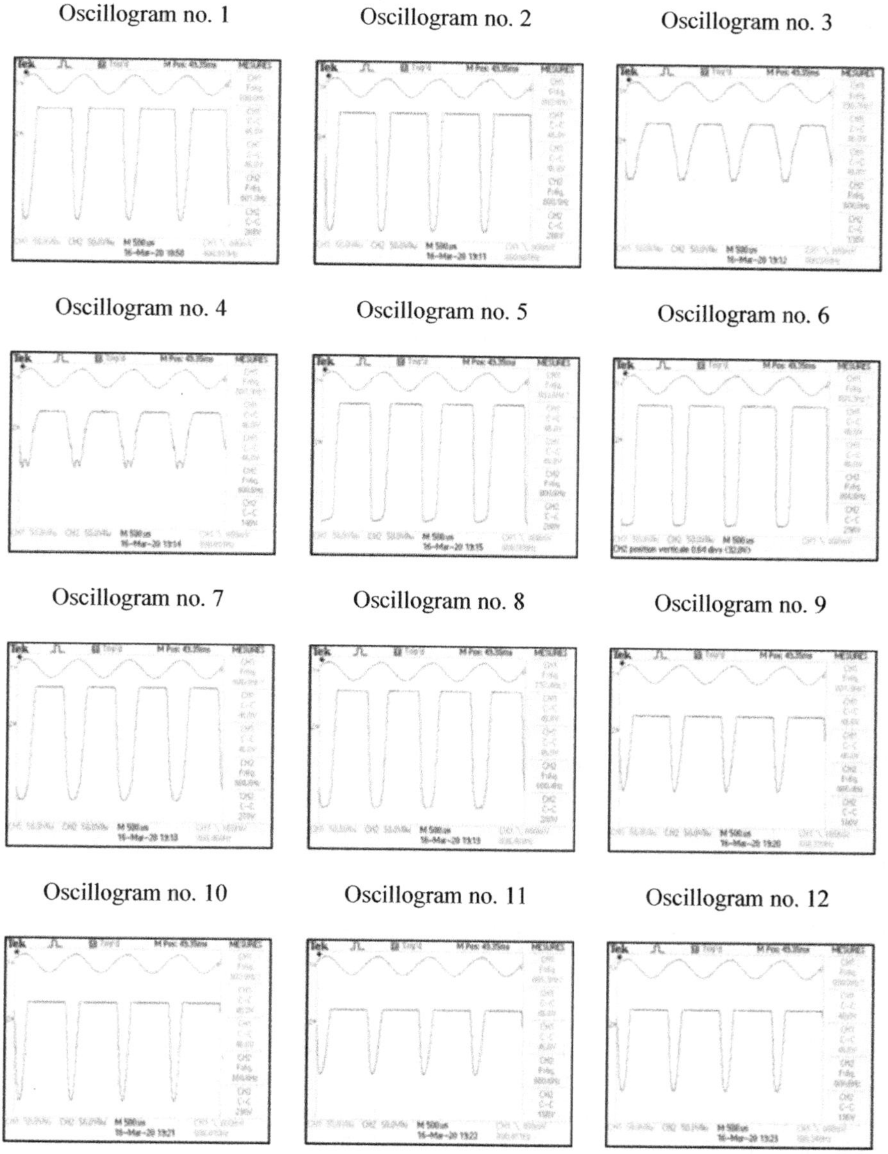

Figure 4.10. *The oscillograms of the choice of timbres (measurements carried out on a model from the author)*

4.3.4. *The T9 timbre*

On the diagrams that we have used, there appears a line marked "timbre reinjection". This involves taking an LF signal at the output of the shaping stage and reinjecting it, when switch T9 is closed through the 350 pF capacitor at the input. A voltage from the mixer's fixed frequency oscillator is applied to this same input.

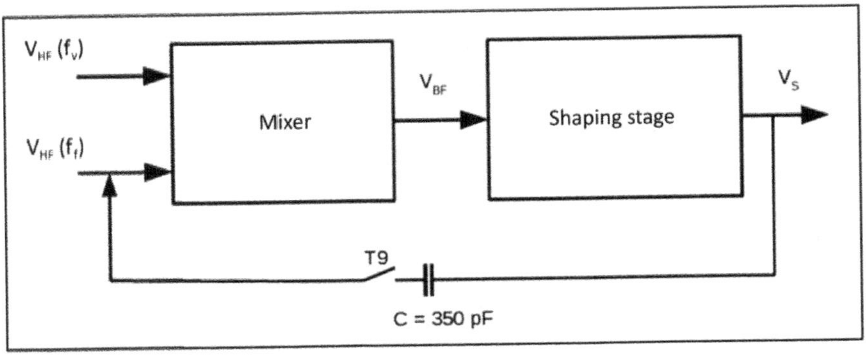

Figure 4.11. *Block diagram of the T9 timbre*

It is therefore a looped system, but this one is particular insofar as it comprises nonlinear elements in the direct chain, and involves a mixture of signals of very different frequencies. The mathematical analysis of this operation is extremely complex and is beyond the scope of this work. The major risk taken when looping a system is the destabilization of this system. It is probable that Mr. Martenot's initiative is purely experimental. We have limited the measurements to the two cases retained by Mr. Martenot for Ondes no. 208: polarized screen grid with the non-decoupled 1 MΩ resistor (no. 3) and polarized screen grid with the 140 kΩ/200 kΩ resistor voltage divider decoupled by a 0.1 µF capacitor (no. 5).

The mixer is powered by two HF signals: on grid G_1, a sinusoidal signal with a frequency equal to 95 kHz and 5.7 Vdc in amplitude, on grid G_3, a sinusoidal signal with a frequency equal to 94.420 kHz and 46 Vdc of amplitude. The supply voltage of the assembly is 300 V. On the oscillograms, the voltage in yellow corresponds to the input of the clipper tube (first LF); the voltage in blue is the output voltage of the assembly. A strong variation in the amplitudes of the output voltages is observed: from 122 V to 304 V. The closing of switch T9 (cases 2 and 4) leads to operations that present instabilities. The modifications of the shapes of the signals are minor, and the variations of the harmonic richness remain low. This technological option will not be retained by Mr. Martenot on the '53 model.

1) R = 1 MΩ and T9 open.

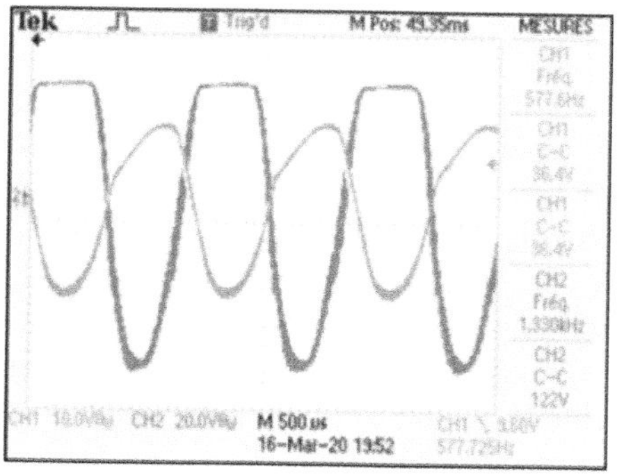

2) R = 1 MΩ and T9 closed.

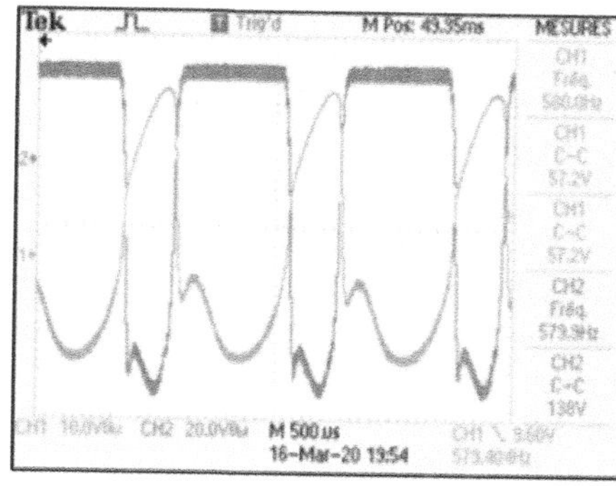

3) Voltage divider and T9 open.

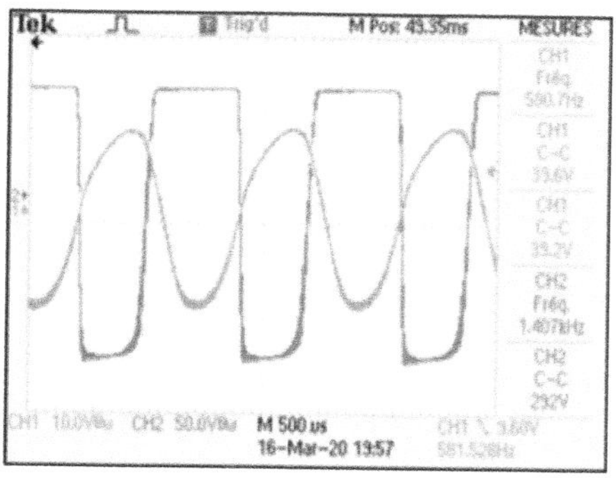

4) Voltage divider and T9 closed.

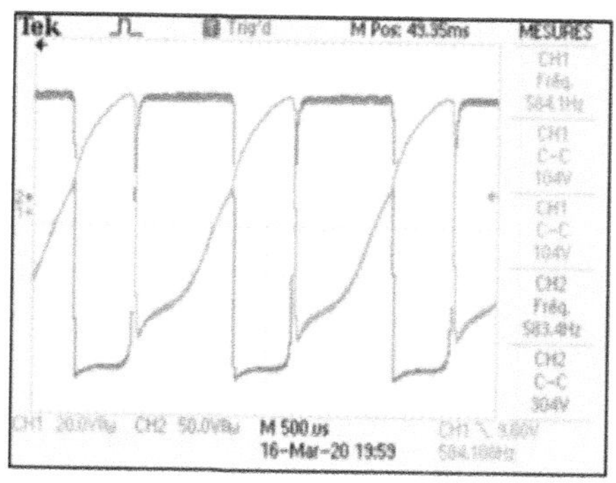

Figure 4.12. *The oscillograms according to the choices for T9 (measurements carried out on a model of the author)*

4.3.5. *LF power stage*

It is built around the LF Octal 6V6 power pentode, this tube, in class A, with a load impedance of 5,000 Ω is capable of delivering a useful power of 4.5 W. The output transformer has three secondaries (with probably an impedance of 5 Ω each), which allows the loudspeaker, the Gong or the Palme to be powered, together or separately. The intensity key is inserted into the cathode circuit whereas in early ondes models the intensity key is in series with the loudspeaker. A foot pedal can be connected either in parallel with the intensity key (volume control) or constituted with the capacitor of 0.1 µF connected to the plate of the tube, a harmonic filter. Three other capacitors – 0.1 µF, 0.05 µF and 0.01 µF – associated with the controls of timbres T1, T2 and T3, allow for the operation of low-pass filters with respective cut-off frequencies 350 Hz, 700 Hz and 3.5 kHz. Note that the screen grid is powered by a 2,500 Ω resistor decoupled by a 16 μF capacitor. This technical choice assumes that the supply voltage of the assembly is greater than 250 V, the study of the characteristics of the 6V6 tube and in accordance with the diagram proposed by Mr. Martenot we can deduce that the supply voltage of the stage is of the order of 300 V. In this case for a key resistance of the order of 200 Ω, the grid voltage is –12.6 V. However, the signal applied at the input has peak-to-peak amplitude of between 138 Vdc and 296 Vdc, which results in a very high saturation of the assembly, the processing of signals prior to this stage is therefore greatly modified or even destroyed. When we performed a simulation of the behavior of the stage using the Ltspice software from linear technology, the intensity key varied from 0.1 Ω to 1 MΩ.

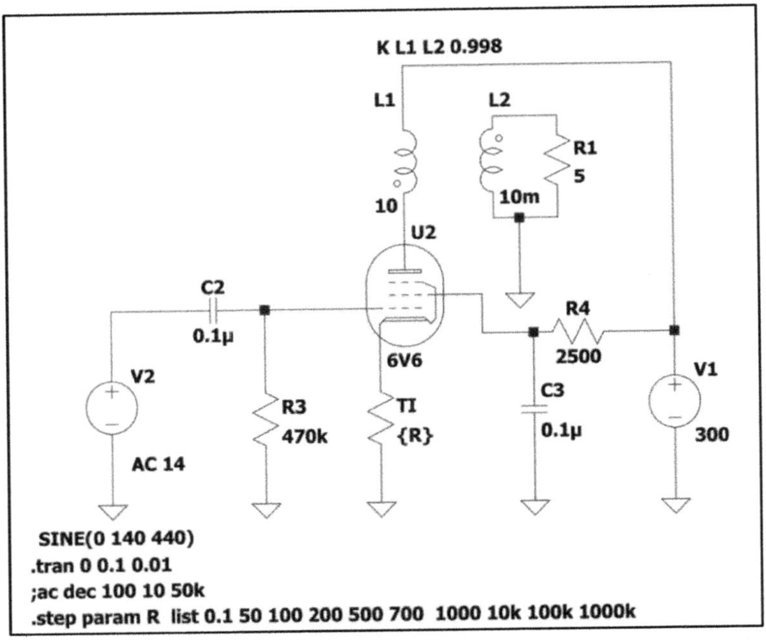

Figure 4.13. *Diagram for Ltspice simulation of the LF power stage*

The results are presented as follows:

– Bode diagram in linear regime;

– output voltage at the terminals of the load, in linear mode and saturated mode (excluding saturation of the LF transformer).

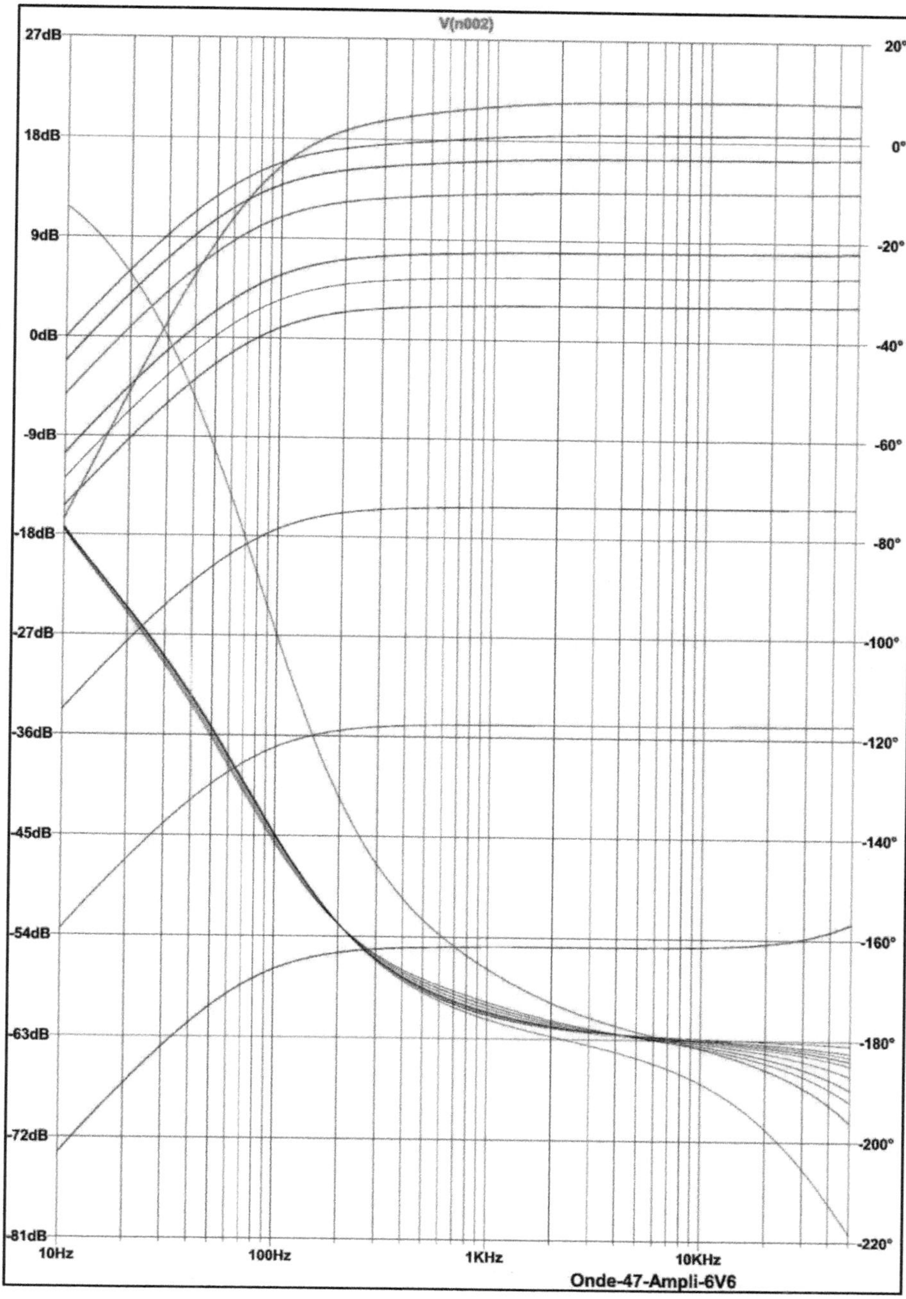

Figure 4.14. *Bode diagram of the power stage*

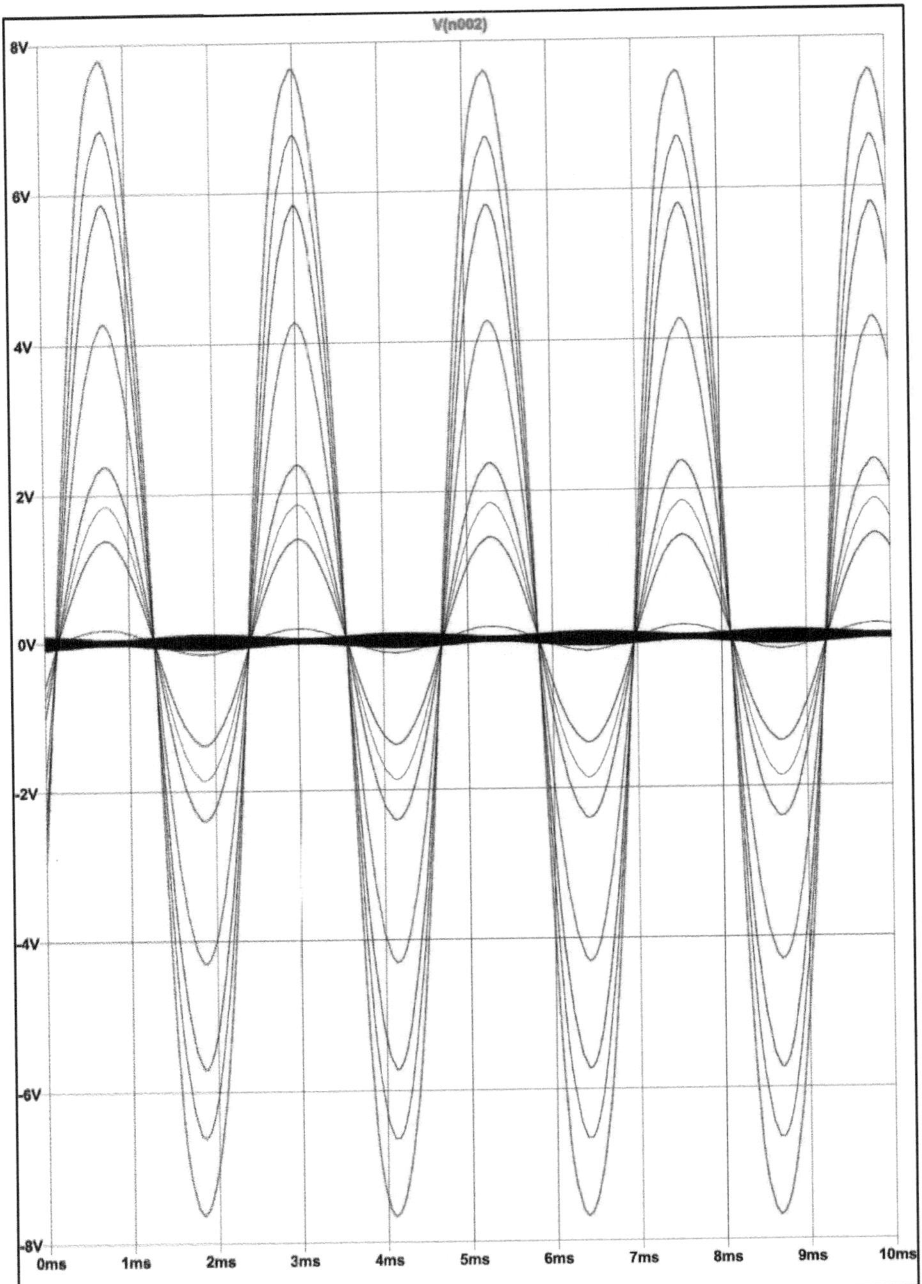

Figure 4.15. *Response of the power stage to a sinusoidal signal*

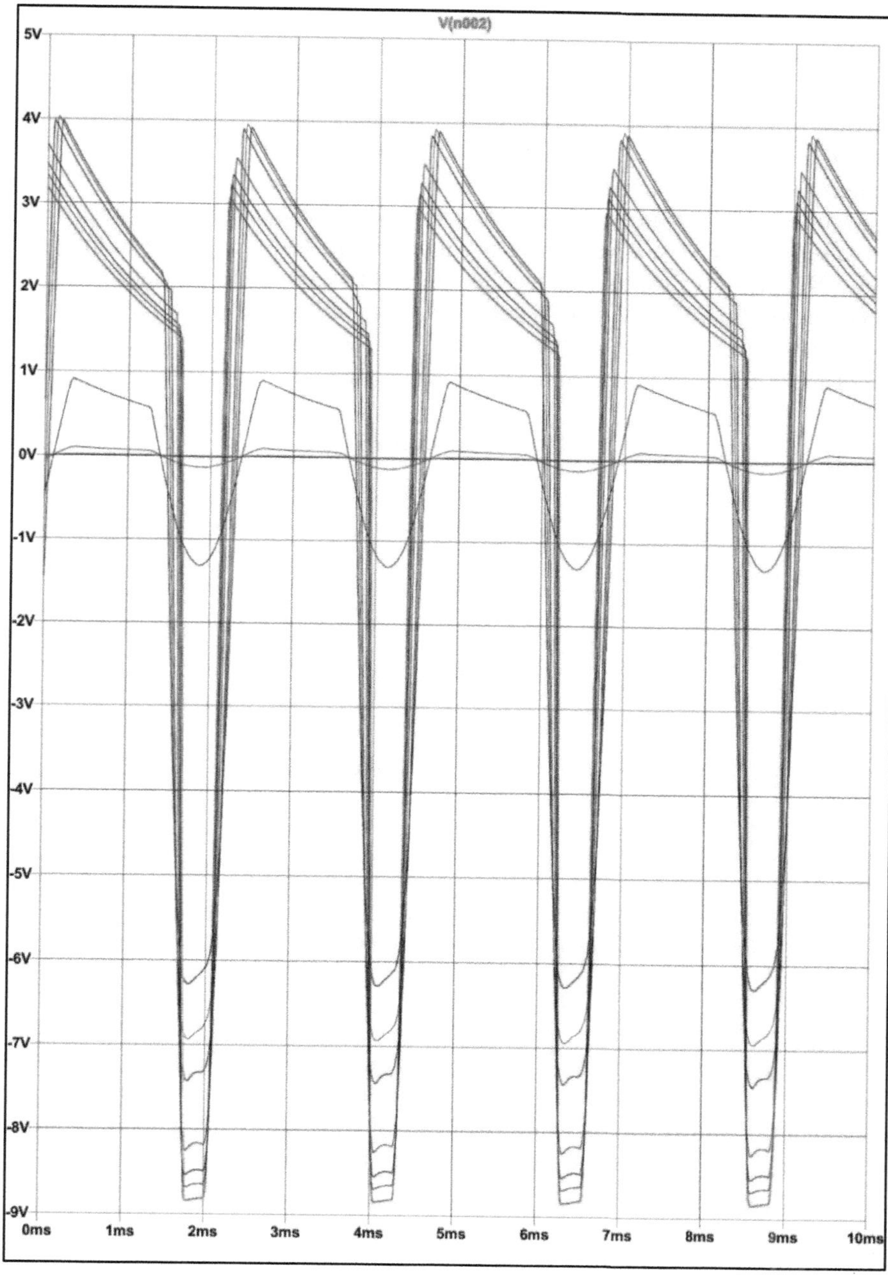

Figure 4.16. *Response in saturation regime of the power stage (excluding saturation of the LF transformer and constant HV supply)*

4.3.6. *Power module*

The diagram proposed by Mr. Martenot for this module is classic, the only unknown is the value of the supply voltage. If we refer to the technical specifications of the 6J7 electronic tubes, the first models proposed by RCA (1935) have a plate voltage of 250 V maximum. Later, Tung-Sol (1941) will raise this voltage to 300 V. If we examine the power stage, we have shown previously that the supply voltage of the 6V6 tube is 300 V. The development of this prototype had taken place in the years 1946–1947, the use of the 5Y3GB rectifier tube, and the preceding remarks, lead us to propose a voltage of 325 V–350 V per HV winding at the secondaries of the power supply transformer, in order to obtain, after rectification and filtering, a DC voltage of the order of 300 V–315 V. A diagram is given in Figure 4.17.

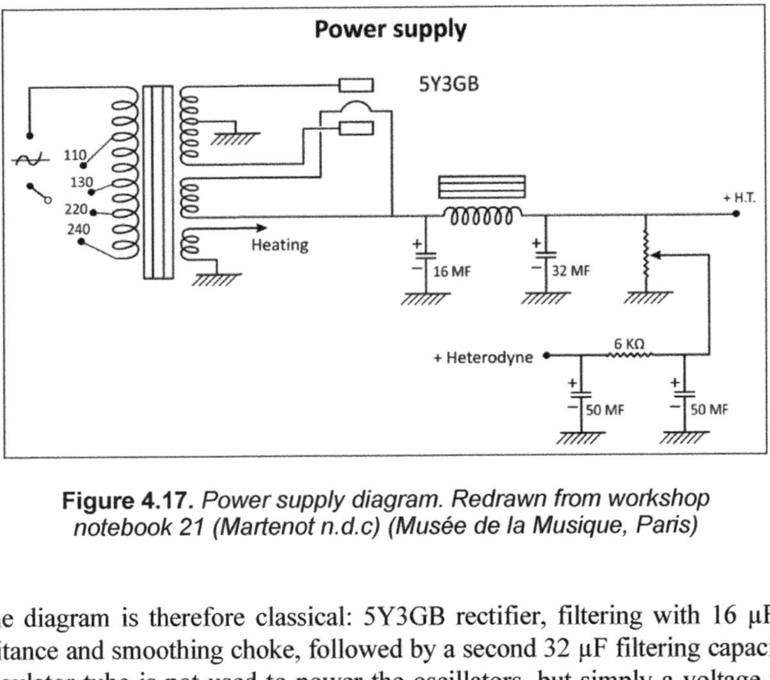

Figure 4.17. *Power supply diagram. Redrawn from workshop notebook 21 (Martenot n.d.c) (Musée de la Musique, Paris)*

The diagram is therefore classical: 5Y3GB rectifier, filtering with 16 μF head capacitance and smoothing choke, followed by a second 32 μF filtering capacitor. A gas regulator tube is not used to power the oscillators, but simply a voltage tap by potentiometer followed by a 50 μF/6 kΩ/50 μF filter cell. This module probably remained in a prototype state without further development.

4.4. The intensity key, ribbon capacitor and keyboard

The drawing of the chassis of this prototype (presented on the next page) shows that the configuration is close to model no. 169, the chassis is rectangular and the power supply is mounted on a separate chassis.

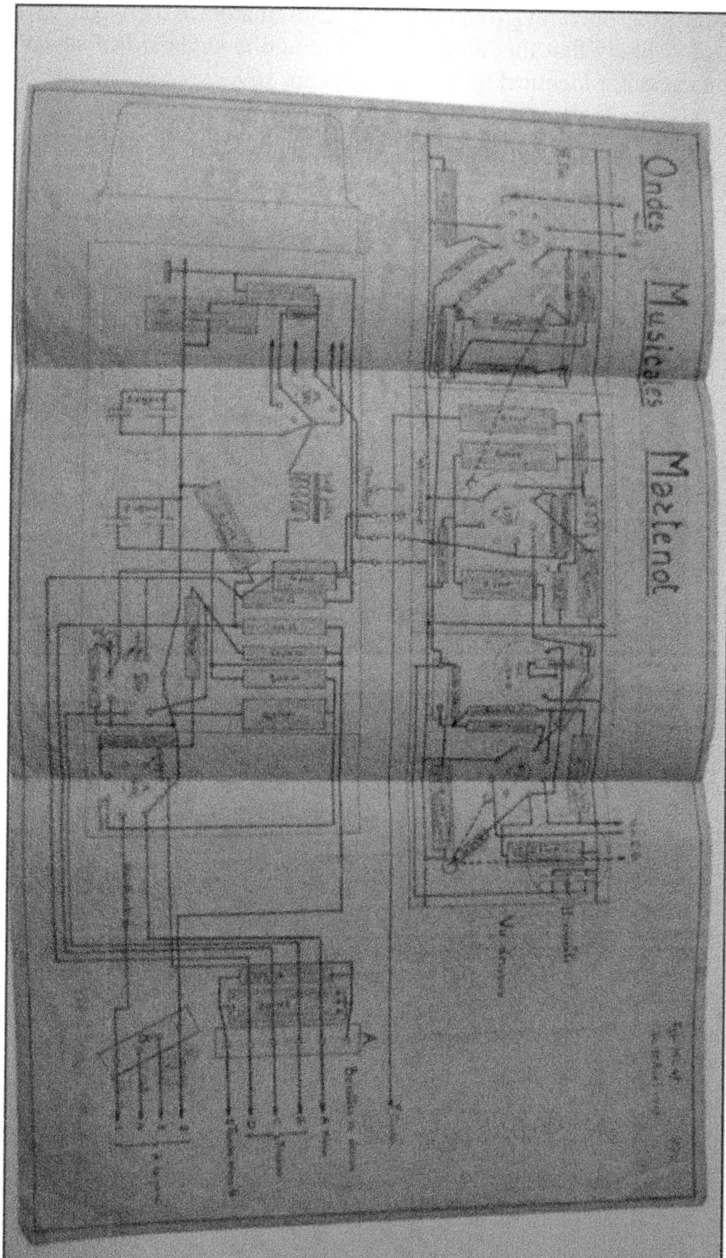

Figure 4.18. *Plan of the chassis of ondes model '47 (Musée de la Musique, Paris)*

In addition, the operating frequency of the oscillators is between 95 kHz and 99 kHz, so it is likely that this prototype had a mobile keyboard of seven octaves, and a ribbon capacitor identical to that of model no. 169.

On the contrary, we have no information relating to the manufacture of the intensity key.

5

Ondes No. 208 (1953)

5.1. Introduction

Following our various works carried out on ondes no. 15 and no. 169, we were interested in the evolution of ondes models which succeeded. By having ondes no. 208, which is representative of the latest electronic tube model, we were able to continue our work analyzing the operation of this type of instrument. In this chapter, we present the results of our research, which we will largely use for the realization of a model.

It is the last ondes model with electronic tubes designed by Mr. Martenot in 1953 and made available to ondists in 1955. By its external aspects, this ondes is presented in the continuity of the preceding models; on the other hand, major innovations are brought to the levels of musical performance and associated play. The first prototype developed by Mr. Martenot was equipped with pentode electronic tubes from the American Octal series, this prototype was indeed manufactured by Mr. Martenot in 1947 (see the previous chapter). On the diagram of this prototype appear all of the innovations that we find on diagram nos. 514 and 541 of the models sold by Mr. Martenot a few years later.

The ondes on which we worked (ondes no. 208) is functional. According to the ondes follow-up notebook from Mr. Martenot's studio, it seems that it was Maurice Martenot's personal ondes. Its wiring strictly conforms to the diagram referenced: 541 Martenot-type "concert" ondes drawn on January 25, 1960.

A detailed examination of this ondes leads us to believe that it is the prototype developed in the workshop.

For the color versions of the figures in this chapter, see www.iste.co.uk/courrier/ondesmartenot.zip.

The range is seven octaves but a switch allows us to play on the low register or on the high register, the keyboard therefore has 72 keys and the *glissando* obtained with the ribbon is spread over six octaves. A vibrato effect is obtained by the mobile keyboard.

The intensity key handles ribbon and keyboard playing, a foot pedal and a knee lever are added, these two controls work on the same principle as the intensity key, the first is in parallel with the manual intensity key, the second acts on a harmonic filter. The following are added: the T0 and T7 timbres, the needle block and the ± 1/4 tone correction.

The instrument has three outputs which allow the connections of the Main loudspeaker or the Resonating loudspeaker, the Metallic gong and the Palme.

But it is in the design of the electronic functions necessary for the functioning of the instrument that the major innovations appear: use of electronic pentode tubes, non-sinusoidal oscillators, electronic mixing of signals from the oscillators, low register/high register switching, timbres T7 and T8 (octaviating effect obtained by a rectifying device).

5.2. Functional descriptions

5.2.1. *The unit*

The unit is an assembly of 18 mm / 20 mm thick oak veneers; the following diagram by Mr. Martenot gives us its general appearance. Two side legs in the same material (thickness ~ 20–22 mm) fit together thanks to "bedstead" type fittings and are blocked by ball pushers. On the upper part, a "piano" type lid with two brass hinges and a closing lock by ball pushers are found. On the front face, on the left, an opening is made for housing the drawer.

When the cover is open in the playing position, the performer can fit into two rectangular (brass) supports inclined towards the back a "music stand", made by an assembly of three strips of plastic material arranged in 1/2 arcs of concentric circles. On the inner face of the lid is also arranged a bevelled plate of engraved plastic material. A plastic accessory of the same nature as the indicative plate allows the storage of the "music holder" before closing the cover.

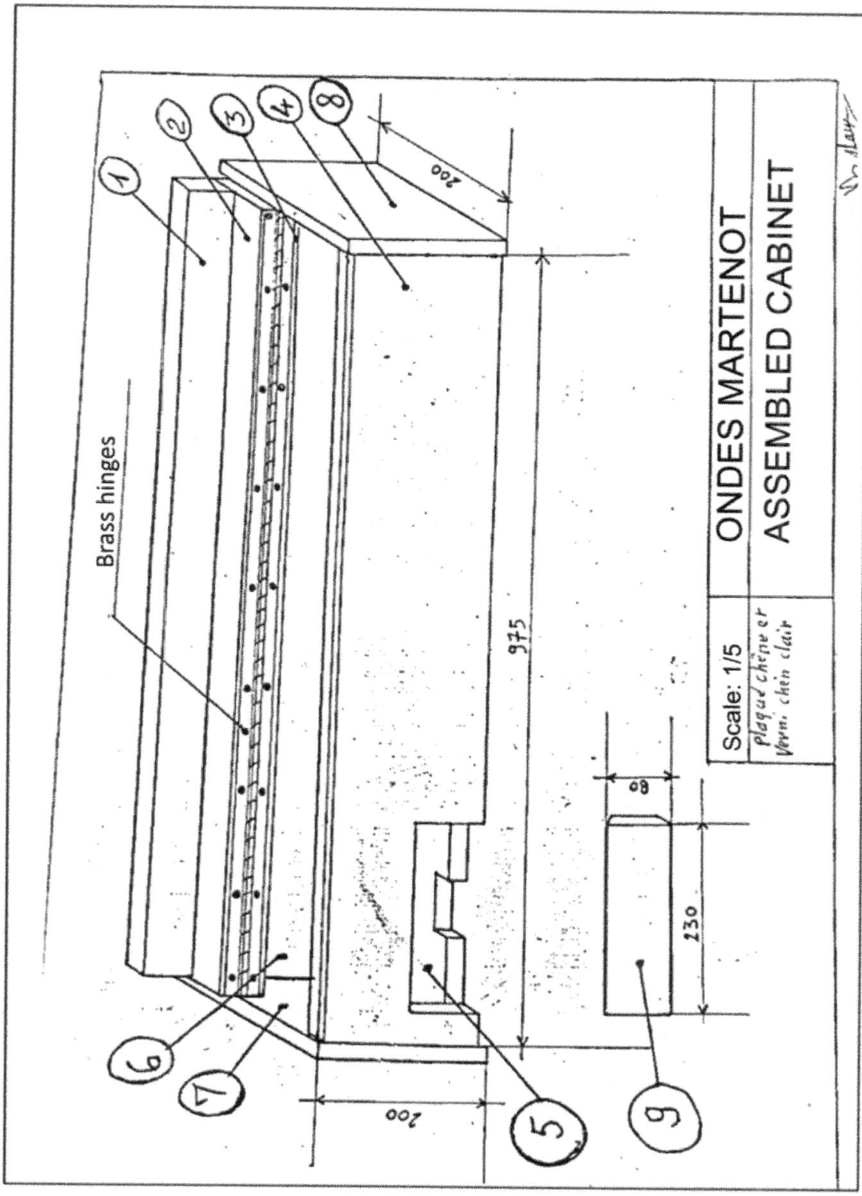

Figure 5.1. *Drawing of the unit (Musée de la Musique, Paris)*

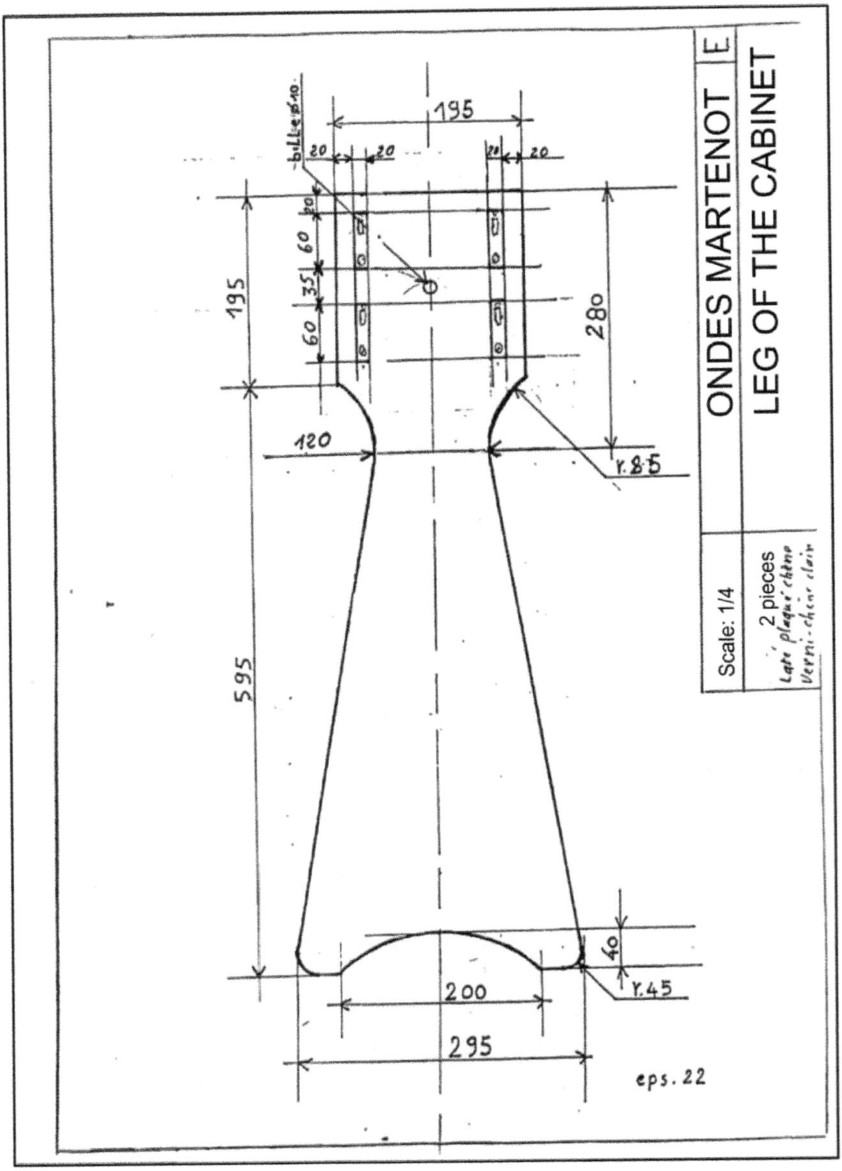

Figure 5.2. *Unit leg (Musée de la Musique, Paris)*

A cover made up of an oak wood strip hides the U-profile which contains the ribbon capacitor. On this stick are placed on the right and on the left two plastic hooks making it possible to hang the ribbon in the case of a play exclusively on the keyboard.

An ebonite (as it seems) strip graduated by note (recess for a white note, a half-spherical brass head for a black note) allows the performer to spatially locate their finger (index) which activates the ribbon. This black strip is semi-recessed on the height as well as on the width in the inner and upper part of the front face.

Finally, to complete this description, the furniture is externally coated with a "Martin" varnish.

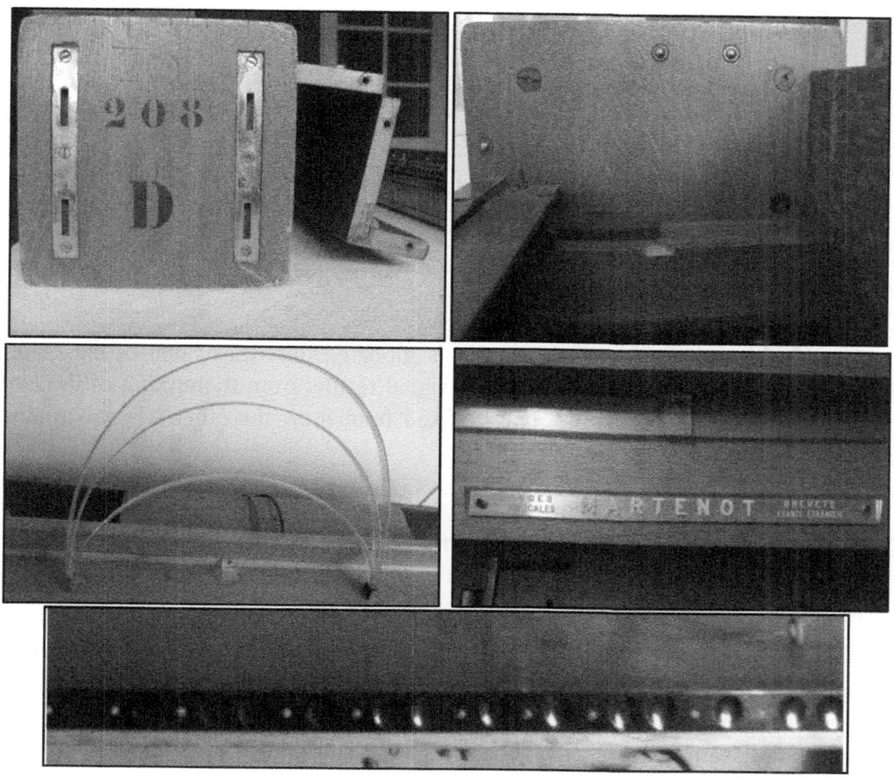

Figures 5.3. *Some descriptive elements of ondes no. 208: (Top) Right side face and inner face. (Middle) Music holder and engraved plate. (Below) Index marking*

5.2.2. *The block diagram*

The synoptic diagram is as follows:

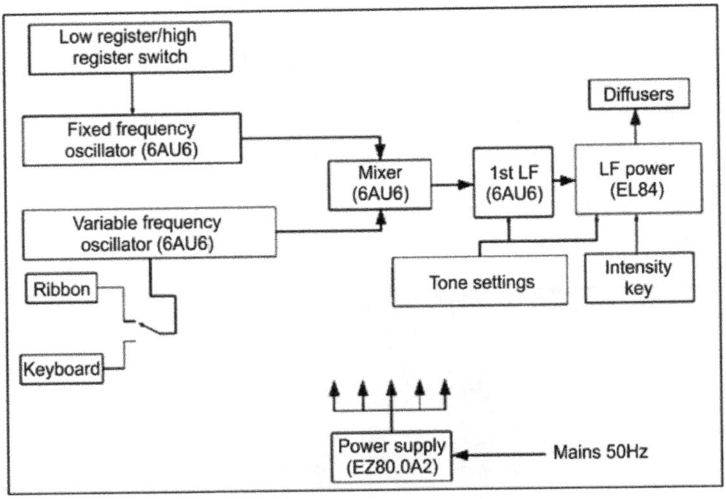

Figure 5.4. *Synoptic diagram of ondes no. 208*

The electrical operation of this ondes is globally identical to that of the ondes built previously by Mr. Martenot, the LF signal results from the mixing of the two signals from two HF oscillators, one of fixed frequency, the other whose variable frequency is controlled by the performer.

5.2.3. *Mechanical/electrical arrangement*

Materially, the instrument is organized into three sub-assemblies:

– The tinned sheet iron chassis which supports: the mixer, the two heterodynes, the LF amplifier, the power supply, the on/off switch, the various adjustment devices: T7 tone, T0 tone, low/high register switch, the various capacitor and inductor tuning settings.

– The keyboard/ribbon sub-assembly which is made up of two interlocking structures: a frame-shaped structure made from aluminum sections and bent sheet metal elements painted black; at the four corners of this structure are placed the ribbon guide pulleys. The U-shaped rear aluminum profile contains all the components making up the ribbon capacitor. The variable fine-tuning capacitor is arranged on one of the walls of the U-section.

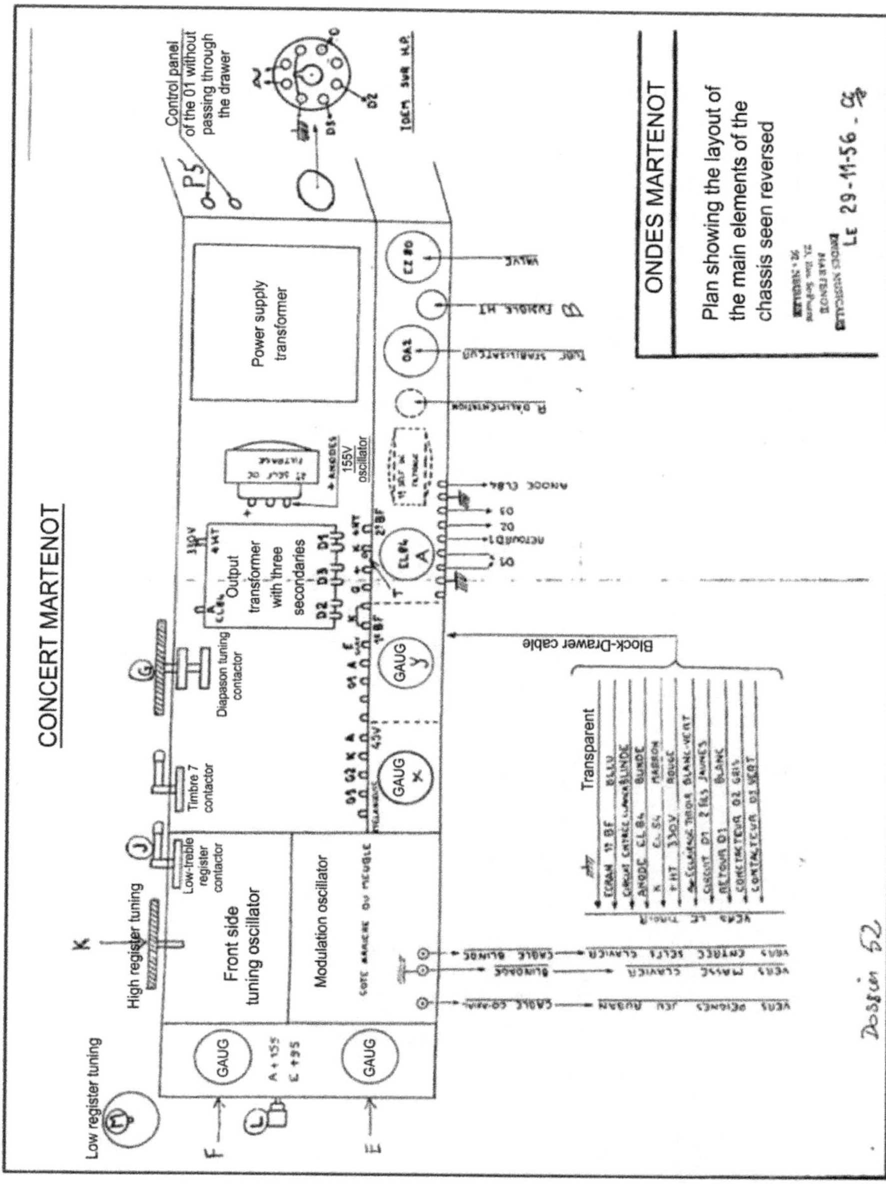

Figure 5.5. *General organization of the chassis, drawing by Mr. Martenot's workshop (Musée de la Musique, Paris)*

Inside this device is suspended by nylon threads a second structure in the form of a frame which can move laterally. This mobile keyboard is also made up of aluminum profiles and Zamak reinforcement parts. The keys are made of white or black plastic material, and they control the moving contacts which activate (or don't activate) the coils which define the musical notes. These coils are aligned and fixed on a hardboard support, itself fixed on the mobile frame. Another subset of coils (whose mode of operation and location we will specify later on), arranged on the mobile structure, allows for the control of the vibrato.

– The drawer consisting of a black painted iron sheet box on which are arranged most of the controls: timbres (1 to 6 and Octaviant), LF power adjustment, selection of diffusers, ribbon/keyboard switch, the needle block, +/-1/4 tone adjustment and intensity key. Also placed on this unit are a miniature incandescent light indicator and the connectors necessary for the operation of the instrument: connection of the pedal, shielded connections to the heterodynes and set of conductors (drawer/chassis connections) grouped together in a single cord.

Figure 5.6. *Drawer, top view*

Figure 5.7. *Drawer, left side*

5.3. Operation analysis

5.3.1. *Oscillators*

Figure 5.8. *Oscillator tubes, low register tuning inductor, fixed-frequency oscillator tubular-piston-type capacitor*

As with all previous models, the oscillators are the key elements in the operation of the instrument designed by Mr. Martenot. When we look at the mechanical/electrical designs, we find the same implementation precautions set out for the design of the '47 model oscillators, namely:

– coils without magnetic core, and strictly identical on the two oscillators;

– importance of the shielding in order to preserve the initial value of the coil after introduction of this one in the shielding;

– moving outside of the shielding of electronic tubes to limit temperature drifts;

– wiring in conductive wire of large section or use of twisted pairs to reduce drifts by mechanical deformations.

Figure 5.9. *Ondes 208 oscillators. The wiring is carried out according to the recommendations cited above. Note (circled in white), the antenna/C33 (see section 5.3.3)*

5.3.2. *The fixed-frequency oscillator*

This oscillator generates two fixed frequencies according to the register we choose: for the low register, the frequency is 80 kHz; for the high register, it is 160 kHz. A diagram is given in Figure 5.11.

The registers are selected by switch J placed on the front face of the chassis. It is an ECO-type oscillator using the 6AU6 pentode tube; the output coupling is electronic and gives this assembly great stability and low influence of the load on the oscillator. The resonance frequency is determined for the high register by the values of S_2, C_{29}, C_{30} and C_{31}. For the low register, capacitors C_{29}, C_{30} and C_{31} are out of circuit and a series structure composed of S_3 and $C_{34}//C_{32}$ is connected in parallel to the terminals of coil S_2. Other links with the oscillator (C_{33}, timbre 7) are ignored at this stage; we will come back to the role of these elements later.

Capacitor C_{30} is of the tubular-piston trimmer type, it allows for the adjustment of the working frequency; capacitor C_{31} of very small capacity (~4 pF) is the fine tuning capacitor of the treble register; it is controlled by a wheel placed in front on the chassis, due to the very low variation in capacity; it was manufactured by Mr. Martenot himself in the first models.

The frequency in the low register is adjusted by the tubular-piston-type capacitor C_{32}, but the fine adjustment is obtained by the movement of a magnetic core which passes through the coil of the inductor S_3. This device is also a manufacture of Mr. Martenot. A core made of small ferromagnetic sheets slides inside the rectangular section coil support. The self-induction coefficient varies approximately from 8 μH to 22 μH. The displacement of the core is obtained by means of an endless screw fixed to a wheel made of plastic material. This device is placed on the front face and on the left of the chassis.

Figure 5.10. *Ondes 208 low register tuning inductance*

Components R_{26}, R_{27} and C_{35} ensure the operation of the oscillator, and capacitor C_{36} decouples grid G_2, which is supplied by a stabilized voltage of the order of 95 V. The plate is biased via resistors R_{28}, R_{29} and R_{30}, by a voltage of 155 V, which is also regulated. Capacitor C_{37} decouples this voltage.

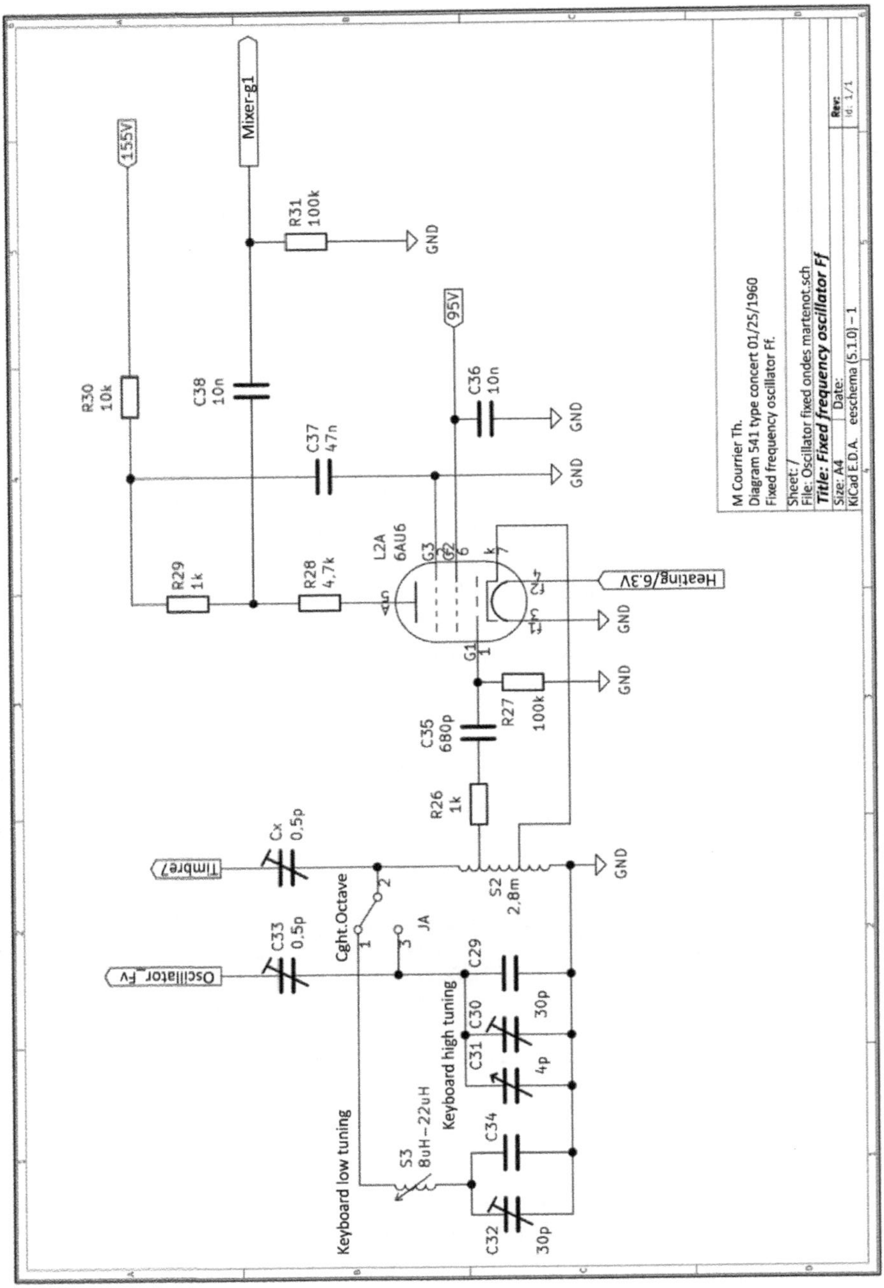

Figure 5.11. *Diagram of the fixed oscillator of ondes no. 208*

The signal is picked up by capacitor C_{38} at the nodal point of the divider formed by resistors R_{28} and R_{29}; it is injected into grid G_1 of the mixer tube. The properties of the pentode tube (steep slope, high output resistance, sharp cutoff) induce a particular shape of the signal at the output of the oscillator. Indeed, the voltage delivered on the plate of the electron tube is not sinusoidal. It is closer to a rectangular signal with a low duty cycle than to a sinusoidal one. In stabilized operation, this voltage has an amplitude of 67 Vcc.

5.3.3. *The variable-frequency oscillator*

This oscillator must generate a high-frequency signal f_v in such a way that the difference with the signal generated by the fixed-frequency oscillator f_f produces the audible frequency signal $f_{musical} = f_f - f_v$

Of course, the performer must have control of this variation; it is obtained through the mobile keyboard or the ribbon. Initially to simplify the study, we place ourselves in the case of a playing on the low register with f_f = 80 kHz, the ribbon capacitor is likened to a simple C_{ribbon} variable capacitor and the keyboard is likened to a variable inductance $S_{keyboard}$. A diagram is given in Figure 5.12.

Overall, it is the same type of oscillator as the previous one; it is structured around inductance S_2 and capacitor C_3, which constitute the main elements of the resonant circuit. Pentode 6AU6 and components R_1, R_2, R_6, R_7, R_8, S_2, C_3, C_4, C_6 and C_7 ensure the operation of the oscillator as in the previous case. But this oscillator must generate a frequency f_v between 76 kHz and 80 kHz, in order to produce by electronic mixing with the signal generated by the oscillator of fixed frequency f_f = 80 kHz, the set of audible musical frequencies which extends globally at 4 kHz, but in this model, the range of audible frequencies is divided into two parts, low register (Do_0 to Si_5) and high register (Do_1 to Si_6). We will return in detail to this operating principle of the instrument.

In ribbon play mode, this operation is obtained by connecting point (B) to ground, placing C_1 and C_2 in parallel with C_3 and $C_{41A/B}$, and commissioning the C_{ribbon} capacitor, which is thus connected in parallel with the previous capacitors.

Capacitors C_{54}, C_{55} and C_{56} constitute the needle block: when the performer touches with their finger one or the needles, which connect the floating poles of these capacitors, they establish a connection to ground, which modifies the resonance frequency of the variable oscillator.

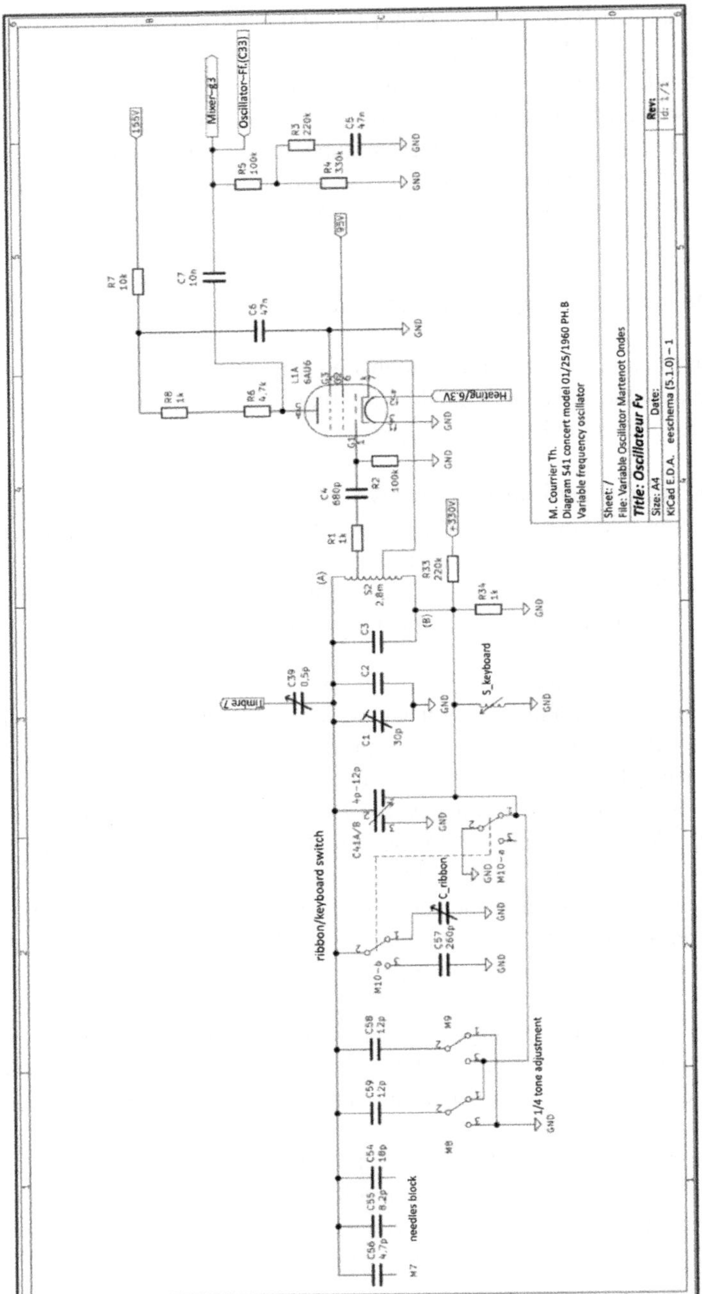

Figure 5.12. *Diagram of the variable oscillator of ondes no. 208*

Capacitors C_{58} and C_{59} are intended (via push buttons M8 and M9) to correct the operating frequency of the oscillator in order to produce an adjustment of +/- 1/4 tone on the musical frequency. They are only operational in keyboard mode.

In keyboard play mode, with switch M10, the ribbon capacitor is isolated and a compensation capacitor C_{57} is paralleled with C_1, C_2 and $C_{41A/B}$.

The keyboard variable inductor is inserted between point (B) and ground, resulting in a more complex series-parallel structure of the oscillating circuit. The nodal point (B) is taken to a very low continuous potential V(B) = 330 * 1/220 = 1.5 V, to minimize switching noise between ribbon play and keyboard play; indeed, by this process, the cathode of the oscillator tube always has a defined potential.

The $C_{41A/B}$ differential capacitor has a particular role. Mr. Martenot explains in his patent no. 841-866 (Martenot 1939) the operation of this device. It makes it possible to obtain transpositions of small importance to adapt to different tuning. Mr. Martenot will name this adjustment device: diapason.

Initially it is carried out using a two-way switch, for each change in the position of the switch, on one way a capacitor is added in parallel, on the other way it is disconnected, so in this way, on a channel the capacitance increases by one step and simultaneously on the other channel, there is a decrease of the same value. The pitch chosen by Mr. Martenot corresponds to the lowest capacitance value available for a capacitor at that time: 2 pF.

The switch has six positions; the total variation is therefore 5 x 2 pF, with an open circuit position for each of the channels.

Subsequently, Mr. Martenot will use the differential variable capacitor. This device is not common, it was found mainly in electronic equipment for aeronautics. It had a major defect: the common point of the two variable capacitors, i.e. the mobile armature, was not electrically isolated but connected to the ground of the capacitor frame. To be usable in the configuration proposed in the diagram of his instrument, Mr. Martenot will have to make on the one hand an insulating support to fix the device to the chassis of the instrument, and on the other hand isolate the control shaft from the user.

Finally to complete this study, the resistors R_4 and R_5 fix the DC grid potential G_3 of the mixer tube to 0 V. When the fixed-frequency oscillator operates on the high register, it produces a frequency of 160 kHz, the variation of the frequencies of the variable oscillator always extends from 78 kHz to 80 kHz and the synthesis of the audible frequency must therefore be done on the harmonic two of these frequencies (156 kHz to 160 kHz). By similarity with the operation of tube radio receivers, we put forward the following hypothesis: the mixing of frequencies on harmonic 2 is favored by the injection of a very small fraction of energy from the fixed oscillator by C_{33} and is injected at the C_7/R_5 nodal point. The choice of the point of the energy sampling is variable according to the ondes constructed; on ondes no. 222, this point is located at the level of the junction of the grid capacitor C = 680 pF and the conductor connected to the grid of the tube oscillator.

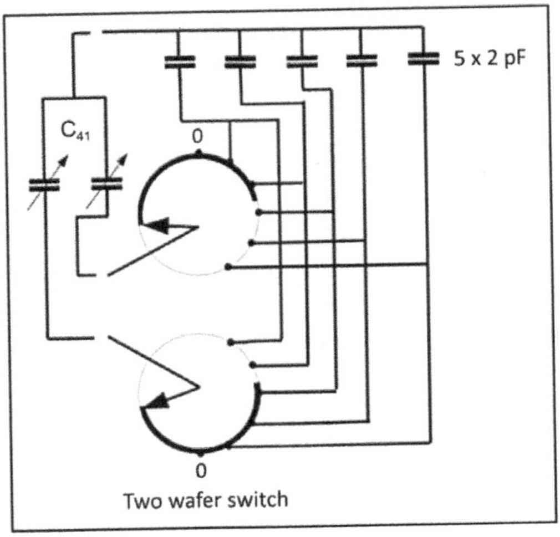

Figure 5.13. *Principle of operation of the C_{41} differential capacitor*

Figures 5.14. *Differential capacitor $C_{41A/B}$: (Left) made from a wafer switch (ondes no. 208). (Right) differential air variable capacitor (ondes no. 222)*

This process is similar to that used in old radio receivers to improve the operation of an HF transformer in a wave range. A pigtail connection was formed by winding on the main conductor connected to the mixer tube, a few turns of insulated conductor, which constituted a capacitor of very low capacity. This conductor is connected to the HF source. We can also think that it is an experiment of Mr. Martenot whose technical objective to achieve is not elucidated.

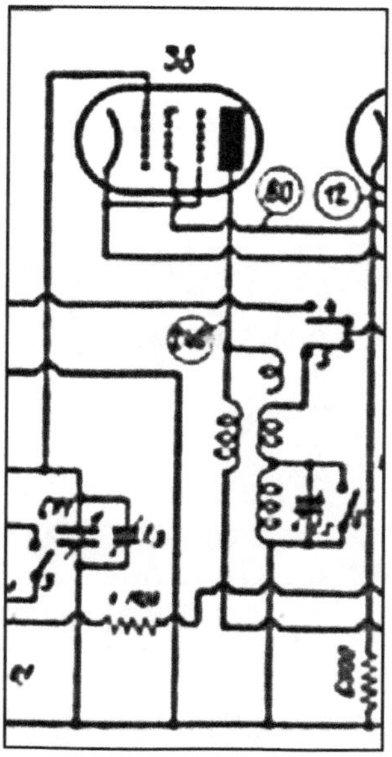

Figure 5.15. *HF coupling, extract from the manufacturer's diagram of the Grammont model 37 receiver*

The R_3/C_5 filter placed in parallel on the R_4 resistor is a remnant of a circuit which in the very first version (Model 1947, see Figure 4.2) consisted of taking from the grid of the oscillator tube a fraction of the signal which is then mixed with the output signal of this same oscillator; this circuit has been abandoned.

5.4. The ribbon capacitor

5.4.1. *General provisions*

The ribbon capacitor built by Mr. Martenot for this instrument conforms to the descriptions he gives in these patents. The development of this device is extremely delicate because the order of magnitude of the capacitance values ranges from one-hundredth of a pF to approximately 60 pF. By way of comparison, the capacitance variation between the hand and a conductive metal rod varies from 0 pF to 1.3 pF, when the hand approaches 20 cm up to approximately 1 cm from this conductor. This property is implemented in the design of musical instruments by

Theremin. Thus, when the protective cover of the ribbon capacitor is removed, the note played on the instrument can be raised by a half tone, just by approaching the hand near the walls of this capacitor.

Figure 5.16. *The ribbon capacitor, a module and the series capacitors*

For this model, the fixed parts are an assembly of six pairs of blades (150 mm * 30 mm) in a 0.5 mm aluminum sheet; they are split over a height of 24 mm. A main blade has as many divisional blades as musical notes per octave; this was not always the case in previous models. In this model of instrument, the variability in the construction of the ribbon capacitor is great, so, for example, the ondes no. 222 has a ribbon capacitor whose walls are made of thin tinned sheet metal, similar to that used to make boxes for storing foodstuffs.

The ribbon identical to those used in the previous models has a conductive part with a height of 10 mm.

Mechanically, to ensure parallel movement of the ribbon, the main blades are grouped and aligned in three pairs. These two groups define a broken line in the middle, the blades are mounted on plastic wedges and these wedges are then fixed to the bottom of a U-shaped shield (l = 35 mm, h = 40 mm, L = 995 mm) and made of 1.5 mm thick aluminum sheet. This device is closed by a removable cover of the same metal.

The ribbon is regularly guided by rods and sliding wedges; it is stretched between four pulleys by a piece of elastic material ribbon sewn to the left end. The remaining end of the elastic ribbon is connected by a hooking system to a ring, as well as the other end of the ribbon. When the operator places the ring on the far left of the instrument, the conductive part is outside the fixed walls of the capacitor, the frequency of the variable heterodyne is the highest and the musical note is the lowest. When the ring is on the very right of the instrument, all of the conductive part of the ribbon is inside of the walls, the frequency of the variable heterodyne is the lowest and the musical note generated is the highest.

In this model, the *glissando* covers six octaves. A switch allows us to choose the low register (octave 0 to 5) or the high register (octave 1 to 6).

Figure 5.17. *The ribbon capacitor, the guide rod and the sliding wedge*

5.4.2. *Sizing*

5.4.2.1. *Problem*

In his patent (Martenot 1947), Mr. Martenot fairly quickly explains the principles used to make a ribbon capacitor. Detailed development is required.

To build the ribbon capacitor, we will use two relationships: the first defines the resonance frequency F(Hz) of a dipole consisting of an inductance L(H) in parallel with a capacitor C(F), and the second defines the value of the capacitor C(F) as a function of the geometric parameters: the surface S(m^2) of the two conductive metal planes, parallel and separated by a distance e(m).

$$\boldsymbol{F} = \frac{1}{2\pi\sqrt{LC}} \text{ and } \boldsymbol{C} = \varepsilon \frac{S}{e}$$

where ε is the permittivity of vacuum or air.

The calculations are simplified by considering only the increase in capacity ΔC(x) that must be obtained to pass from the previous musical frequency to the next. To do this, the capacitor is made up of a set of pairs of blades of conductive metal of width d located at the distance n*Delta(x) (with 0≤ n≤72), from the origin. This origin is defined by the entry of the ribbon by the edges located to the right of the first pair of blades.

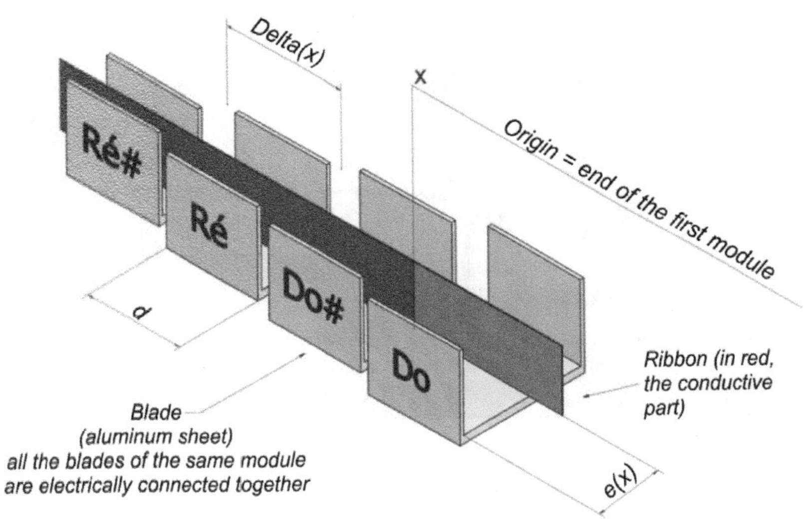

Figure 5.18. *Block diagram of the ribbon capacitor*

The median of the advance step Delta(x) of the ribbon and the median of the walls facing the ribbon are aligned, so the surface S of the capacitor is the product h*d, where h is the height of the ribbon; this product is constant. The capacitor thus constituted has the value:

$$\Delta C(x) = 2\varepsilon \frac{hd}{e(x)}$$

Thus, in this way, the value of the deviation e(x) of the strip relative to a wall of the pair of blades is calculated step by step. Note that the total size is 2e(x) + the thickness of the ribbon.

We have chosen a blade width equal to 12.4 mm because it corresponds to 1/12 of the total length of the module, taking into account an insulation space of the order of 0.11 mm. We will see at the end of this part that the accuracy of the calculations is sufficient.

Let us recall the data that define the ribbon capacitor and the associated oscillators:

– The variable capacitor C_v:

- fixed armature d = 12.4 mm, grouping by octave, 12 blades per octave over 150 mm;

- movable armature (ribbon), h = 10 mm, elementary surface S = h*d = 124 mm^2;

- vacuum or air permittivity $\varepsilon = 8.85E\text{-}12$ F.m^{-1};

- step width (to move from one key to another) Delta(x)= 12.5 mm;

- total length 72 *12.5 = 900 mm → $0 \leq x \leq 900$.

– The oscillating circuit is an inductor L_0 = 2.8 mH in parallel with a fixed capacitor C_0 # 1412 pF.

– The frequency of the fixed oscillator is f_f= 80 kHz.

A diagram is given in Figure 5.19.

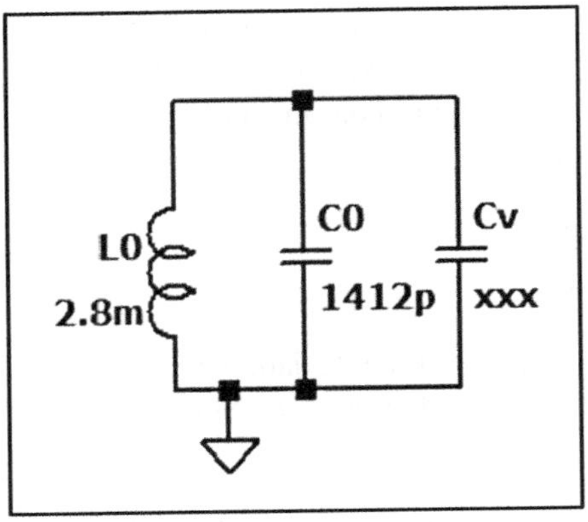

Figure 5.19. *Ribbon capacitor circuit diagram*

Let us recall the values of the musical frequencies (Hz) on the first six octaves:

Octave	0	1	2	3	4	5
Do	32.70	65.41	130.81	261.63	523.25	1046.50
Do#	34.65	69.30	138.59	277.18	554.37	1108.73
Ré	36.71	73.42	146.83	293.66	587.33	1174.66
Ré#	38.89	77.78	155.56	311.13	622.25	1244.51
Mi	41.20	82.41	164.81	329.63	659.26	1318.51
Fa	43.65	87.31	174.61	349.23	698.46	1396.91
Fa#	46.25	92.50	185.00	369.99	739.99	1479.98
Sol	49.00	98.00	196.00	392.00	783.99	1567.98
Sol#	51.91	103.83	207.65	415.30	830.61	1661.22
La	55.00	110.00	220.00	440.00	880.00	1760.00
La#	58.27	116.54	233.08	466.16	932.33	1864.66
Si	61.74	123.47	246.94	493.88	987.77	1975.53

Table 5.1. *Frequency (Hz) of tempered scale notes (La_3 =440 Hz)*

The values of the total capacities $C_0 + C_v$ (in nF) are given in Table 5.2.

Octave	0	1	2	3	4	5
Do	1.41468	1.41584	1.4181	1.4228	1.4322	1.4512
Do#	1.41475	1.41597	1.4184	1.4234	1.4333	1.4535
Ré	1.41482	1.41612	1.4187	1.4240	1.4345	1.4560
Ré#	1.41490	1.41627	1.4190	1.4246	1.4358	1.4586
Mi	1.41498	1.41644	1.4194	1.4252	1.4371	1.4613
Fa	1.41506	1.41661	1.4197	1.4259	1.4385	1.4642
Fa#	1.41516	1.41680	1.4201	1.4267	1.4400	1.4673
Sol	1.41525	1.41699	1.4205	1.4275	1.4416	1.4706
Sol#	1.41536	1.41720	1.4209	1.4283	1.4433	1.4741
La	1.41547	1.41742	1.4213	1.4292	1.4451	1.4778
La#	1.41558	1.41765	1.4218	1.4301	1.4470	1.4818
Si	1.41571	1.41789	1.4223	1.4311	1.4491	1.4860

Table 5.2. *Total capacitance $C_0 + C_v$ (in nF i.e. 10^{-9}F)*

We deduce the increases ΔC of the capacitor C_v in pF in Table 5.3.

Octave	0	1	2	3	4	5
Do	0.0647	0.1300	0.2606	0.5241	1.0579	2.1566
Do#	0.0690	0.1378	0.2763	0.5551	1.1222	2.2904
Ré	0.0729	0.1460	0.2927	0.5886	1.1900	2.4325
Ré#	0.0771	0.1545	0.3103	0.6244	1.2624	2.5838
Mi	0.0817	0.1641	0.3288	0.6617	1.3398	2.7448
Fa	0.0867	0.1737	0.3485	0.7015	1.4211	2.9165
Fa#	0.0920	0.1840	0.3696	0.7436	1.5079	3.0997
Sol	0.0973	0.1951	0.3915	0.7890	1.6002	3.2945
Sol#	0.1030	0.2068	0.4148	0.8360	1.6984	3.5028
La	0.1094	0.2189	0.4399	0.8870	1.8025	3.7245
La#	0.1158	0.2321	0.4662	0.9403	1.9135	3.9617
Si	0.1229	0.2460	0.4942	0.9974	2.0314	4.2141

Table 5.3. *Increases ΔC of capacitor C_v (in picofarad or 10^{-12}F)*

We deduce the gap e(x), in mm, between the ribbon and the fixed blade, according to the note played, as well as according to the distance x.

Octave	0	1	2	3	4	5
Do	33.923	16.883	8.421	4.188	2.075	1.018
Do#	31.810	15.926	7.943	3.954	1.956	0.958
Ré	30.110	15.035	7.497	3.729	1.844	0.902
Ré#	28.450	14.205	7.074	3.515	1.739	0.849
Mi	26.847	13.374	6.674	3.317	1.638	0.800
Fa	25.310	12.635	6.297	3.129	1.544	0.753
Fa#	23.848	11.927	5.938	2.952	1.455	0.708
Sol	22.545	11.252	5.606	2.782	1.372	0.666
Sol#	21.303	10.613	5.291	2.626	1.292	0.627
La	20.060	10.026	4.989	2.474	1.218	0.589
La#	18.953	9.456	4.708	2.334	1.147	0.554
Si	17.859	8.922	4.441	2.201	1.080	0.521

Table 5.4. *Gap e(x) between the ribbon and the fixed wall of the slats (in mm)*

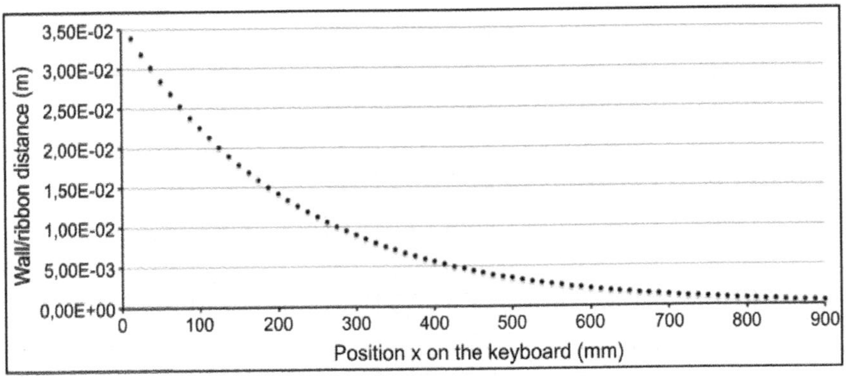

Figure 5.20. *Wall/ribbon distance for the entire keyboard*

It is observed that the fixed wall has a flared shape. This shape results, on the one hand, from the discontinuity which appears when the ribbon is positioned at the entrance to the first pair of blades. Indeed to the left of the point x = 0 (on the graph), the capacitance of the ribbon capacitor must be zero, which results in e (x = 0) tending towards + ∞. On the other hand, the capacitance variation is weak on the low-frequency side and strong on the high-frequency side. In practice, the ratio of the two dimensions has the value:

$$\frac{e(x=0)}{e(x=L)} \approx 65$$

And note that the total size in width is E = 2 * 3.39 = 6.78 cm neglecting the thickness of the ribbon. At the other end, e(x = L) is measured in several 1/10 of a millimeter, so this capacitor is not technically feasible!

A first necessary intervention consists of reducing the increase in the value C at high musical frequencies so as to obtain values of e(x) of the order of a millimeter. To do this, Mr. Martenot proposes to associate with the main resonant circuit L_0C_0, an auxiliary resonant circuit L_1C_1 with a higher resonant frequency than that of the main circuit. The following diagram shows this.

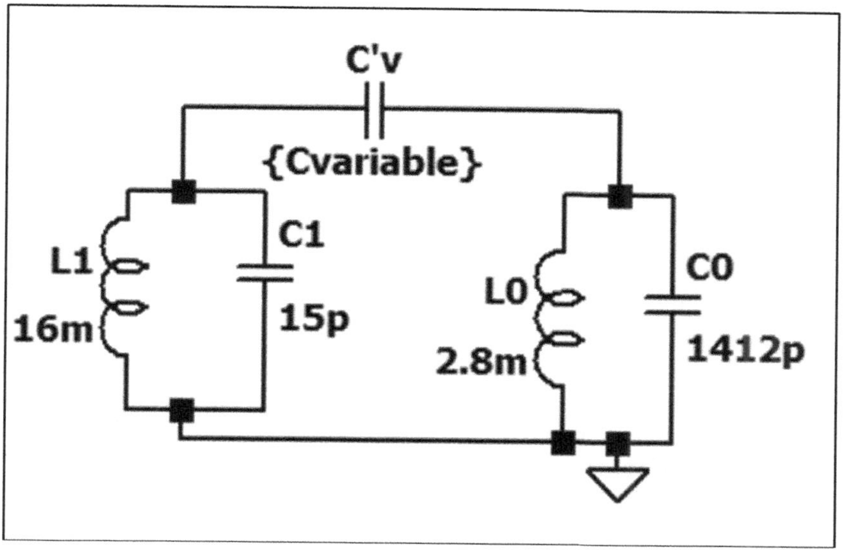

Figure 5.21. *Diagram of the auxiliary circuit of the ribbon capacitor*

In this specific case, the resonance frequency of the auxiliary circuit is of the order of 315 kHz. The laws that govern this circuit are those of coupled circuits and give this diagram the following properties:

– the natural frequency of each resonant circuit is unchanged;

– the variation in capacitance ΔC_v of the capacitor C_v for the same range Δf of the resonance frequencies of the main circuit is reduced.

We carried out a simulation using the Ltspice software in order to determine the variations induced by the settings of the L_1 or C_1 parameters. The initial variable (ribbon) capacitor is C_v, and with the auxiliary resonant circuit, its new value is C'_v.

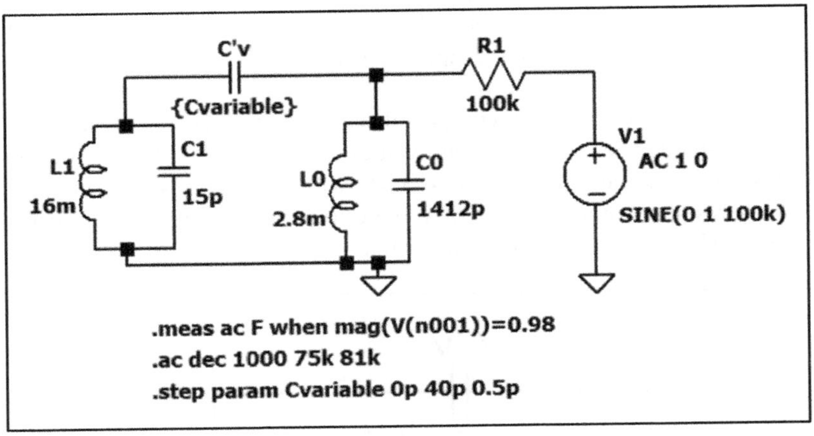

Figure 5.22. *Diagram and parameters for Ltspice simulation of the ribbon capacitor with the addition of the auxiliary circuit*

We get the following results:

– Figure 5.23 shows that the desired effect is indeed obtained for the highest musical frequencies (the notes between Si_4 and Si_5), i.e. at the lowest frequencies of the oscillator. It is also observed that an increase in the value of inductance L_1 reduces the range of variation of C'_v. Thus, for the pair of values (16 mH/15 pF) at the frequency of 78,012 Hz, the value of C'_v is 55.5 pF, whereas without an auxiliary resonant circuit, the value of C_v is approximately 70.5 pF.

– We also resumed these simulations by fixing the value of L_1 and modifying that of C_1, it appears that the effects obtained are much less, and that the possibility of adjustment of the capacitor C_1 is reserved for a fine adjustment.

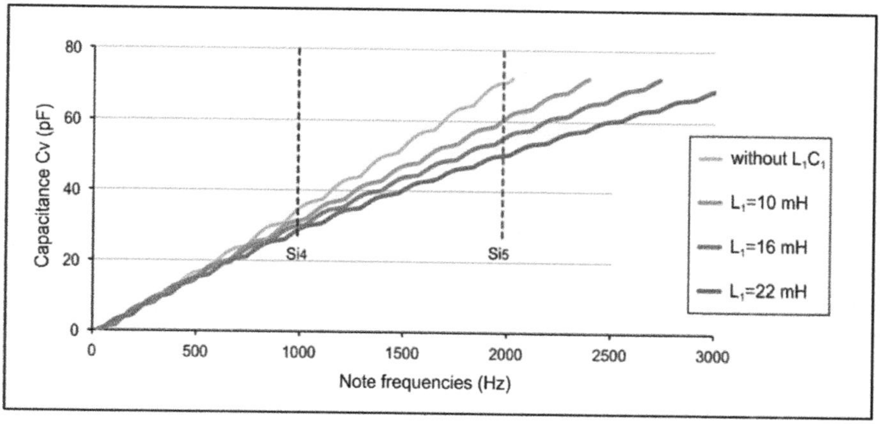

Figure 5.23. *Effect on the capacitance C_v of the addition of the auxiliary circuit (Ltspice simulation; C_1 = 15 pF and L_1 = 10 – 16 – 22 mH)*

The second interesting intervention adopted by Mr. Martenot consists of placing in series with a subset of blades intended for an octave, a series capacitor. He applies this process to octaves 1 to 5, the objective of which is to increase the overall value of the ribbon capacitor in order to reduce the difference e(x). For this, it is necessary to electrically and separately supply each of the sections, by using, for example, impedances of high values, we then find, for each sector, the flared shape caused by the discontinuity of capacitance value which occurs at the input of each module and the frequency evolution law. Now we connect to the main resonant circuit each sector via a series capacitor. To simplify, we define an operating point at a fixed frequency such that the ribbon is in the middle of the section; the overall capacitance value seen by the oscillating circuit is C_{eq}, which verifies:

$$\frac{1}{C_{eq}} = \frac{1}{C_{sec}} + \frac{1}{C_s}$$

where C_{sec} is the value of the capacity of the section at the point previously defined.

In order to keep the same overall value of capacitance C_{eq} seen by the oscillating circuit, it is necessary to increase the overall value of the ribbon capacitor C_{sec}, which is equivalent to reducing the distance e(x) between the walls and the ribbon. Another favorable effect of this structure is the linearization of the values of the capacities of the sector, as we will show in calculations a little later in this part.

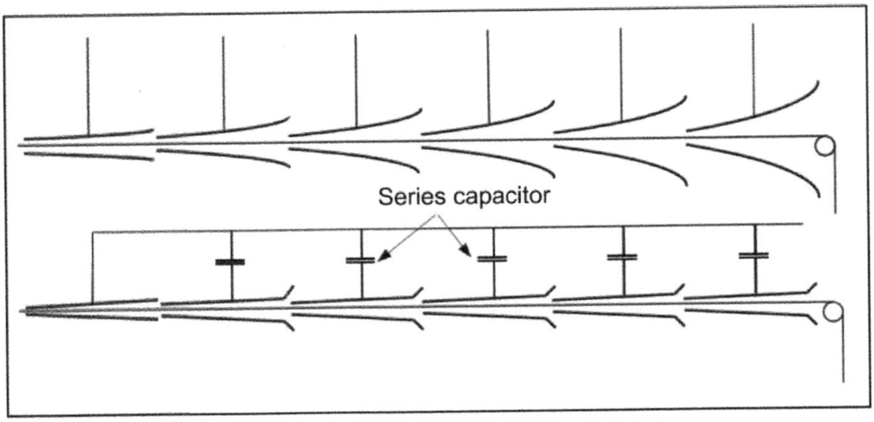

Figure 5.24. *Evolution of the shapes of the armatures of the ribbon capacitor: splitting into modules by octaves and placing the capacitors in series (for a better understanding, the shapes have been accentuated)*

5.4.2.2. *Resolution*

In the previous part, we exposed the various difficulties which appear during the realization of the ribbon capacitor, we also gathered certain data necessary for the dimensioning and we will reuse them in the following calculations.

5.4.2.2.1. Capacitance calculations of series capacitors for octaves 0 to 4

For each octave, we determine the optimal value of C_S, i.e. the value which makes it possible to obtain a gap e(x) at the end of the sector ideally varying from 10 mm to 1 mm. The calculation that we give below is an approximate calculation that takes into account the following observations:

– The total variation in capacitance of the ribbon capacitor is the same, whether it is made up of continuous walls over the entire length of the keyboard or whether it is made up of six successive sections. The cancellation of the capacitance value comes from the "electro-mechanical" break that is introduced at the entrance of each section.

– But very quickly the values of e(x) of a section *n* merge with the values e(x) obtained from a single capacitor.

In this way of proceeding, it will be noted that the value of e(x) at the input of each module must be corrected by adjustment at the time of production.

Octave 0, C_{s0} = 1.30 E-13 F.

	ΔCv(x) *(pF)*	ΔC'v(x) *(pF)*	e'(x) *(mm)*	x *(mm)*
Do	0.0647	0.1290	17.02	12.5
Do#	0.0690	0.1470	14.93	25.0
Ré	0.0729	0.1659	13.23	37.5
Ré#	0.0771	0.1898	11.57	50.0
Mi	0.0818	0.2203	9.96	62.5
Fa	0.0867	0.2604	8.43	75.0
Fa#	0.0920	0.3151	6.96	87.5
Sol	0.0974	0.3877	5.66	100.0
Sol#	0.1030	0.4966	4.42	112.5
La	0.1094	0.6909	3.18	125.0
La#	0.1158	1.0602	2.07	137.5
Si	0.1229	2.2498	0.98	150.0

Table 5.5. *Result of ribbon capacitor calculations for octave 0*

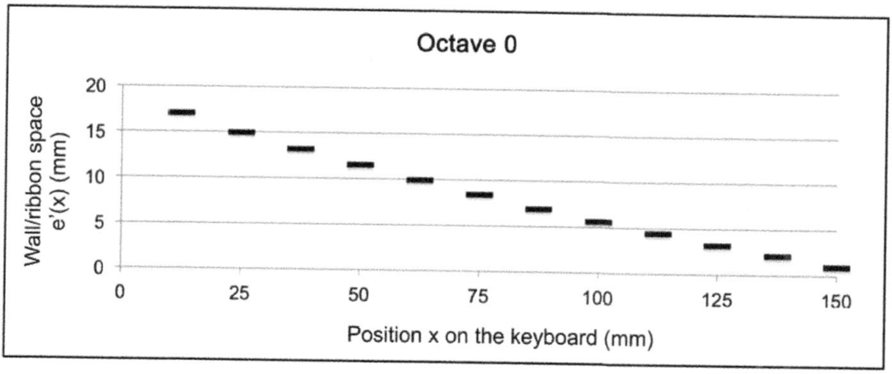

Figure 5.25. *Wall/ribbon distance e'(x) for octave 0*

Octave 1, C_{s1} = 1.0 E-11 F.

	ΔCv(x) *(pF)*	ΔC'v(x) *(pF)*	e'(x) *(mm)*	x *(mm)*
Do	0.1300	0.1317	16.66	162.5
Do#	0.1378	0.1397	15.71	175.0
Ré	0.1460	0.1481	14.82	187.5
Ré#	0.1545	0.1569	13.99	200.0
Mi	0.1641	0.1668	13.16	212.5
Fa	0.1737	0.1768	12.42	225.0
Fa#	0.1840	0.1875	11.71	237.5
Sol	0.1951	0.1989	11.03	250.0
Sol#	0.2068	0.2112	10.39	262.5
La	0.2189	0.2238	9.81	275.0
La#	0.2321	0.2376	9.24	287.5
Si	0.2460	0.2522	8.70	300.0

Table 5.6. *Result of ribbon capacitor calculations for octave 1*

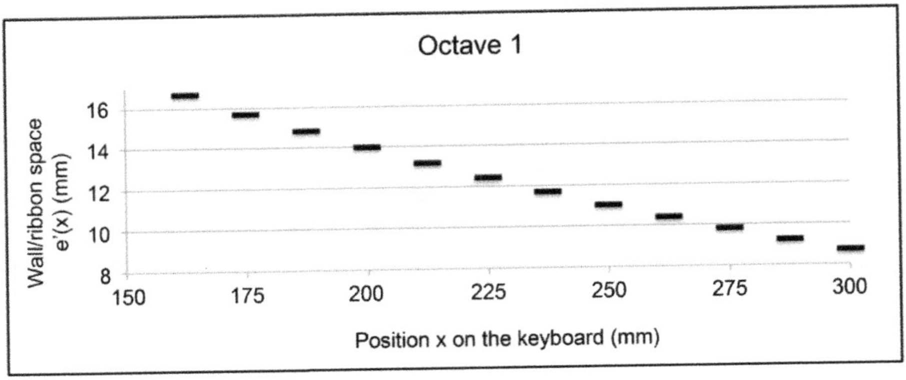

Figure 5.26. *Wall/ribbon distance e'(x) for octave 1*

Octave 2, C_{s2} = 2.70 E-11 F.

	ΔCv(x) *(pF)*	ΔC'v(x) *(pF)*	e'(x) *(mm)*	x *(mm)*
Do	0.2606	0.2632	8.34	312.5
Do#	0.2763	0.2792	7.86	325.0
Ré	0.2927	0.2960	7.42	337.5
Ré#	0.3103	0.3139	6.99	350.0
Mi	0.3288	0.3329	6.59	362.5
Fa	0.3485	0.3531	6.22	375.0
Fa#	0.3696	0.3748	5.86	387.5
Sol	0.3915	0.3973	5.52	400.0
Sol#	0.4148	0.4213	5.21	412.5
La	0.4399	0.4472	4.91	425.0
La#	0.4661	0.4744	4.63	437.5
Si	0.4942	0.5034	4.36	450.0

Table 5.7. *Result of ribbon capacitor calculations for octave 2*

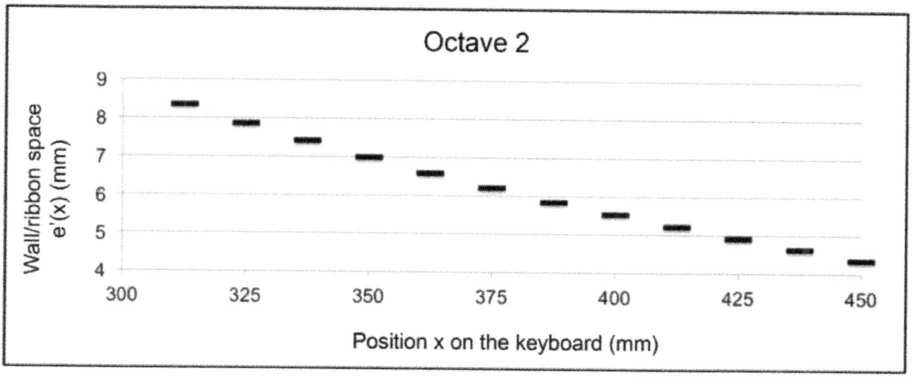

Figure 5.27. *Wall/ribbon distance e'(x) for octave 2*

Octave 3, C_{s3} = 4.70 E-11 F.

	ΔCv(x) *(pF)*	ΔC'v(x) *(pF)*	e'(x) *(mm)*	x *(mm)*
Do	0.5241	0.5300	4.14	462.5
Do#	0.5551	0.5617	3.91	475.0
Ré	0.5887	0.5961	3.68	487.5
Ré#	0.6244	0.6328	3.47	500.0
Mi	0.6617	0.6711	3.27	512.5
Fa	0.7015	0.7121	3.08	525.0
Fa#	0.7436	0.7556	2.90	537.5
Sol	0.7890	0.8025	2.74	550.0
Sol#	0.8360	0.8511	2.58	562.5
La	0.8870	0.9041	2.43	575.0
La#	0.9403	0.9595	2.29	587.5
Si	0.9974	1.0190	2.15	600.0

Table 5.8. *Result of ribbon capacitor calculations for octave 3*

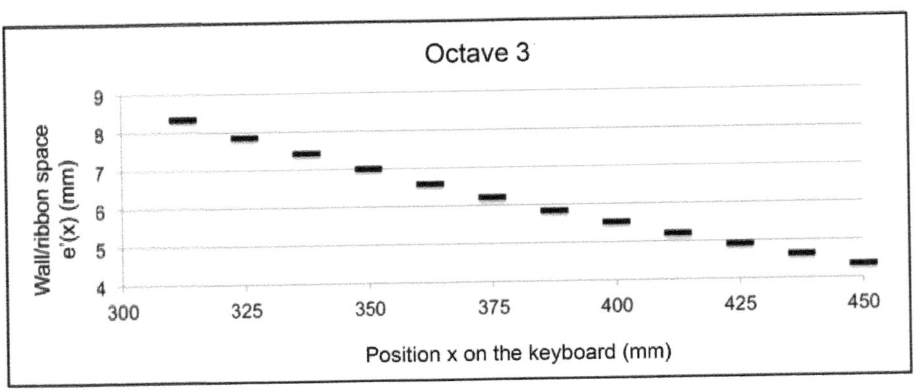

Figure 5.28. *Wall/ribbon distance e'(x) for octave 3*

Octave 4, C_{s4} = 1.50 E-10 F.

	ΔCv(x) *(pF)*	ΔC'v(x) *(pF)*	e'(x) *(mm)*	x *(mm)*
Do	1.0579	1.0654	2.06	612.5
Do#	1.1222	1.1307	1.94	625.0
Ré	1.1900	1.1995	1.83	637.5
Ré#	1.2624	1.2731	1.72	650.0
Mi	1.3398	1.3519	1.62	662.5
Fa	1.4211	1.4347	1.53	675.0
Fa#	1.5079	1.5232	1.44	687.5
Sol	1.6002	1.6174	1.36	700.0
Sol#	1.6983	1.7178	1.28	712.5
La	1.8025	1.8245	1.20	725.0
La#	1.9135	1.9382	1.13	737.5
Si	2.0314	2.0593	1.07	750.0

Table 5.9. *Result of ribbon capacitor calculations for octave 4*

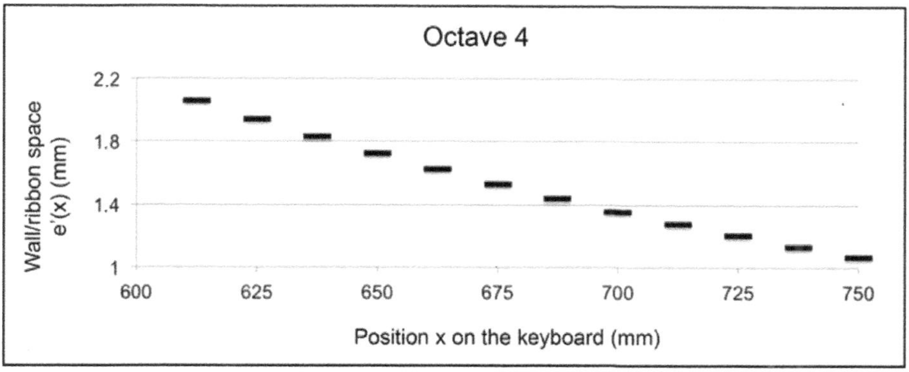

Figure 5.29. *Wall/ribbon distance e'(x) for octave 4*

5.4.2.2.2. Calculation for octave 5

We have shown that for high musical frequencies, to obtain a moderate increase in the values of the ribbon capacitor, it is necessary to give for L_1 values between 10 mH and 22 mH, associated with a capacitance C_1 = 15 pF. In most ondes of this type, capacitor C_1 is a tubular-piston-type trimmer capacitor with an adjustment range of 7.5 pF–25 pF. The inductance is wound in a honeycomb on a mandrel equipped with an adjustable ferrite core, and the adjustment amplitude is quite high; by choosing the average value of 17 mH, we give ourselves a wide adjustment amplitude.

	ΔCv(x) *(pF)*	ΔC'v(x) *(pF)*	e(x) (en mm)	e'(x) *(mm)*	x *(mm)*
Do	2.21	1.50	0.99	1.46	762.5
Do#	2.32	1.62	0.95	1.36	775
Ré	2.46	1.72	0.89	1.28	787.5
Ré#	2.60	1.82	0.84	1.21	800
Mi	2.76	1.93	0.80	1.14	812.5
Fa	2.92	2.04	0.75	1.08	825
Fa#	3.10	2.16	0.71	1.02	837.5
Sol	3.28	2.29	0.67	0.96	850
Sol#	3.48	2.43	0.63	0.90	862.5
La	3.68	2.57	0.60	0.85	875
La#	3.90	2.72	0.56	0.81	887.5
Si	4.13	2.88	0.53	0.76	900

Table 5.10. *Result of calculations of the ribbon capacitor for octave 5*

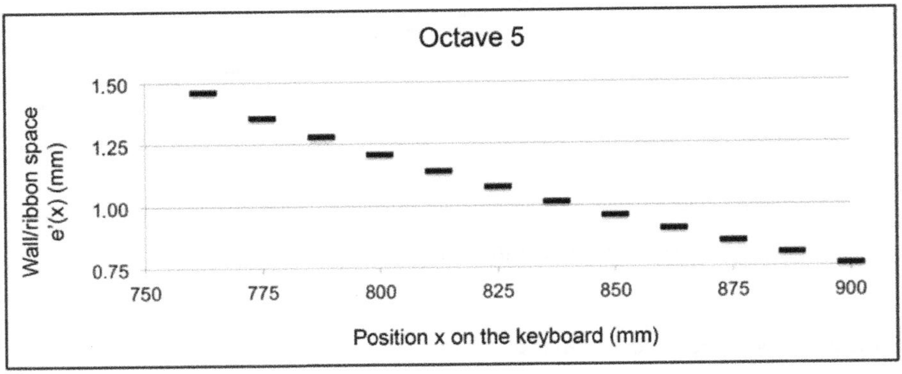

Figure 5.30. *Wall/ribbon distance e'(x) for octave 5*

5.4.3. *Comments*

Before giving the conclusions on these results, we checked that the step (½ tone) of calculation adopted for an octave leads to results of an acceptable precision. For this, we have taken the calculations for octave 2 with a step corresponding to 1/16 tone. The data are therefore as follows: first calculation, 12 intervals d = 12.4 mm, second calculation, 48 intervals d = 3 mm, the height is the same in both cases: h = 10 mm. We get the following results:

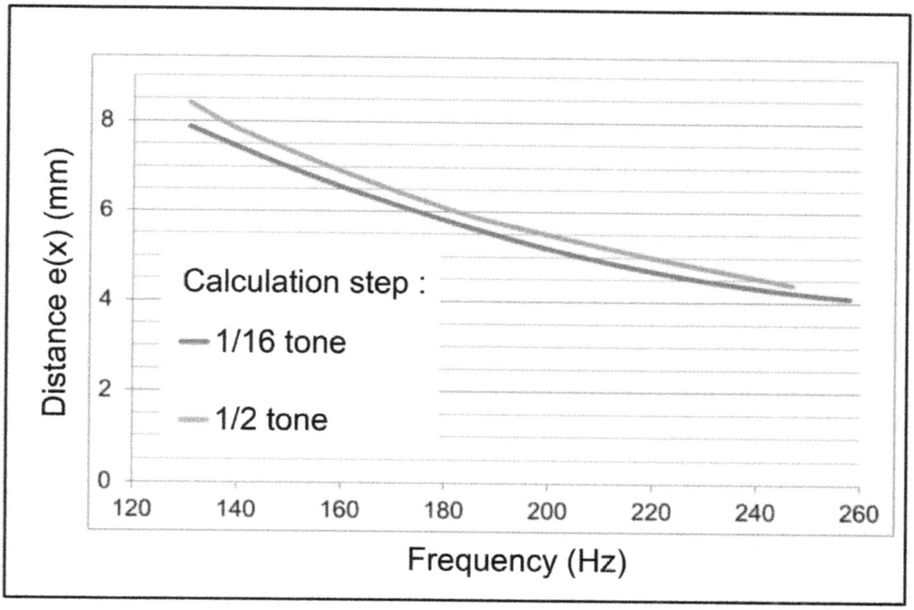

Figure 5.31. *Comparison of calculations of e(x) for octave 2*

The error on the distance e(x) is less than 5/10 mm over the whole octave. Taking into account the mechanical precision which it is possible to obtain by a traditional production of the ribbon capacitor, we deduce that the precision of the calculations is largely sufficient. Tilting the blades parallel to the average line of evolution of the variation of e(x) will allow us to obtain a satisfactory result.

For the correction of the ribbon capacitor of octaves 1 to 4, we obtain values close to those obtained by M. Martenot; for octave 0, the value of C_{s0} chosen by Mr. Martenot is higher than that which we have determined. It should be noted that for this octave, the values of the tuning capacitor are very low, and therefore the space e(x) is very large, and the relation $C = \varepsilon\frac{S}{e}$ is no longer strictly applicable; moreover, many interfering capacitances are no longer negligible.

For the correction of the ribbon capacitor of octave 5, we obtain at the output of the sector a space e(x) = 0.76 mm; this value is low and the risk of contact between the wall and the ribbon is significant, although at the exit, the ribbon is guided by the pulley. It will therefore be necessary to adjust the value of L_1 on the device installed, because here too the interfering capacitances are not negligible.

In addition, all the modules are installed in an aluminum metal shield; capacitive effects are created between the walls of the module and the sides of the shield. Local influences are reinforced because each module is powered via a series capacitor, near the openings of each of the modules which are flared, near the guide rods and near the line break of the ribbon between the modules of the octaves 2 and 3. During the calculations, we disregarded the discontinuity of the capacitance values at the input of each module, resulting in an additional adjustment of the first blade which can sometimes be removed. The realization of the ribbon capacitor and its tuning requires adding here and there local capacitive compensations; thus, we observed the installation of a plane parallel to a wall of the module of the octave 5 connected to the mass, and on another ondes (no. 257) we noted an additional capacitive compensation located under the central staple which fixes the lid of the shielding. Adjusting the ribbon capacitor to obtain tuning is an extremely delicate, long and repetitive operation (more details in section 6.11).

5.5. Keyboard inductors

5.5.1. *General provisions*

The monodic and mobile keyboard associated to a ribbon play (model 5), appears for the first time in the '37 model; let us recall the principle: a set of inductors is put in series using contact actuated under each key of the keyboard, with the inductance specific to the main oscillating circuit. From the outset, the keyboard is mobile and a vibrato system controlled by the lateral displacement of the entire keyboard will be proposed by Mr. Martenot for this model. It is simply a matter of causing the movement of a moving magnetic core in the coil which produces the first note of each octave. This device will be reproduced in model no. 208.

The technical realization does not pose any particular difficulties. Each coil is individually wound on a mandrel equipped with a ferrite core; the value of the self-induction coefficient is adjustable over large amplitude. The coils are arranged and aligned on the underside of the keyboard on an "Isorel" support. They are electrically in series, a conductor connects the common point between two successive inductors to the fixed contact of the key and the moving contact then establishes the connection to ground. The vibrato coils are offset and arranged horizontally in such a way that the mandrel is secured to the mobile frame supporting the keyboard. The sliding magnetic core inside the chuck is mechanically linked to the fixed keyboard frame by a rigid metal wire. In most models, these magnetic cores are made from sheet metal cut into a rectangle 25 mm long and 6 mm wide and shaped into a more or less closed U whose diameter is of the order of 2 mm. The sheet metal, as in the case of the walls of the ribbon capacitor, probably comes from a white metal box (tinned) used for the preservation of foodstuffs.

Figure 5.32. *Central part of the keyboard of ondes no. 208, bottom view*

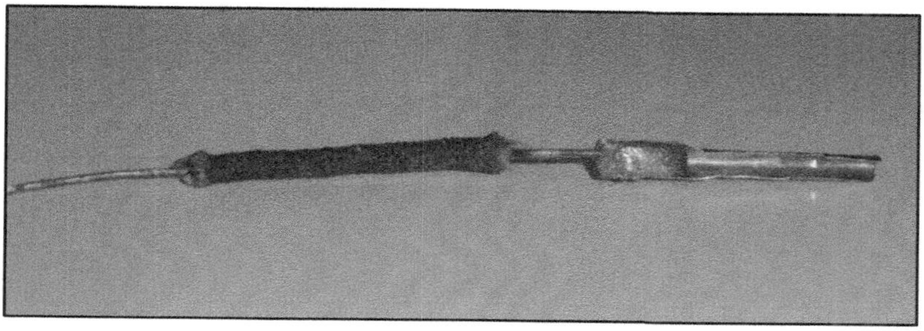

Figure 5.33. *Core of the vibrato inductor, ondes no. 208*

5.5.2. *Sizing*

5.5.2.1. *Problem*

The main oscillating circuit is identical to that used for the ribbon capacitor, for the record:

L_0 = 2.8 mH, C_0 = 1,412 pF, fixed oscillator frequency f_f = 80 kHz.

The corresponding diagram is given in Figure 5.34. By applying the relation which establishes the resonant frequency of a circuit (L_0+L_v)// C_0, we obtain Table 5.11 of the values of the variable inductance L_v.

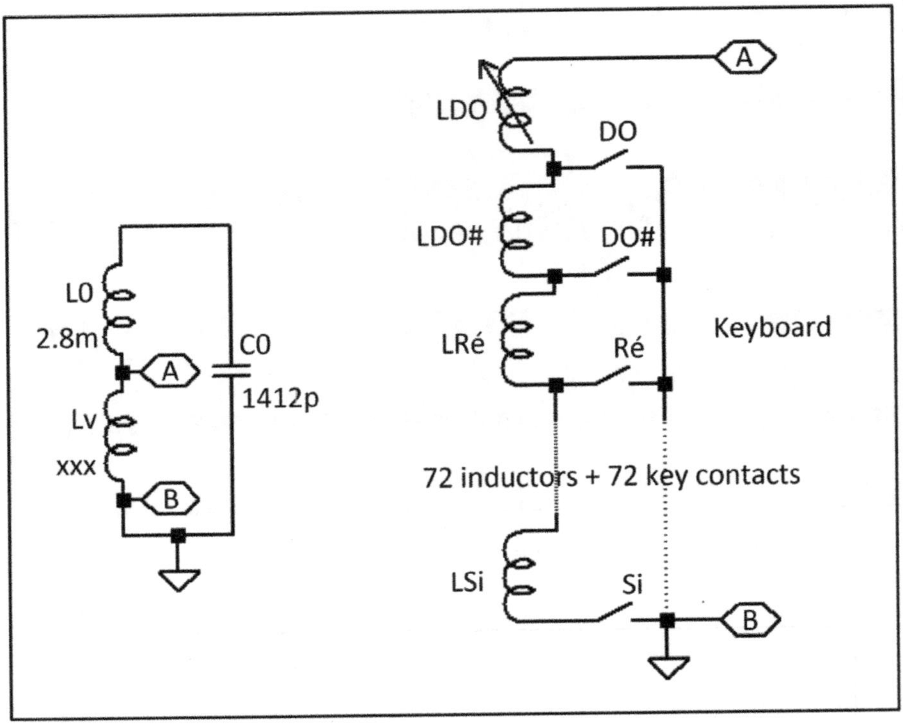

Figure 5.34. *Diagram of the oscillating circuit of the keyboard*

Octave	0	1	2	3	4	5
Do	2.8023	2.8046	2.8092	2.8184	2.8370	2.8747
Do#	2.8024	2.8049	2.8097	2.8195	2.8392	2.8793
Ré	2.8026	2.8052	2.8103	2.8207	2.8416	2.8841
Ré#	2.8027	2.8055	2.8109	2.8219	2.8441	2.8892
Mi	2.8029	2.8058	2.8116	2.8232	2.8467	2.8946
Fa	2.8031	2.8061	2.8123	2.8246	2.8495	2.9004
Fa#	2.8032	2.8065	2.8130	2.8261	2.8525	2.9066
Sol	2.8034	2.8069	2.8138	2.8277	2.8557	2.9131
Sol#	2.8036	2.8073	2.8146	2.8293	2.8591	2.9200
La	2.8039	2.8077	2.8155	2.8311	2.8626	2.9274
La#	2.8041	2.8082	2.8164	2.8329	2.8664	2.9352
Si	2.8043	2.8087	2.8174	2.8349	2.8705	2.9436

Table 5.11. *Calculated values of the variable inductances Lv (mH) of the keyboard*

To go from the previous musical note to the next (i.e. from one key to another on the keyboard), the increment of values is:

Octave	0	1	2	3	4	5
Do	0.128	0.257	0.516	1.038	2.096	4.272
Do#	0.137	0.273	0.547	1.100	2.223	4.537
Ré	0.144	0.289	0.580	1.166	2.357	4.818
Ré#	0.153	0.306	0.615	1.237	2.501	5.118
Mi	0.162	0.325	0.651	1.311	2.654	5.437
Fa	0.172	0.344	0.690	1.390	2.815	5.777
Fa#	0.182	0.364	0.732	1.473	2.987	6.140
Sol	0.193	0.386	0.776	1.563	3.170	6.526
Sol#	0.204	0.410	0.822	1.656	3.364	6.939
La	0.217	0.434	0.871	1.757	3.571	7.378
La#	0.229	0.460	0.923	1.863	3.790	7.848
Si	0.243	0.487	0.979	1.976	4.024	8.348

Table 5.12. *Values of the increase in inductances Lv (µH) of the keyboard*

Thus, the total variation of L_v is 141 μH with a minimum value of 0.128 μH and a maximum value of 8.348 μH. These inductors are not easily realizable. The evolution of Lv values is given in Figure 5.35.

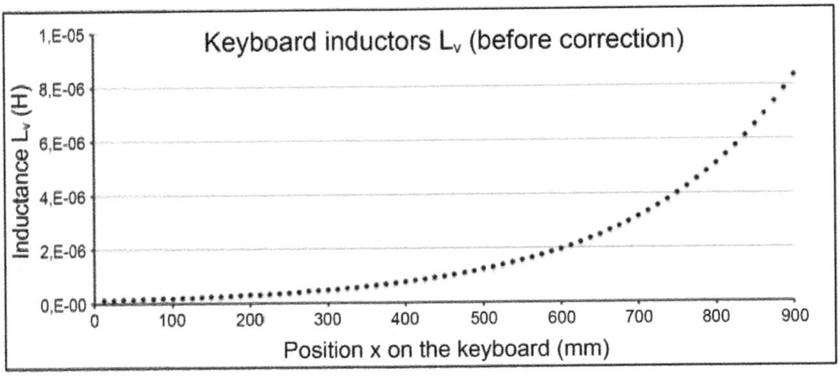

Figure 5.35. *Calculation of the inductances L_v between each key of the keyboard*

5.5.2.2. *Resolution*

Technically, given the relatively low frequencies, it is possible to wind adjustable coils from 0.30 μH. To obtain this value, it is necessary to wind four to five turns of conductor around the mandrel. If the number of turns is too low, the magnetic core no longer plays its role. To increase the values of the inductors, Mr. Martenot uses the following diagram:

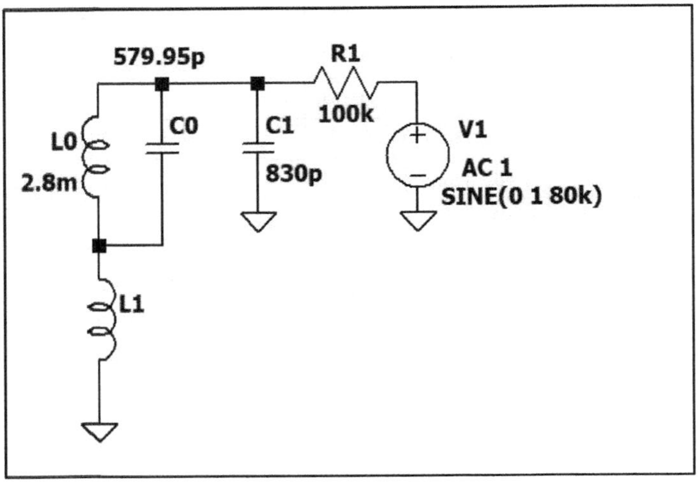

Figure 5.36. *Correction circuit for the keyboard inductors*

Around the main oscillating circuit L_0C_0, it connects in series the new inductor L_1 (variable representative of keyboard inductances) and all put in parallel with the capacitor C_1. To determine the new values of the capacitors C_0 and C_1, we give ourselves a range of total variation of the inductance L_1; on ondes no. 208, this value is approximately 400 μH. Then, we determine the values of C_1 and C_0 so as to cover the range of necessary frequencies; the settings of C_0 and C_1 are not independent, but quickly by successive adjustments, we obtain the following values, C_0 ~ 580 pF and C_1 ~ 830 pF.

A simulation using the Ltspice software gave the table of frequencies according to the values of inductance L_1 as follows:

Sector 1 *7.15 μH <L_1<15.8 μH* ΔL_1 = 0.60 μH		**Sector 2** *15.8 μH < L_1< 30.5 μH* ΔL_1 = 0.80 μH		**Sector 3** *30.5 μH < L_1< 54.0 μH* ΔL_1 = 1.60 μH	
L_1	**f_v**	**L_1**	**f_v**	**L_1**	**f_v**
7.15	79967.2	15.8	79939.7	30.5	79874.7
7.75	79965.3	16.6	79937.1	32.1	79861.0
8.35	79963.5	17.4	79934.4	33.7	79842.2
8.95	79961.6	18.2	79931.7	35.3	79813.4
9.55	79959.7	19.0	79928.9	36.9	79805.9
10.15	79957.8	19.8	79926.1	38.5	79800.9
10.75	79955.9	20.6	79923.2	40.1	79795.8
11.35	79954.0	21.4	79920.2	41.7	79790.8
11.95	79952.1	22.2	79917.2	43.3	79785.7
12.55	79950.2	23.0	79914.0	44.9	79780.7
13.15	79948.3	23.8	79910.8	46.5	79775.7
13.75	79946.4	24.6	79907.4	48.1	79770.6
14.35	79944.5	25.4	79903.8	49.7	79765.5
14.95	79942.5	26.2	79900.1	51.3	79760.4
15.55	79940.6	27.0	79896.1	52.9	79755.1
15.80	79939.7	27.8	79891.9	54.0	79751.4
		28.6	79887.3		
		29.4	79882.3		
		30.2	79876.9		
		30.5	79874.7		

Sector4 *54.0 μH < L_1< 105.0 μH* ΔL_1 = 4.50 μH		Sector5 *105.0 μH < L_1< 210.0 μH* ΔL_1 = 7.50 μH		Sector6 *210.0 μH < L_1< 400 μH* ΔL_1 = 15.0 μH	
L_1	f_v	L_1	f_v	L_1	f_v
54.0	79751.4	105.0	79500.3	210.0	78983.7
58.5	79735.4	112.5	79433.4	225.0	78879.1
63.0	79716.3	120.0	79409.6	240.0	78829.4
67.5	79689.5	127.5	79385.1	255.0	78733.7
72.0	79634.4	135.0	79356.6	270.0	78670.7
76.5	79613.8	142.5	79310.5	285.0	78610.4
81.0	79599.6	150.0	79248.4	300.0	78510.9
85.5	79585.4	157.5	79224.6	315.0	78459.5
90.0	79570.7	165.0	79199.7	330.0	78351.3
94.5	79554.8	172.5	79170.1	345.0	78302.5
99.0	79536.4	180.0	79119.7	360.0	78236
103.5	79511.6	187.5	79063.6	375.0	78143.7
105.0	79500.3	195.0	79039.7	390.0	78090
		202.5	79014.5	400.0	78024.5
		210.0	78983.7		

Table 5.13. *Calculations, for the simulated inductances L_v (μH), of the frequencies f_v (Hz) of the oscillating circuit of the keyboard*

The graphical representation of this table of values is given in Figure 5.37.

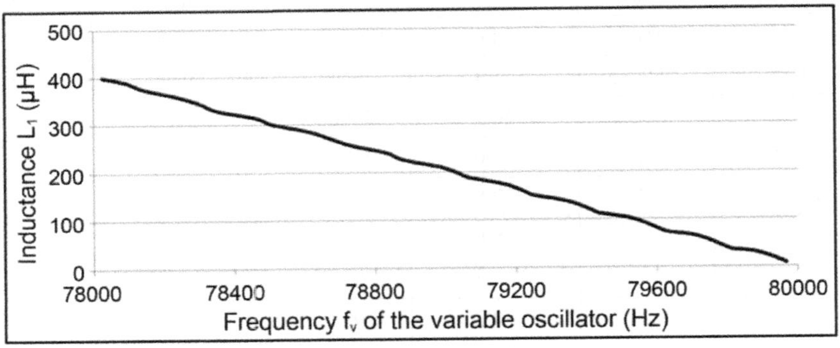

Figure 5.37. *Variation of inductance L_1 as a function of frequency f_v*

Thus, knowing the fixed frequency of the oscillator (80 kHz) and the frequencies of the musical notes (Table 5.1), we then build Table 5.14 giving the values of L_1 according to the musical frequencies and then that giving the increments of L_1.

Octave	0	1	2	3	4	5
Do	0.34	0.74	1.48	2.96	5.92	11.84
Do#	0.39	0.78	1.57	3.13	6.27	12.55
Ré	0.42	0.83	1.66	3.32	6.64	13.29
Ré#	0.44	0.88	1.76	3.52	7.04	14.08
Mi	0.47	0.93	1.86	3.73	7.46	14.92
Fa	0.49	0.99	1.98	3.95	7.90	15.81
Fa#	0.52	1.05	2.09	4.19	8.37	16.75
Sol	0.55	1.11	2.22	4.44	8.87	17.74
Sol#	0.59	1.18	2.35	4.70	9.40	18.80
La	0.62	1.24	2.49	4.98	9.96	19.91
La#	0.66	1.32	2.64	5.27	10.55	21.10
Si	0.70	1.40	2.79	5.59	11.18	22.35

Table 5.14. *Calculation of increments of L_1 (μH) according to the notes*

The graphical representation of Table 5.14 is given in Figure 5.38.

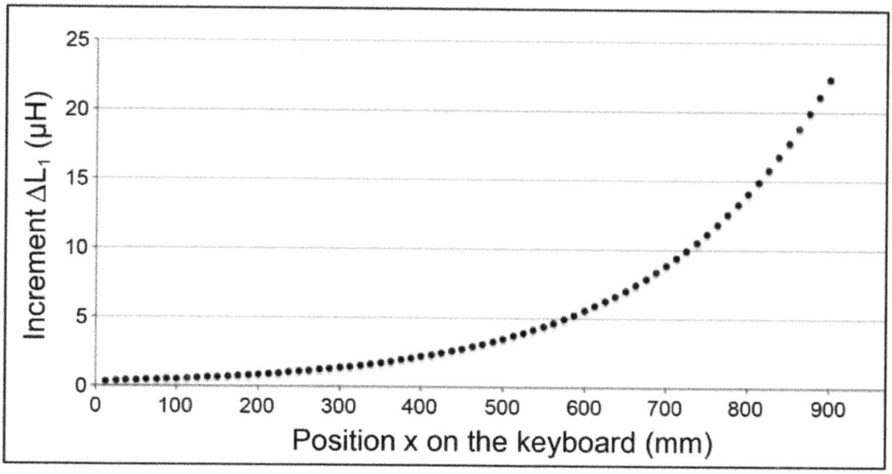

Figure 5.38. *Increase ΔL_1 between each key on the keyboard*

The total variation is 400 μH, with a minimum inductance of 0.34 μH and a maximum value of 22.35 μH. With the exception of the first two octaves, the values of the inductances obtained by these calculations correspond to the values recommended by Mr. Martenot. For these octaves 0 and 1, the values proposed by M. Martenot are lower than those obtained by calculations, but these values are very low, the wiring inductances are no longer negligible, as are the mutual inductances.

Note: it may happen that after the careful adjustments to obtain the musical tuning, there remains a slight shift. Mr. Martenot will add in series with the keyboard coils, an adjustable coil (1 μH–8 μH) which by its adjustment will make it possible to obtain the global setting of the keyboard.

5.5.2.3. *Comments*

In the same way as for the sizing of the ribbon capacitor, the operating principle adopted by Mr. Martenot to obtain a variation of the musical frequencies from a set of inductors encounters technical difficulties. By the judicious choice of a diagram, he manages to make the keyboard. Note that, unlike the ribbon capacitor calibration adjustment, the inductor calibration adjustment is quite simple, but it is cumulative (from Do_0 to Si_5), i.e. it is necessary to adjust the value of the rank n inductor before setting the value of that of row n + 1.

5.5.3. *The register change*

In this ondes, the ribbon or the keyboard covers six octaves from 0 to 5 for the low register and from 1 to 6 for the high register. The analysis that we have just carried out for the realization of these devices shows that the problem is even more delicate for a design of an instrument which relates to a single range of seven octaves. In the study of ondes no. 169 and its model, we pointed out that it was difficult to rigorously tune such an instrument.

It is also probable that Mr. Martenot opted for a system with a change of register to reduce the dimensions of the instrument and make it more compact.

A register extension would suppose in a first approach a setting up of new values of the ribbon capacitor and therefore a module and keyboard inductors. But even better than this: Mr. Martenot will use the second harmonic of the frequencies generated by the variable-frequency oscillator; indeed, it is enough to bring the frequency of the fixed oscillator to the value of 2 * 80,000 = 160 kHz. The mixer will produce the $f_f - f_v$ frequencies as well as $f_f - 2f_v$ and many other multiples. But only frequencies of the types $f_f - 2f_v$ are audible.

Let us give an example: the frequency of La_4 is 880 Hz; in the "low register" position, the oscillator must produce the frequency 80,000 – 880 = 79,120 Hz.

In the "treble register" position, the frequency of the fixed oscillator is doubled. We then obtain the frequencies 160,000 – 79,120 = 63,120 Hz and 160,000 – 880 = 159,120 Hz, which are both inaudible. Only the frequency 160,000 – 2 * 79,120 = 1,760 Hz will be audible, which is the frequency of La_5. So in this way, by keeping the same devices as the ribbon capacitor and the keyboard with switchable inductors, the instrument can cover all seven octaves thanks to a simple switch!

This supposes that the oscillators deliver HF signals rich in harmonics; this is the case of the ECO-type oscillators implemented by Mr. Martenot. When switching from "low register" to "high register", each note is shifted one octave to the right.

The simplified diagram of the oscillator and register switch is as follows:

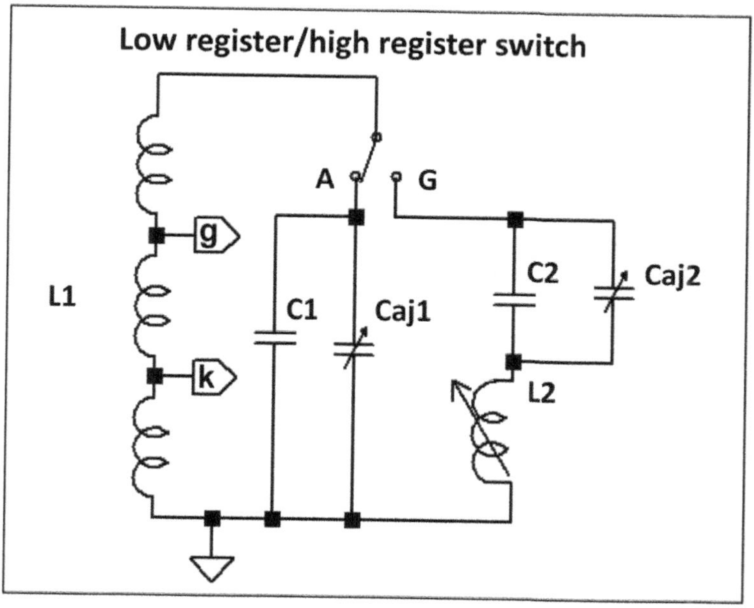

Figure 5.39. *Diagram of the register switch circuit*

When the switch is in the high register position, the resonant circuit is $L_1//(C_1+C_{aj1})$. Capacitor C_{aj1} consists of the parallel connection of a tubular-piston-type adjustable capacitor with a second adjustable capacitor (manufactured by M. Martenot) and controlled by the wheel placed on the front of the chassis. This is the treble register tuning.

When the switch is in the low register position, a series circuit L_2 + (C_2//C_{aj2}) is connected in parallel to the inductor L_1, the resonant frequency is lowered and the basic adjustment is made by an adjustable tubular-piston-type capacitor C_{aj2}, which can be finely adjusted by the adjustable inductor L_2, whose adjustment knob is located on the front of the chassis. This is the low register tuning setting. In operation, we observe a drift of musical frequencies over time, and the ondist, during play, must often retune the instrument.

5.6. The mixer

The role of this circuit is to produce a signal whose musical frequency verifies:

$$F_{musical} = f_f - f_v$$

A diagram is given in Figure 5.40.

Let us leave aside the timbre T7 control device made up of the elements M_2, C_{12}, C_{39} and C_x; we will come back to it later.

The assembly revolves around the 6AU6 pentode which has three accessible grids. The cathode is biased around 2 V by R_9 and C_9; HF filtering is provided by capacitor C_{10}. Grid G_2 is supplied continuously via resistor R_{10} decoupled by capacitor C_{11}. The plate load is formed by the series connection of resistors R_{18} and R_{11}, which creates a voltage divider for the output signal, this one is taken from the nodal point of resistors R_{11} and R_{18}; a capacitor C_{13} of low value shunts the high-order harmonics. The resistors which ensure the continuous grid potentials are not represented because they are physically placed on the plates of each of the oscillators.

The signals coming from each of the oscillators are respectively applied to the grid G_1 for those coming from the fixed-frequency oscillator, and to the grid G_3 for those coming from the variable-frequency oscillator. The mixing of the signals is multiplicative. Capacitor C_8 connected to grids G_1 and G_3, of low value, plays the same role as C_{13}.

We have noted the performance of this assembly in real operation. On grid G_1, a signal of frequency 79.571 kHz and amplitude 7.84 V is noted, and on grid G_3, a signal of frequency 79.095 kHz and amplitude 67.2 V is noted; the oscillograms obtained confirm the proper functioning of this device, and the LF signal has an amplitude of 28 V and a frequency of: 79.571 – 79.095 = 476 Hz.

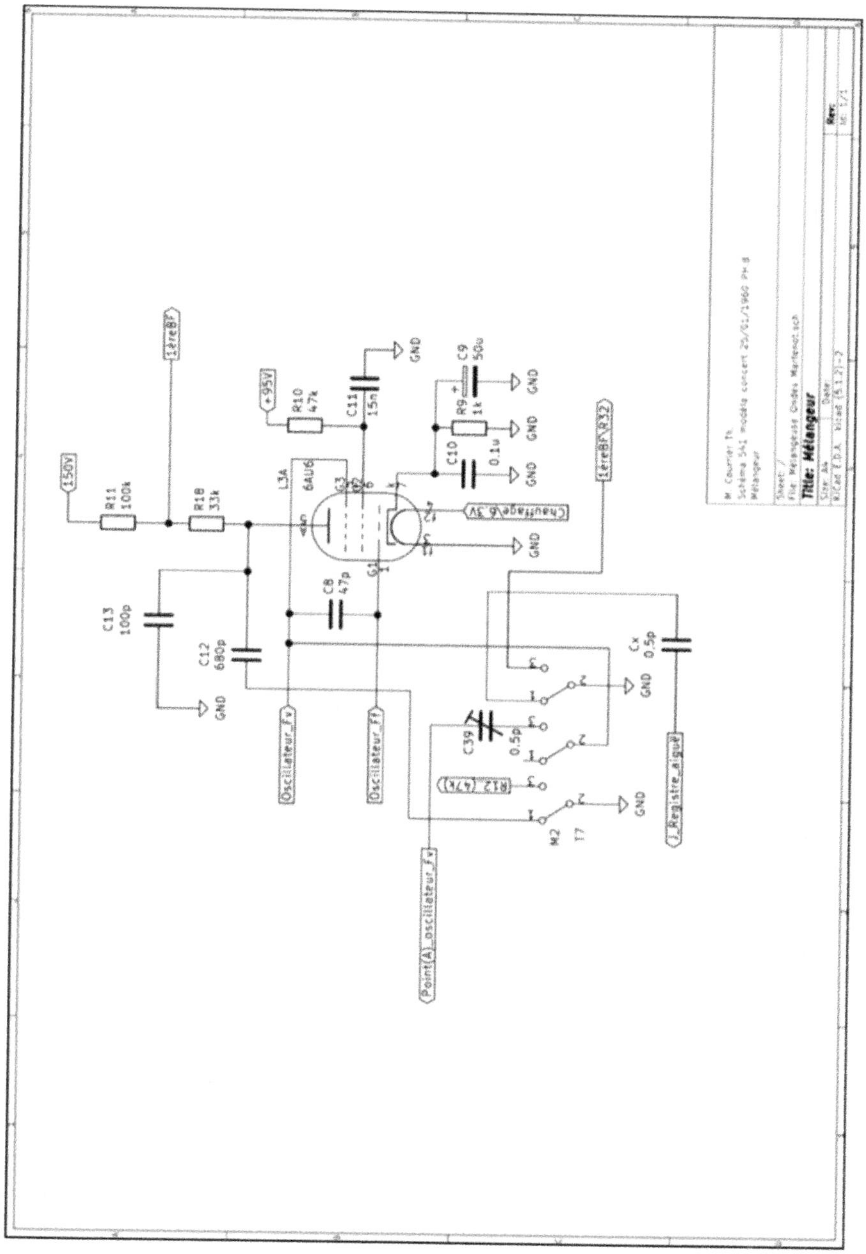

Figure 5.40. *Diagram of the mixer (grids are represented by capital Gs)*

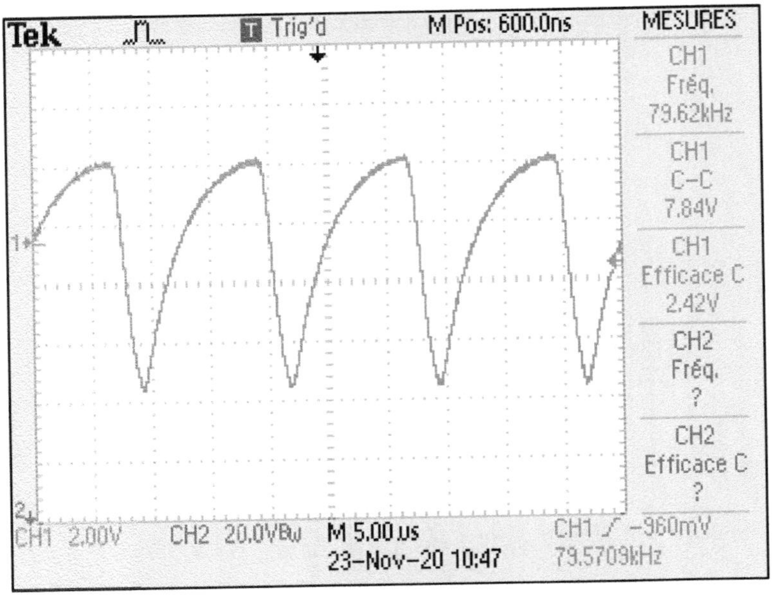

Figure 5.41. *Oscillogram: measurement on the G1 grid of the mixer*

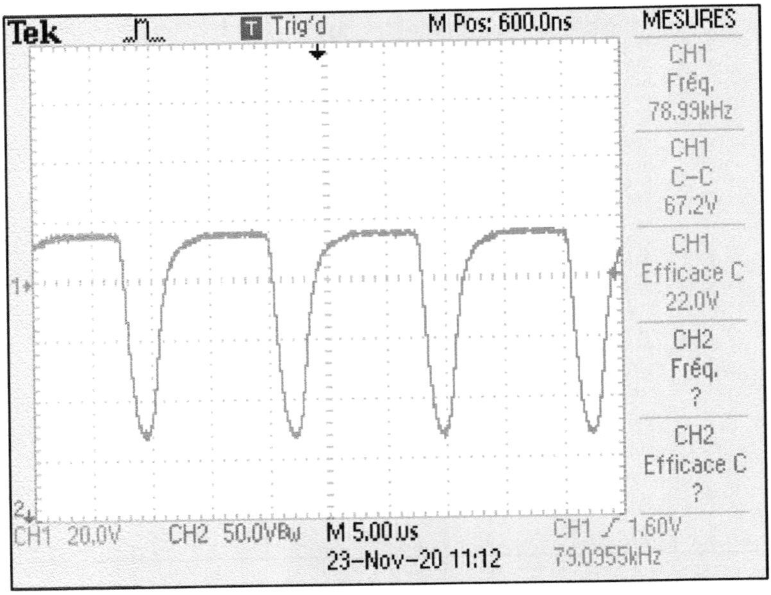

Figure 5.42. *Oscillogram: measurement on the G3 grid of the mixer*

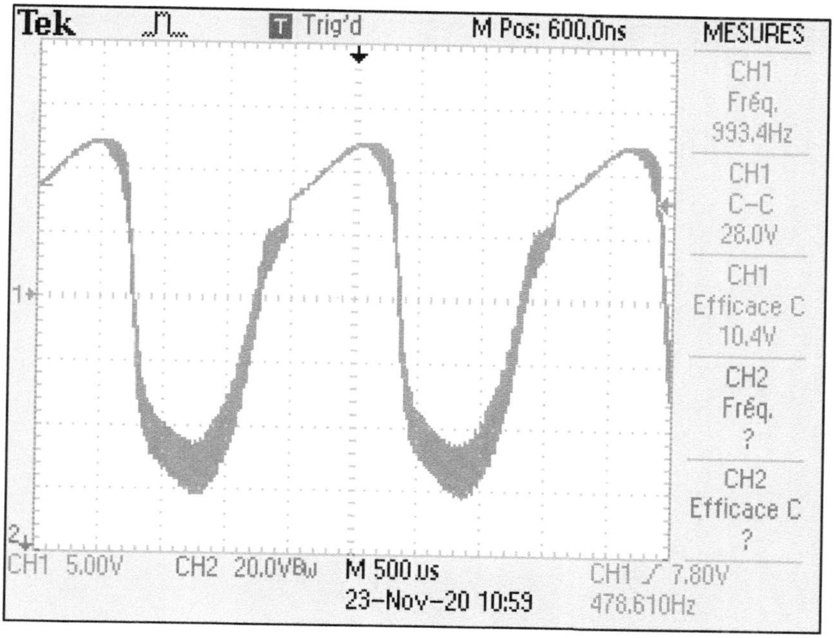

Figure 5.43. *Oscillogram: measurement on the mixer tube plate*

NOTE.– The signals obtained at the terminals of the oscillators are not sinusoidal and contain many harmonics; these are essential for the functioning of the ondes in the high register. The electronic mixture of these signals will also generate a strong presence of harmonics, but these do not contribute in any way to the richness of the musical signal, because on the one hand they are of too high frequencies to be audible, and on the other hand, in that the first LF tube operates as a comparator: the signal obtained at the output of this stage is the source of the LF harmonics of the signal before amplification by the final pentode.

5.7. 1st LF

In conventional systems, the stage preceding the power output circuit is generally referred to as an LF pre-amplifier, the gain of this assembly is high, for signals applied at the input of a few hundred millivolts in amplitude, and we generally obtain in outputs signals up to 10 V to 15 V in amplitude. This is not the case here; the stage operates identically to that studied in model 47.

A diagram is given in Figure 5.44.

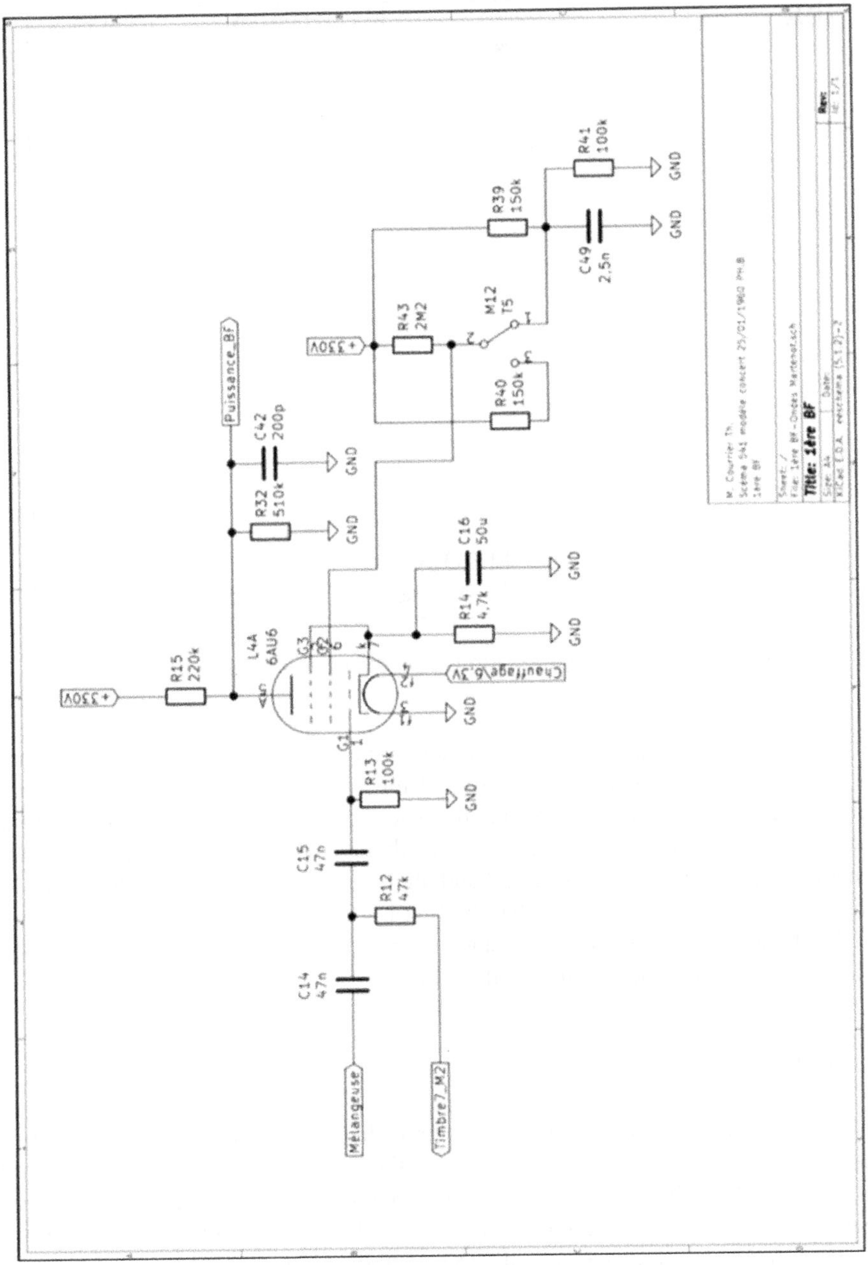

Figure 5.44. *Diagram of the pre-amplifier circuit*

Let us analyze the operation of the circuit in sinusoidal mode: the signal applied at the input is of high amplitude, the tone switch T5 is in the "inactive" position, the potential of the screen grid is 140 V, it is fixed by the divider network R_{39} + (R_{41}//C_{49}), the grid voltage from which the anode current is zero, is relatively low, the pentode operates in saturated mode and we obtain a rectangular-shaped signal at the output.

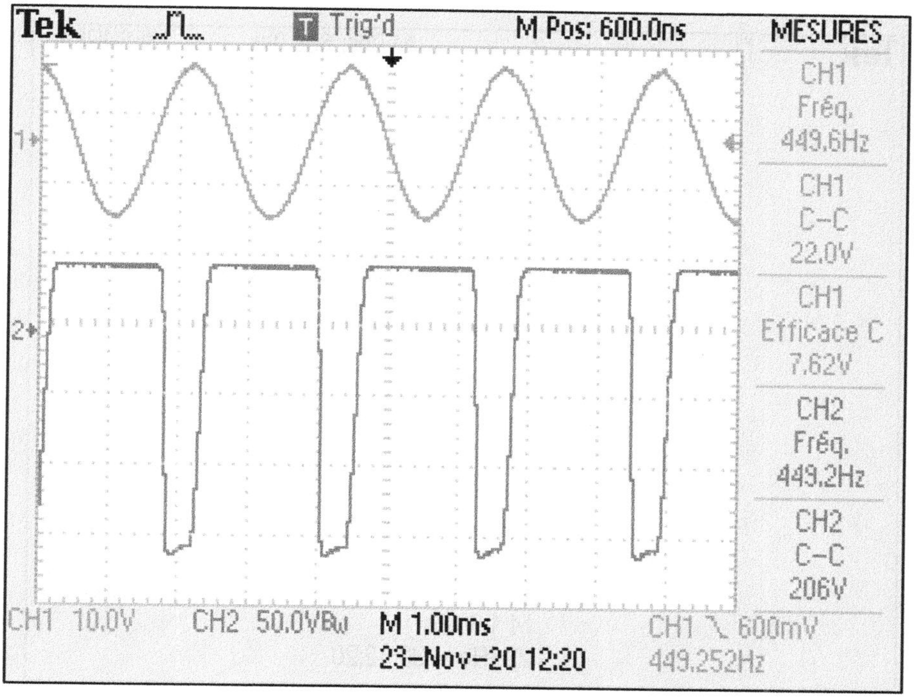

Figure 5.45. *First LF, switch T5 inactive: tube with screen powered by a voltage divider*

When the tone switch T5 is in the "active" position, the screen grid is brought to a potential close to 300 V by the resistor R_{40}, the resulting clipping on the signal is highly asymmetrical.

Resistor R_{43} reduces switching noise, and capacitors C_{14} and C_{15} are connection capacitors with the previous assembly. Resistor R_{12} is grounded when timbre T7 is "on", so two high-pass filters (C_{14}/R_{12}) and (C_{15}/R_{13}) are brought into service in the signal path. The usefulness of this circuit remains to be demonstrated, but we found that this system had been abandoned on the following models. The low-capacitance capacitor C_{12} shunts high-order harmonics. The other components fix the operating point of the pentode. Resistor R_{32} shunted by capacitor C_{42} is a remnant of the circuit which participated in the operation of an old version of the T7 timbre, but it is no longer useful.

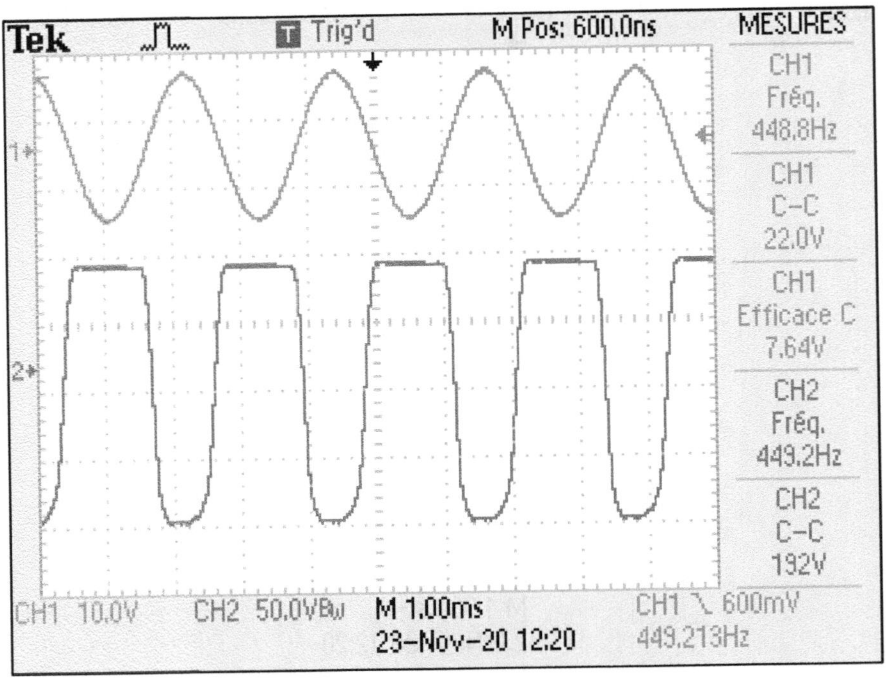

Figure 5.46. *First LF, switch T5 active: screen connected to HT by a resistor*

5.8. LF power stage

It delivers the musical signal to the diffusers; it is built around the EL84 pentode which is capable of delivering a power of the order of 6 W in class A. The diagram is given in Figure 5.47.

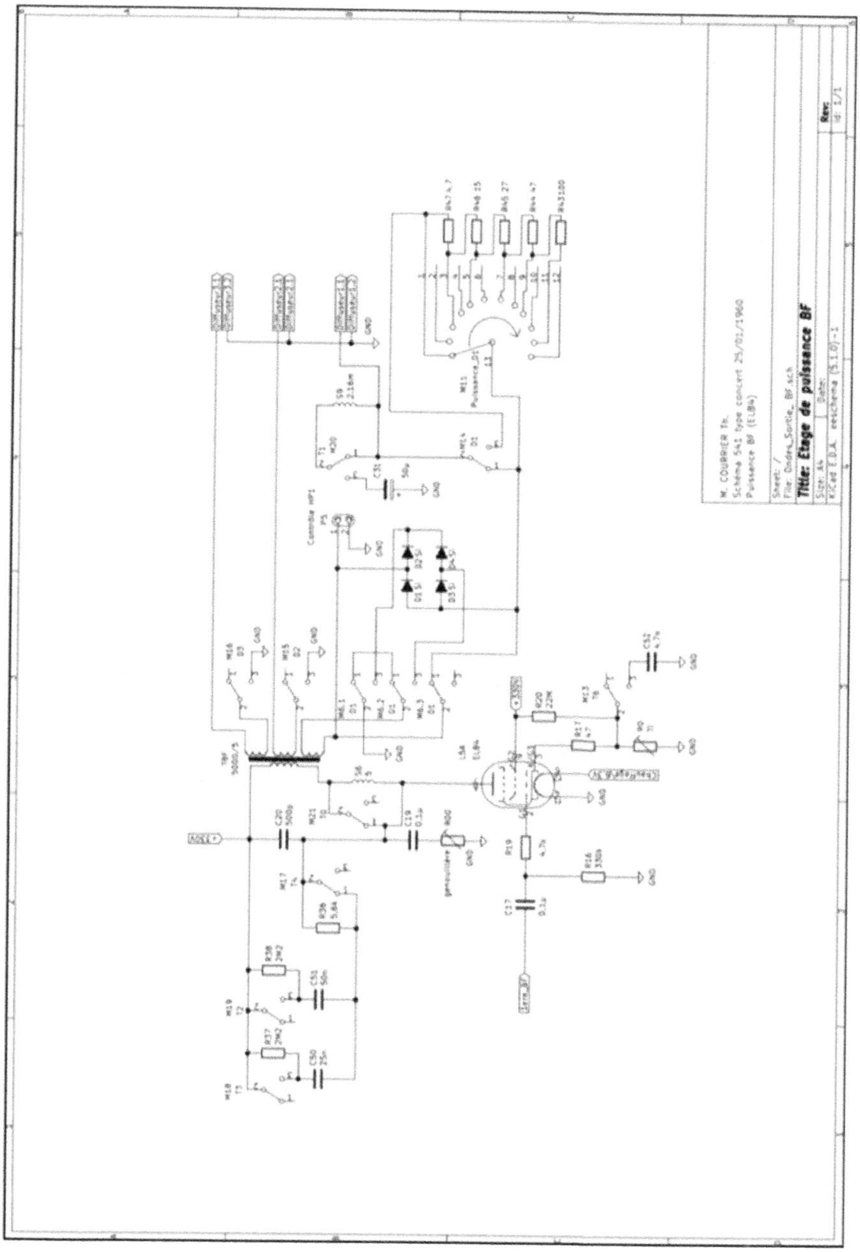

Figure 5.47. *Diagram of the power stage*

If we disregard the various timbre control circuits (they are all shown in the "inactive" position), the pentode is biased at the cathode by resistor R_{17}, no decoupling capacitor is used, when the intensity key is fully pressed, the maximum power is obtained, note that the tube is used at the maximum of these capacities, in fact, for a high voltage close to 300 V under load and an average cathode resistance (resistance of the intensity key + R_{17}) close to 120 Ω, we reach the maximum value of the admissible plate current (65 mA), in other words, during the play, the maximum admissible limits by the tube can be exceeded by transiently (staccato), which will certainly cause accelerated aging of the electron tube. Resistor R_{16} fixes the grid potential at 0 V. Resistor R_{19} eliminates the risk of HF oscillations.

Capacitor C_{17} ensures the LF link with the previous stage. The load consists of a LF transformer: 5,000 Ω/5 Ω. It has the particularity of having three independent secondaries which supply the diffusers and which can be put into service by the switches D1, D2, D3, separately or together. By inserting one or more resistors (R_{43} to R_{47}) in series with the loudspeaker, an M11 rotary wafer switch makes it possible to adjust the power at its terminals, independently of the powers delivered to the other diffusers (Gong, Palme). Resistor R_{20} is placed to apply a high voltage to the cathode when the intensity key is raised. In this case, the cathode is no longer at a floating potential, and the plate and the cathode are at the same potential. This process makes it possible to electrically place the electronic tube in a blocking situation: any risk of leakage is eliminated.

The adjustment of the gain of the amplifier stage is carried out by the intensity key which we assimilate to a pure resistance inserted in the circuit of cathode and in series with the resistance R_{17}. To obtain the gain of the amplifier stage, we vary the resistance by discrete value 0.1 Ω, 100 Ω 1 kΩ, 10 kΩ, 100 kΩ and 1 MΩ; beyond that we can consider that the cathode circuit is open and that the tube is blocked. The other adjustment parameters are specified a little later during the study of the timbres. We obtain the Bode diagram of the response as a function of the value of the resistance of the intensity key (see Figure 5.49: this is the graph corresponding to the case "no timbre in use"). We observe the remarkable efficiency of the intensity key whose dynamic exceeds 60 dB.

5.9. Timbre filters

It remains to study the filters of the timbres, and there are many of them: T0, T1, T2, T3, T4, T6, Octaviant (T8) and knee lever. With the exception of the Octaviant, they all consist of passive harmonic filters. In order to avoid giving too long of an explanation, we present for each timbre the harmonic response of the assembly, this one was simulated on the software Ltspice. The Bode diagrams show the response of the circuit to a sinusoidal voltage, the voltage is measured across the terminals of the 5 Ω load assimilated to a pure resistance (in solid lines the amplitude, in dotted lines the phase).

The operating parameters of the EL84 tube are fixed at the maximum values: $V_{a0} = 300$ V, $I_{k0} = 65$ mA, with a sinusoidal input voltage of 5 V amplitude. The LF transformer has an average performance. We have voluntarily extended the frequency ranges of the simulations beyond the musical frequency spectrum in order to highlight overvoltages which could possibly create instabilities.

The simulation diagram is recalled in Figure 5.48.

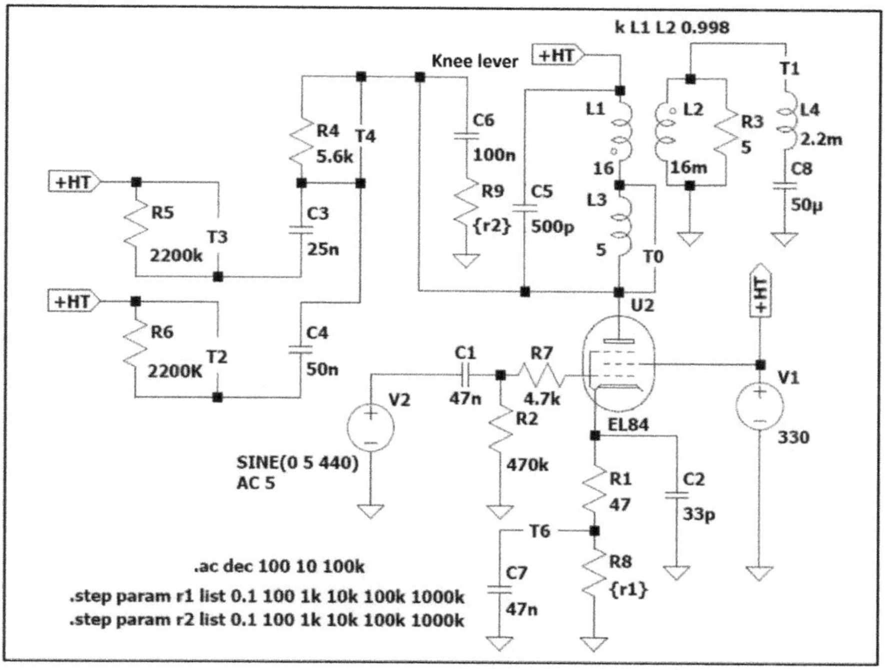

Figure 5.48. *Simulation diagram (Ltspice software) for the study of timbres. R_8 is the intensity key, R_9 is the variable resistance of the knee lever. Timbres switches are represented by open circuits*

No timbre in use:

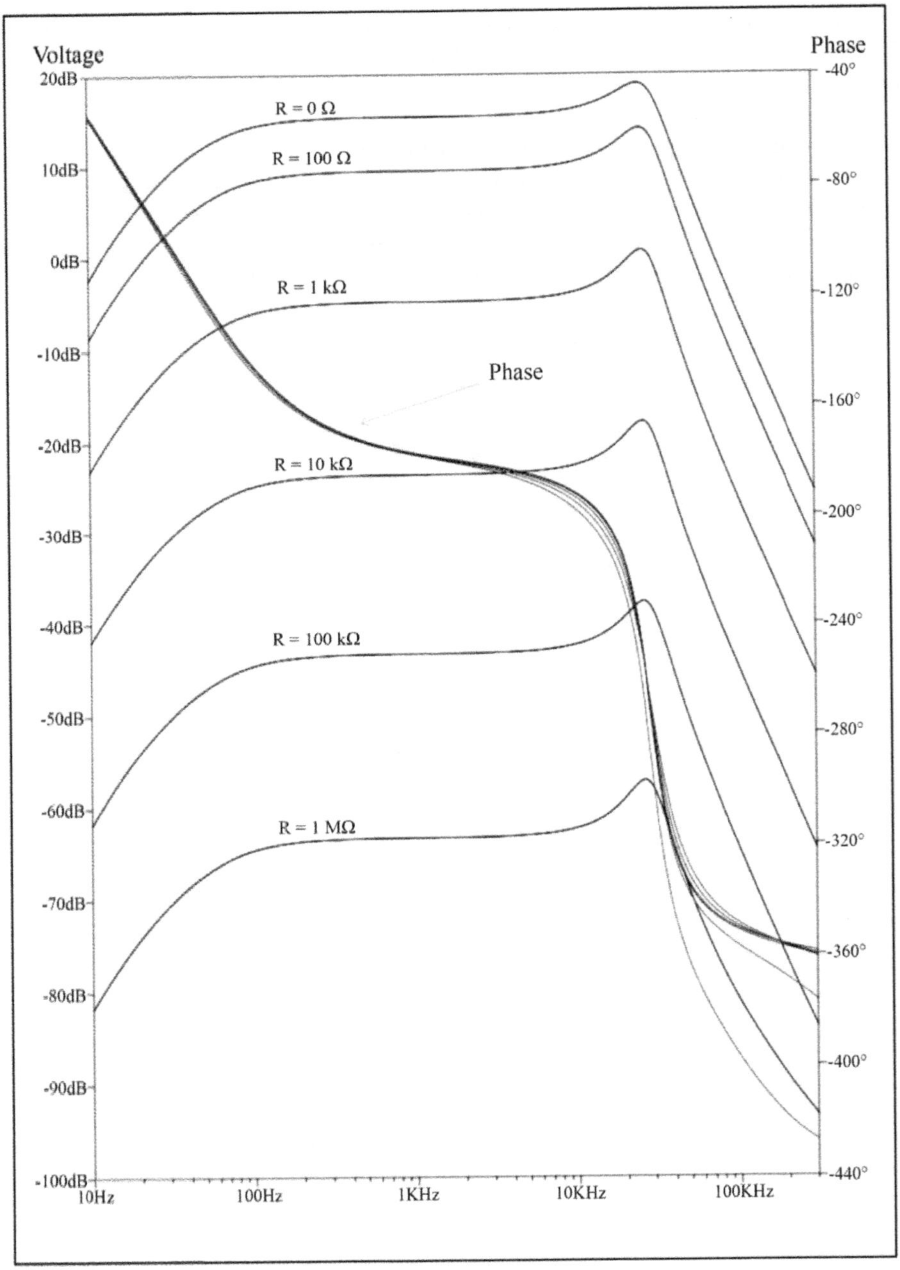

Figure 5.49. *Bode diagram following intensity key values*

T0 timbre in use:

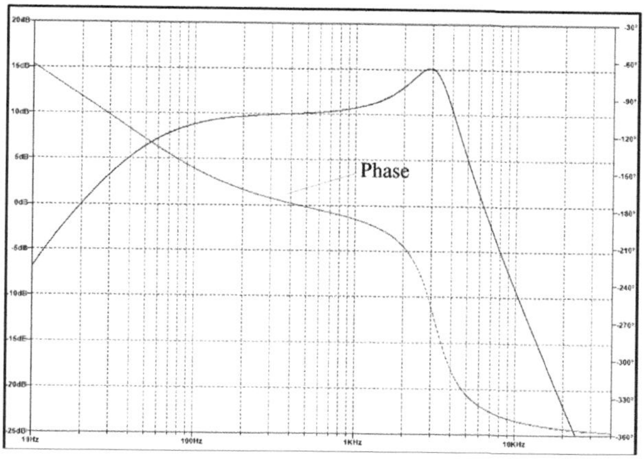

Figure 5.50. *Bode diagram: T0 on*

The insertion of an inductor (L_3) in series with the primary of the transformer modifies the high cut-off frequency which drops around 3 kHz; beyond that, the slope is -40dB/decades.

T1 timbre in use:

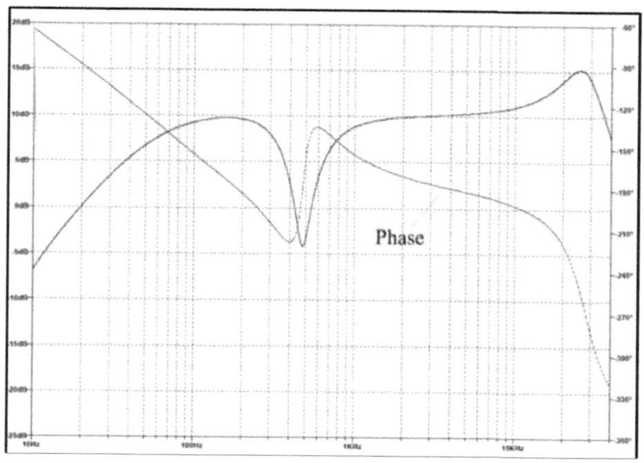

Figure 5.51. *Bode diagram: T1 on*

This timbre (T1) only applies to signals delivered to the loudspeaker (diffuser D1). This response may seem surprising since it cuts the audio band into two ranges around the frequency of 500 Hz, but it is justified by the choice of the electrical configuration of the circuit: an inductor (2.2 mH) in series with a capacitor (50 μF).

It is an anti-resonant circuit or frequency trap; the notes played around this frequency have low amplitude.

Timbres T2, T3 and T4 work in a particular way: timbres T2 and T3 insert a capacitor in parallel with the primary of the LF output transformer, but these capacitors can be put in series with a resistor of 5.6 kΩ using the T4 timbre switch. The different configurations show the modifications of the different responses that obey the laws that govern the RC filters.

T2 timbre in use:

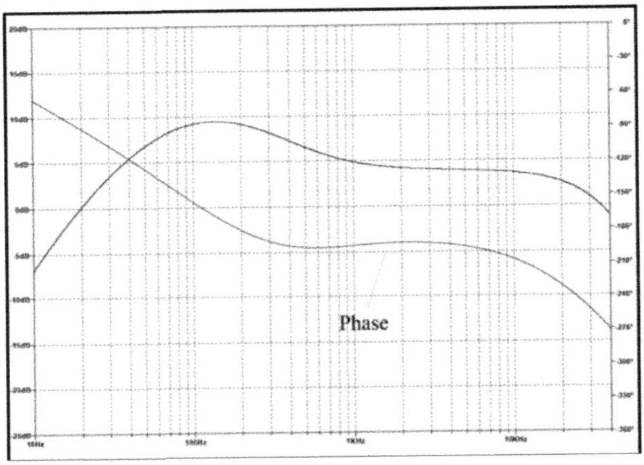

Figure 5.52. *Bode diagram: T2 in use and T4 open*

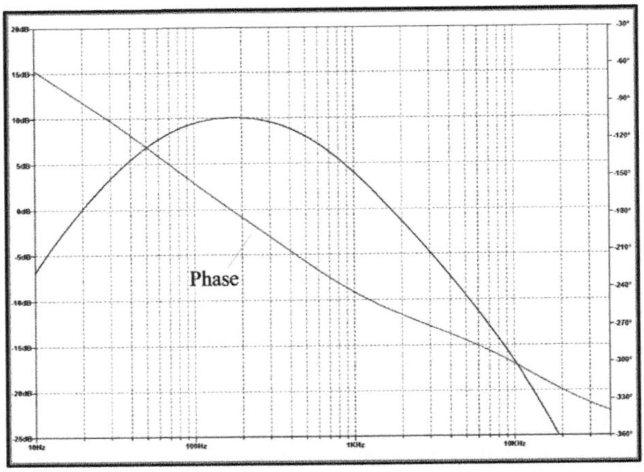

Figure 5.53. *Bode diagram: T2 in use and T4 closed*

T3 timbre in use:

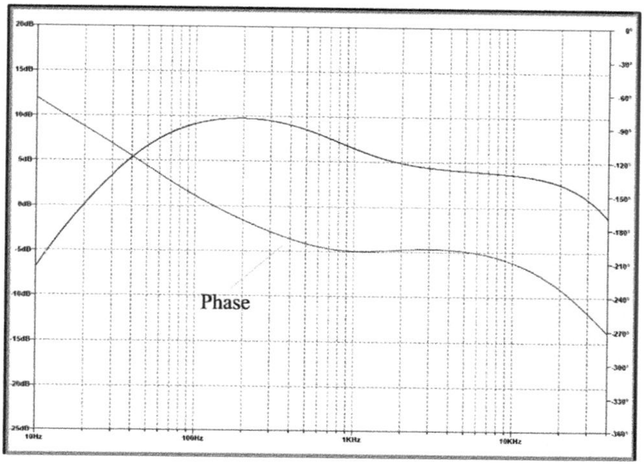

Figure 5.54. *Bode diagram: T3 in use and T4 open*

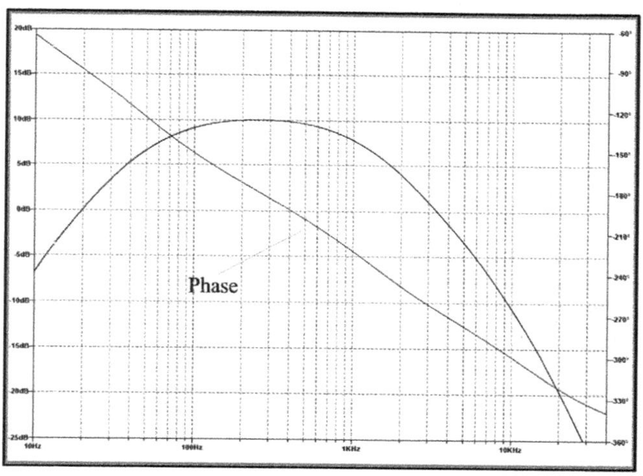

Figure 5.55. *Bode diagram: T3 in use and T4 closed*

T6 timbre in use:

This one is special, since it is achieved by placing the capacitor C_{52} in parallel with the intensity key; we obtain a gain variation combined with a bandwidth variation, the effect is particularly notable for an intensity key resistance between 1 kΩ and 100 kΩ.

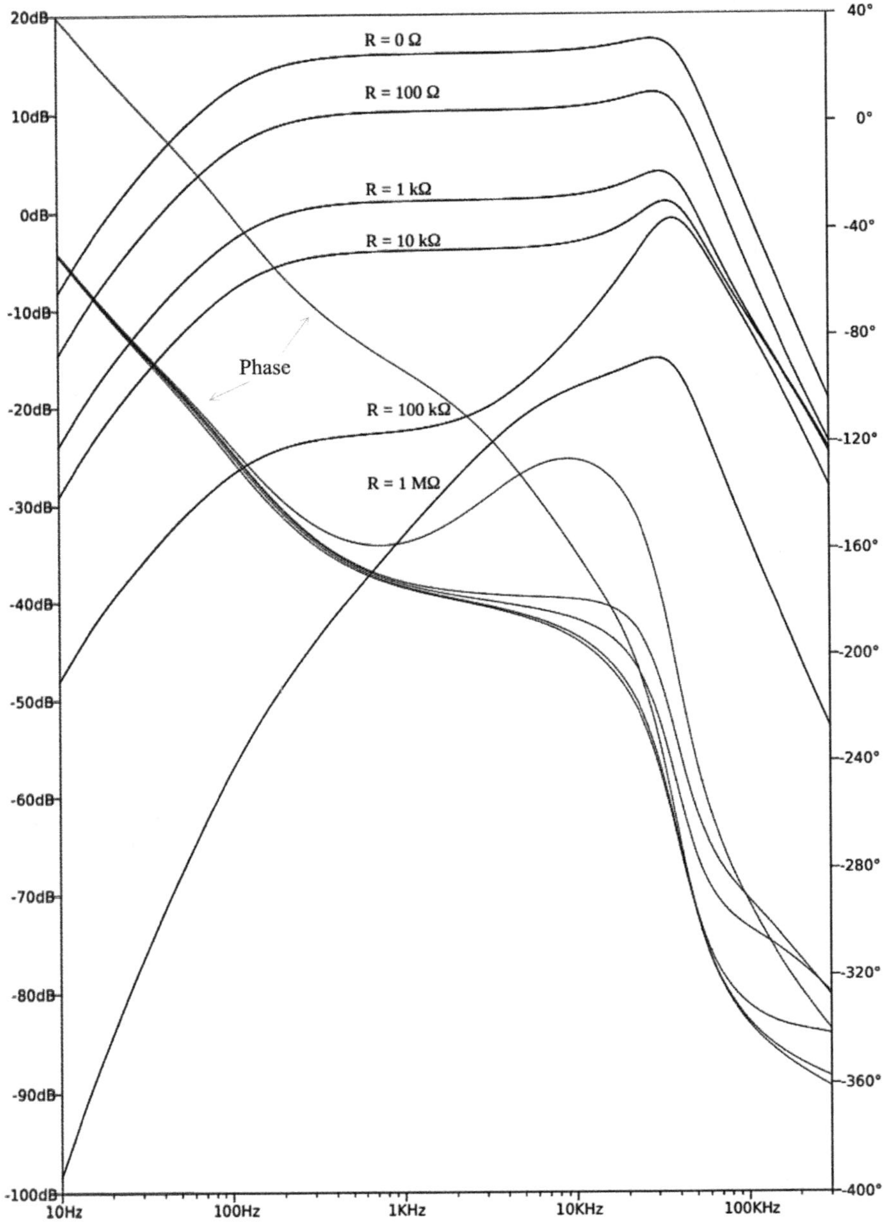

Figure 5.56. *Bode diagram: T6 in use, according to the values of the intensity key*

T7 timbre:

This timbre works in a very particular way. We encountered five different configurations, there are probably others, two of these versions come from Mr. Martenot's diagrams, but the parts of the diagrams relating to the T7 timbre are illegible, because these are more or less erased, apparently with many handwritten changes. The analysis under these conditions is subject to discussion, verifications are necessary when using other sources of information.

We will only describe the version observed on ondes no. 208. The timbre T7 is constructed with the components C_x, C_{12}, C_{39} and R_{12}, the three-way switch M2, the mixer pentode L_3 and the oscillator pentode L_1; see the diagram of the mixer given in section 5.6 and that of the variable-frequency oscillator in section 5.3.3. Capacitors C_x and C_{39} have very low capacities (<1pF); they are manufactured by Mr. Martenot using a conductive wire mounted as a hairpin or with two parallel conductive wires for which the spacing can be adjusted.

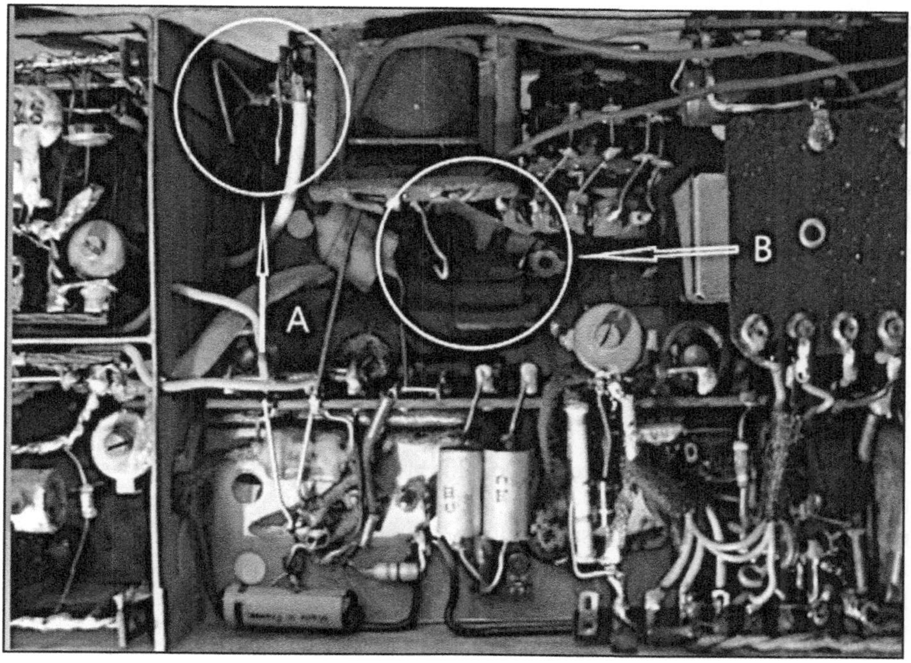

Figure 5.57. *Cx capacitor with parallel conductors (A) and C39 hairpin capacitor (B). Ondes no. 208*

In the "inactive" position, C_{12} is in circuit and is connected to R_{12}; capacitor C_{39} is off, capacitor C_x is grounded and is therefore in parallel with C_{29}, C_{30} and C_{31}. In the "on" position, C_{12} is off, resistor R_{12} is grounded, operating two first-order high-pass filters in the LF signal path. Capacitor C_{39} reinjects at point (A) of the resonant circuit, a very small fraction of the signal coming from the oscillator and taken from grid G_3 of mixer tube L_3. Capacitor C_x is taken out of service. We have not succeeded in determining the role of each component; the only element which seems decisive to us is the reinjection of the signal from the oscillator tube L_1 on the upper terminal of the winding which constitutes the inductor of the resonant circuit. Thus, the signal delivered by the oscillator has new characteristics. Capacitor C_x is taken out of service. Thus in an experimental way Mr. Martenot obtains a new form of signal, which it is difficult to classify in the conventional types: sine, triangle, square, sawtooth.

Octaviant timbre:

This tone is achieved by the commissioning of a bridge rectifier, which is then connected to the terminals of the secondary of the LF transformer; it delivers a "rectified" voltage to the loudspeaker. The first rectifiers were selenium; later the device will be replaced by silicon diodes. If the signal is purely sinusoidal with frequency F, a new signal is obtained, which has the form of a rectified sinusoid whose frequency is 2F and whose harmonic decomposition contains only even harmonics. If the signal is arbitrary, the signal obtained retains amplitude of the fundamental but is accompanied by even harmonics, harmonic 2 has large amplitude, and this is the octaviant effect.

However, this technique has the disadvantage of reducing the amplitude of the signal in significant proportions and generates a DC voltage component which is applied to the coil of the loudspeaker. As for the T1 timbre, it only applies to signals delivered to the D1 diffuser.

Note: the preceding analyzes assume linear operation of the amplifier stage, but this is not the case since the voltage applied to the input grid of the final tube is much higher than the necessary voltage. The final tube works in saturation, it follows the idea that the harmonic decomposition of the signal sent to the loudspeaker cannot be determined by the original shape of the signal and its filtering.

Knee lever:

Figure 5.58. *Knee lever leather bag. Ondes no. 257*

This RC filter is placed between the plate output of the power tube and ground; the variable resistance is obtained as in the case of the intensity key. But the control is done with the knee via a lever located under the cabinet. A nylon thread connects the lever to the mobile end of the bag.

The curves obtained assume a resistive behavior of the powder bag. The resistance variation for this simulation is as follows: 0.1 Ω – 100 Ω – 10 kΩ – 100 kΩ – 1 MΩ.

These results show the diversity of sounds that can be achieved through the use of timbre controls. Remember that the final amplifier of this instrument works in saturated mode. The harmonic signal processing at the level of the power circuit generates significant losses and variations in the output level. The load of the final tube varies in significant proportions, depending on the diffusers in service. It also modifies its operating points. Finally, operation in (very) saturated mode fatigues the electronic tube and the output transformer which are subjected to significant voltage and current stresses.

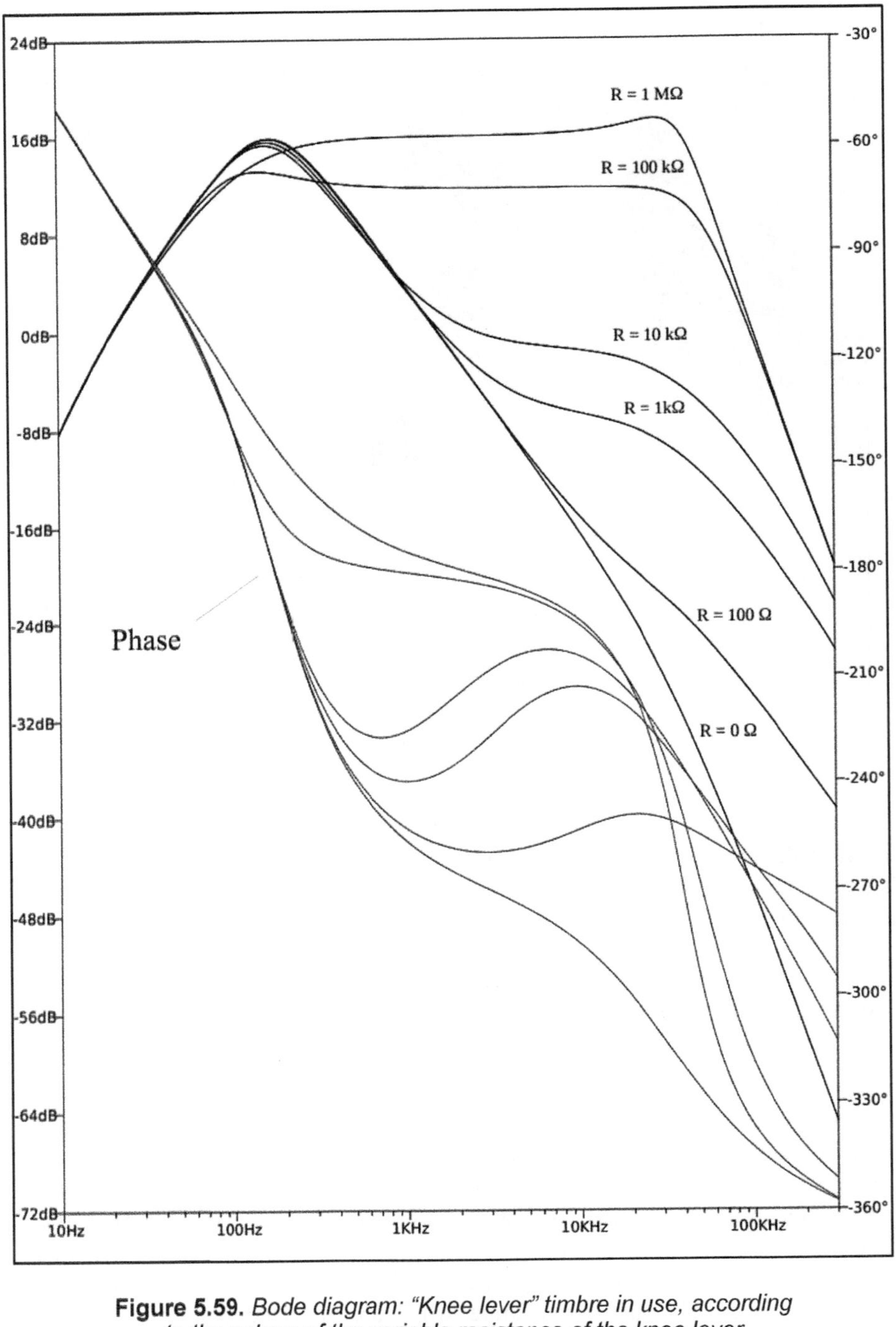

Figure 5.59. *Bode diagram: "Knee lever" timbre in use, according to the values of the variable resistance of the knee lever*

5.10. Diffusers

The analysis of the diagram of the power stage shows that this amplifier has three independent outputs; we can therefore connect to this ondes the three diffusers that Mr. Martenot had built the loudspeaker, the Palme and the Gong. The latter two are driven by modified electromechanical loudspeaker devices. The detailed study of these diffusers is beyond the scope of this document.

5.11. Power supply

The design of this device is entirely consistent with the know-how of that time. The power supply transformer delivers 2 * 350 V for a flow of 75 mA, two heating windings (6.3 V) are provided, one for the valve power supply and the other for the heating circuit other electron tubes. In the primary, a winding with multiple taps makes it possible to adapt to the various voltages of the networks which coexisted at the time. Rectifying is provided by the EZ80 rectifier tube. A 6.3 V incandescent bulb placed between the midpoint of the HV and ground acts as a fuse. HV filtering is provided by two cascaded LC cells with a lead capacitor. Subsequently, Mr. Martenot will remove a filtering cell and will increase the values of the filtering capacitors.

A 0A2 gas regulator tube is used to generate two stabilized voltages 155 V and 95 V for the supply of electronic oscillator and mixer tubes.

In some models which have undergone power supply troubleshooting, the EZ80 tube is replaced with a silicon diode bridge rectifier and the gas tube with Zener diodes.

The power supply diagram is given in Figure 5.60.

Important notes relating to the safety of persons against electrical risks: the connection of the mains to supply the ondes is carried out by a connector called octal cap; on this male connector are also arranged the connections of the diffusers. However, the other end of the power cord is the mains socket, it may happen that this is connected to the network while the octal plug is in the hands of the operator, and the risk of electrocution is certain.

Note also that the DC voltages present at different parts of the chassis are very high: only a qualified person can carry out an intervention on this type of equipment.

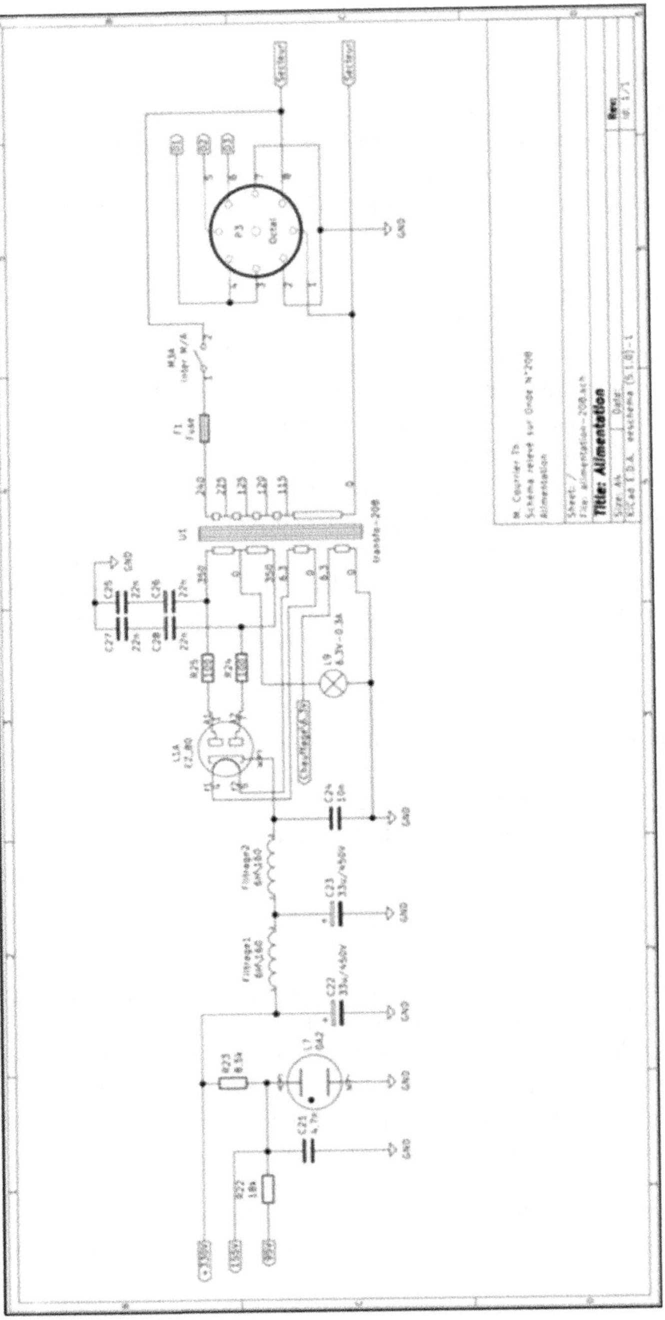

Figure 5.60. *Diagram of the power supply of ondes no. 208*

Figure 5.61. *The power cord and octal connector*

6

Building an Ondes with Vacuum Tubes

6.1. Introduction

All of the work that we have presented has been carried out either on real instruments, or on models or using simulation tools, when simulated models of vacuum tubes exist. This work has essentially focused on analyses of the electronic functioning of the various models of instruments, but we have only very rarely approached the manufacturing aspect. However, Mr. Martenot built his instruments himself with the help of a technician and few parts were subcontracted to outside companies.

We propose the construction of an ondes in order to specify the different technical solutions that Mr. Martenot and his technicians have adopted. The reference model will be ondes no. 208; the copy will be no. 2208. A certain number of elements will be included in order to differentiate in the future an ondes built by Mr. Martenot from this one.

6.2. The electrical schematic

We will adopt without any modification the electrical schematics studied in detail in Chapter 5. The basic nomenclature of the current components is defined by the values indicated on the diagrams. For other components, in particular the capacitors or the inductors which define the frequencies of the oscillators, the values are determined at the end of the tuning of the instrument, it is therefore necessary to look for these values in the last section. The components specially manufactured by us are described in the following sections.

For the color versions of the figures in this chapter, see www.iste.co.uk/courrier/ondesmartenot.zip.

6.3. The chassis

The chassis is one of the parts subcontracted by Mr. Martenot, and it is made by forming a tinned iron sheet of 10/10 mm thick. The various folds and drillings required to form the frame require specialized tools.

A sketch before folding is shown in Figures 6.1–6.4. The frame consists of two volumes linked at the level of a partition. The first contains the two oscillator circuits; the second includes the mixer, the LF chain and the power supply. Before folding, two sheets are therefore used; after forming, the two volumes are joined by screws and welded. The volume of the part containing the oscillators is separated by a sheet into two identical volumes in the direction of the greatest length of the chassis.

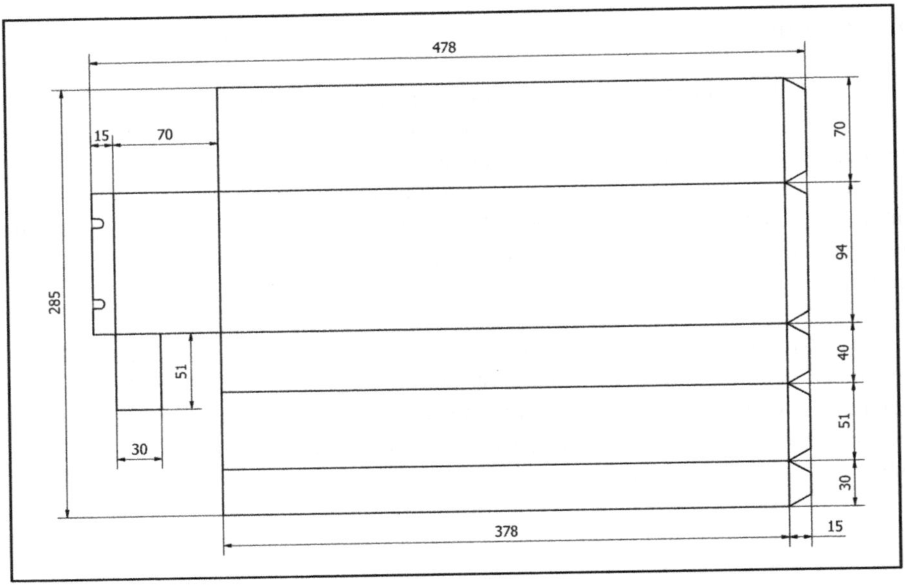

Figure 6.1. *Electronic tube block: 10/10 mm tinned metal sheet*

The various holes for placing the tube supports, the inductors, the power supply transformer and the various control or connection accessories were made visually from ondes no. 208. Figures 6.5 and 6.6 show an overview of the chassis after drilling.

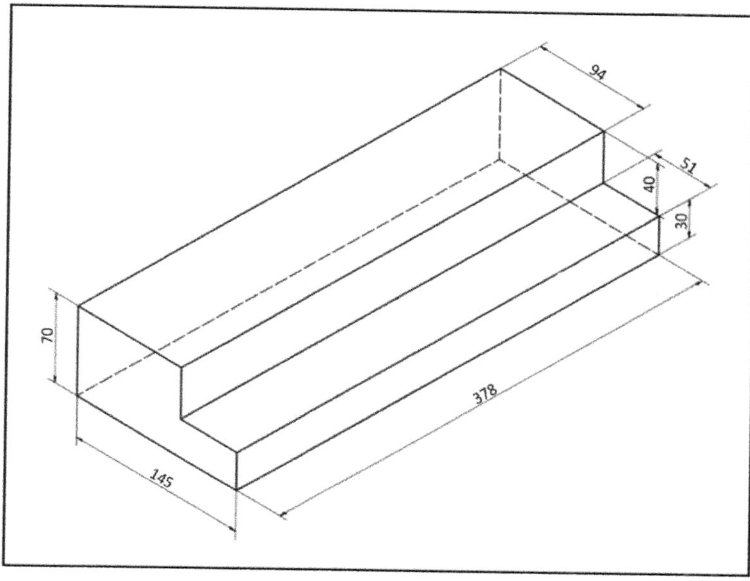

Figure 6.2. *Folded electronic tube unit*

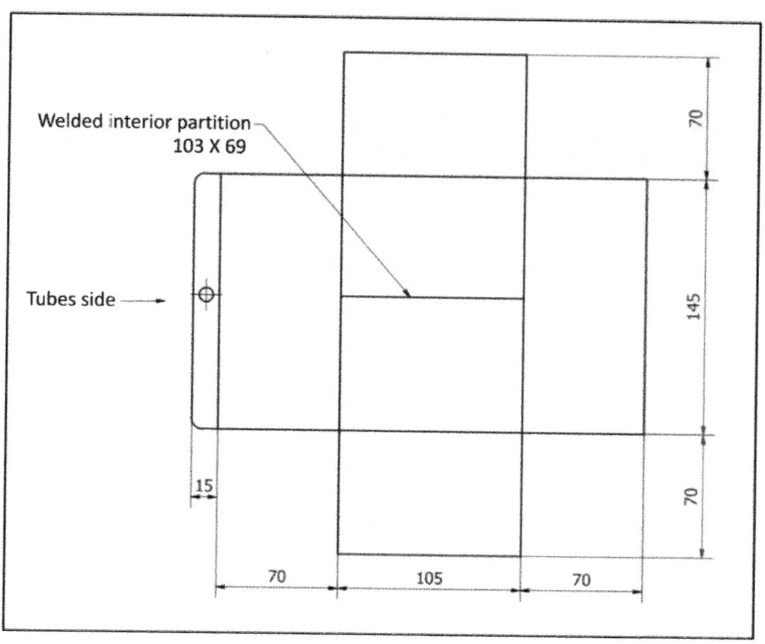

Figure 6.3. *Unfolded oscillator block*

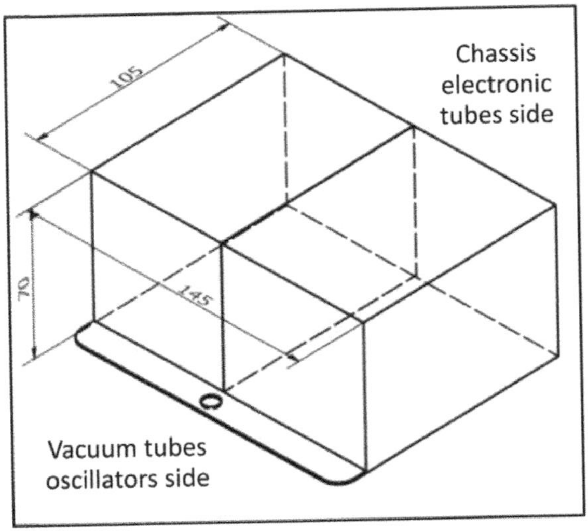

Figure 6.4. *Block-oscillators. Tinned sheet 10/10mm*

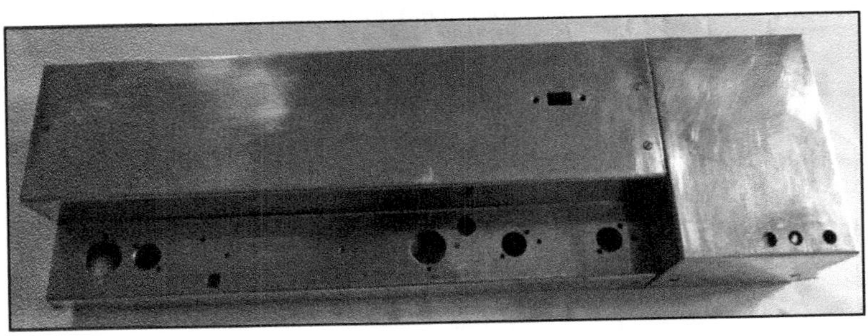

Figure 6.5. *Chassis, seen from above*

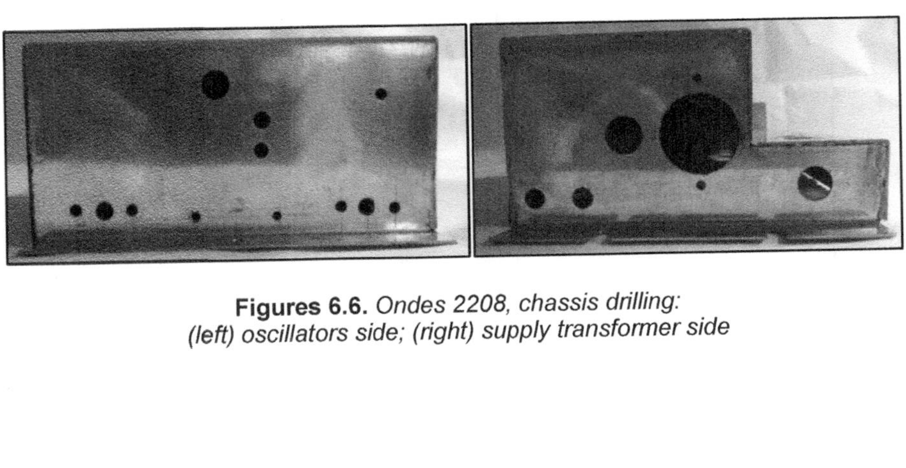

Figures 6.6. *Ondes 2208, chassis drilling: (left) oscillators side; (right) supply transformer side*

6.4. The wiring

The wiring of the components uses a mixed technique; some components are welded, for example, on the vacuum tube supports, and others are welded on terminal boards. Mr. Martenot has made three terminal boards, two of them support the components necessary for the operation of the oscillators and a third, larger one, supports the components necessary for the operation of the LF chain. A large wiring system is used to connect these sub-assemblies as well as the adjustment devices: switches, adjustable capacitors, transformers, fuses, etc.

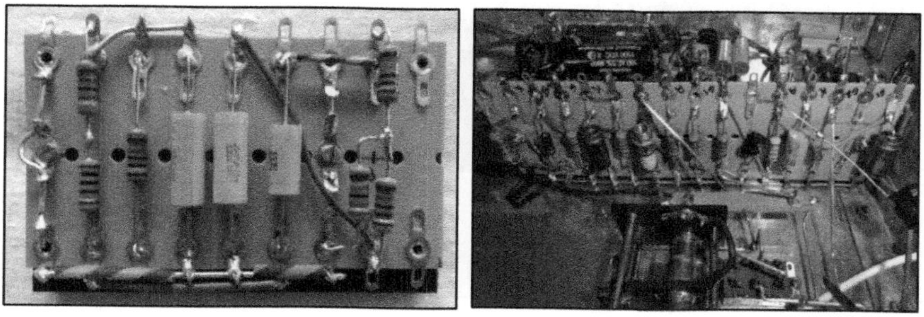

Figures 6.7. *Ondes 2208, wiring examples: (left) oscillator stage; (right) LF terminal board in place*

6.5. Special devices

To ensure the different settings of the ondes, Mr. Martenot imagined different devices. Due to the electromechanical characteristics that define them, these devices are not commercially available, so they must be manufactured.

6.5.1. *The high register tuning capacitor*

To achieve tuning in the high register, Mr. Martenot made a variable capacitor with very low capacitance variation. It is operated by a large knurled wheel. It is attached to a threaded rod which presses on a leaf spring. The M4 threaded rod is guided by a 5 mm thick Bakelite shim with dimensions: L = 60 mm, l = 20 mm. The shim is also the rigid support for the entire device. The slat supports the group of mobile parallel walls, opposite is the other group of parallel walls which is fixed. The mutual interpenetration of these two groups defines capacitance. Figure 6.8 shows the realization.

Figure 6.8. *Ondes 2208, variable capacitor, for high register*

6.5.2. *The diapason capacitor*

Figure 6.9. *Ondes 2208, differential capacitor*

The pitch tuning is achieved using a differential variable capacitor; this very particular device remains difficult to find. One of the solutions that Mr. Martenot had adopted for the first units of the 1953 model was to use an inverted wafer contactor; this is no longer available. We then designed and manufactured a device that fulfills its role perfectly. The walls of this capacitor are cylindrical and concentric; a threaded rod allows the coaxial movement of the outer cylinders along the inner cylinders, and an insulating plastic tube is inserted between the mobile wall and the fixed wall. The system comprises two identical devices controlled by the threaded rod. When the facing surfaces increase on one of the inner cylinders with the wall of the outer cylinder, the facing surfaces decrease on the other cylinder with the corresponding wall. Figure 6.9 shows its realization.

6.5.3. *The variable tuning inductor of the low register*

The low register tuning is obtained by a variable inductor; this device does not exist commercially. In Mr. Martenot's productions, the variation of the inductance is obtained by a coaxial displacement of a small sheet of ferromagnetic material inside a winding with contiguous turns of a conducting wire. This coil has a rectangular section. The ferromagnetic core is cut from a sheet metal used for the construction of transformers. We have reproduced this system using a ferromagnetic ferrite core of cylindrical shape.

Figure 6.10. *Ondes 2208, low register tuning coil*

6.5.4. *Capacitors for timbre 7*

Timbre 7 brings two capacitors into circuit (see Figure 5.40). The first, C_x, is put in parallel with capacitor C_{31} which corrects the tuning of the high register when the T7 timbre is in use. It is composed of two parallel conductors, one is the circular section (diameter 2 mm), and the other is made of a copper blade 20 mm long and 3 mm wide. The spacing at the fixed points is 20 mm. The two conductors are soldered at one end to lugs fixed to an insulating support. The second, C_{39}, is made from a terminal board with two lugs (40 mm apart), a hairpin-mounted conductor, is soldered onto a lug; the capacitance is defined by the distance between the other lug and this conductor.

Figures 6.11. *Ondes 2208, the capacitors of timbre 7: (left) capacitor C_x; (right) capacitor C_{39}*

6.5.5. *HF inductors*

6.5.5.1. *The coils of heterodyne circuits*

These devices were probably outsourced. If we observed a variability of presentation in the components which equip the ondes of the same model, this one is related to the different dates of supply. On the other hand, in all the ondes of the 1955 model that we have studied, the HF inductors intended for the heterodynes are all identical. These are air-core inductors, in order to eliminate the drifts caused by the ferrite cores. They are wound using the honeycomb technique. In section 4.3.1, we gave the parameters taken from a handwritten document by M. Martenot. Let us recall these parameters:

– total inductance value: L = 2.8 mH;

– total number of turns N = 380, wound on a mandrel 15 mm in diameter and 35 mm in length;

– honeycomb technique, 15/100 mm thread insulated under two layers of silk, the width of the winding is e = 5 mm;

– a 110 turn tap (cathode connection), a 330 turn tap (grid connection), then 50 turns (connection to the tuning capacitor).

To make these inductors, we called on a collector of old radios, who had a winding machine. The know-how of the winder and the use of available conductor will determine the final dimensions of the winding as well as its external appearance. But two parameters must absolutely be respected: the value of L (2.8 mH) and the ratio of the number of cathode/ground turns to the number of grid/ground turns. This ratio r must be between 0.29 and 0.37 to ensure the operation of the ECO oscillator. In the case of the inductance of ondes no. 208, this value is r = 110/330 ~ 0.33.

In order to adjust to the right value, with good precision, the inductor is wound with a number of turns slightly higher than necessary, on the last turns. Collodion is deposited in order to fix them reversibly. Then, we measure the value of the inductance on an LRC measurement bridge, and we unwind turn by turn until the exact value of L to be obtained has been so. With this method, we obtained L = 2.801 mH for 405 turns and a ratio r = 117/351 ~ 0.33; the width of the coil is e = 6 mm. Before mounting in the chassis, the inductance is impregnated with an HF varnish.

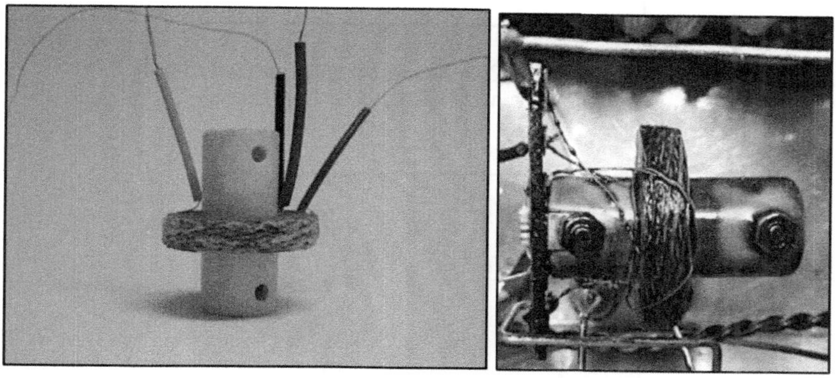

Figures 6.12. *Ondes 2208, HF coil of the heterodyne: (left) end of winding; (right) in place*

6.5.5.2. *Keyboard inductors*

The turning on a keyboard inductor determines the corresponding musical note, and to tune the note to the exact frequency requires some means of adjusting the value of the inductor. These coils therefore have a core of ferromagnetic material, which is a ferrite. We saw in section 5.5.2 that these inductors had very low values; they range from 0.30 μH to 24 μH. All of these coils are wound on a mandrel of the same type, the first have a few turns, the last a higher number distributed over two layers. It is advisable not to exceed more than two layers because the winding which is manual becomes complicated to achieve. The mandrel used by Mr. Martenot has a total height of 13 mm; for a useful height of 8 mm, the external diameter is 7 mm. At the upper part are arranged two diametrically opposite lugs; they allow the fixing of the conductive wire at the start and end of the winding. The magnetic core is a threaded ferrite cylinder of 4.5 mm in diameter and 6 mm in height; it has the particularity of having a very fine thread pitch. Ferrites used in old radios made around the same time generally have a larger pitch. These coils are immobilized by insertion at about a third of the height. They are arranged in a succession of aligned holes and drilled in a 4 mm thick hardboard plate. They are fixed at the bottom to another very thin bakelite plate (5/10 mm), using plastic screws whose pitch is identical to that of the magnetic cores.

Figure 6.13. *Ondes 208, keyboard coil, from left to right: magnetic core, wound mandrel, fixing screws*

Figure 6.14. *Ondes 208, assembly of keyboard coils: left side view with a mirror, right top view*

The disadvantage of this arrangement is that the plastic screw takes up practically about a third of the useful height of the chuck. The magnetic core has the remaining two-thirds. It therefore often happens that the magnetic core is only retained by a few threads of the chuck. However, the position of the keyboard is such that the core undergoes its weight, and to this is added the vibrato movement of the keyboard. To immobilize the magnetic cores as well as possible, all of the coils will be covered with a piece of adhesive, which at the time is often a piece of diverted "Sparadrap" medical tape.

Figure 6.15. *Ondes 2208, keyboard coils, 0.32 μH (left) and 22.6 μH (right)*

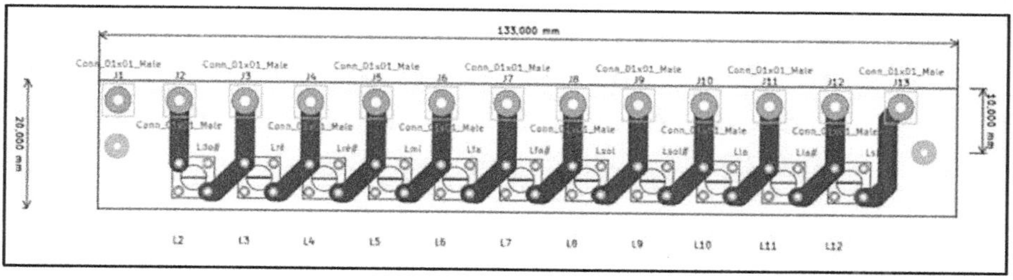

Figure 6.16. *Ondes 2208, a board fastening the 12 inductors of an octave*

We used an equivalent system. The chucks (external diameter 5 mm, useful height 10 mm) currently available do not have ears but have pins to be welded. We therefore manufactured a printed circuit-type support. For octaves 1–5, the coils are grouped by octave on the same support, for the 1st octave, the final tuning-coil must be added.

Figure 6.17. *Ondes 2208, coils fastening board, octaves 1–5*

To these inductors of fixed values, 11 units per octave, it is necessary to add vibrato inductors. They are not arranged on the printed circuit, they are fixed using small aluminum brackets of 8/10 mm (L_1 = 14 mm, L_2 = 10 mm, l = 10 mm). The conductive wire is wound on a mandrel with a length of 24 mm for a diameter of 5 mm. This coil is provided with a square base of 10 mm * 10 mm; this base is fixed using rivets on the smallest plane of the square. The other plane is fixed by M3 screws inside the U-shaped angle of the fixed frame. These inductors determine the Do note of each octave, but the magnetic core, made up of a formed sheet, slides inside the mandrel according to the lateral displacement of the mobile keyboard.

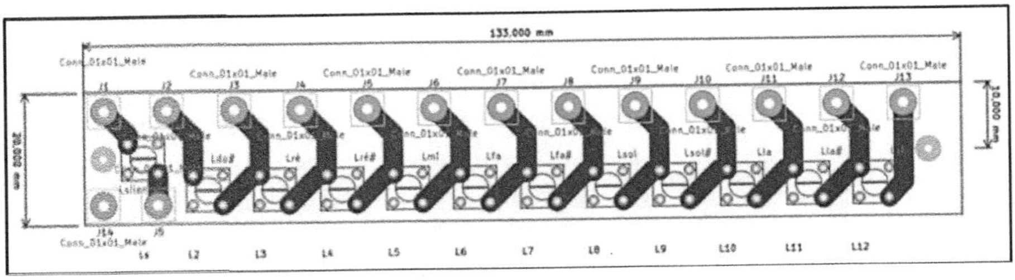

Figure 6.18. *Ondes 2208, inductance support board, octave 0*

Figure 6.19. *Ondes 2208, vibrato coil*

6.6. The LF output transformer

This LF transformer which allows the adaptation of the impedances between the loudspeaker and the electronic amplifier tube has the following characteristics: $P_s = 6$ W, $Z_p = 5000\ \Omega$, Zs = 5 Ω. It has three secondaries that allow the power to be distributed to the three diffusers: the loudspeaker, the Palme and the Metallic gong. This device does not exist commercially, and we wound this transformer using 35/100 mm grain-oriented EI 75 sheets, the iron section is 6.25 cm². After winding, the performances measured are: $L_p = 16$ H, $R_p = 209\ \Omega$, $R_{s(average)} = 3 * 1.08\ \Omega$ and m = 31.629.

6.7. Generic components

Most electronic components, resistors, high-voltage capacitors, smoothing schokes, power transformers, switches, terminal strips, high-voltage wiring wires, shielded wires, tube supports, vacuum tubes, etc., are available from electronic component suppliers. Thus for the power supply transformer, we used a toroidal transformer whose specifications are: P = 80 VA, 2 * 350 V/0.08 A, 6.3 V/3.5 A, its dimensions are: diameter d = 77 mm, height h = 46 mm for a weight of 0.8 kg. Smoothing chokes: 8 H / 0.1 A / 260 Ω and timbre inductor: 5 H / 75 mA / 135 Ω are also available from many suppliers.

6.8. The drawer

6.8.1. *Introduction*

The drawer is an essential device of the ondes Martenot; it supports the intensity key as well as all the timbre controls. This functionality is associated with a strong aesthetic component: the black painted case, the wing switches, with markings engraved in red and white, as well as the Plexiglas key, are in everyone's mind.

The ear switches, manufactured by Mr. Martenot, are unreliable devices; the last instruments of this model will be equipped with standard switches.

6.8.2. *The box*

The box is a parallelepiped formed from a 10/10 mm sheet metal; its external dimensions are: overall width 197 mm, upper face width 186 mm, length 155 mm, height 42 mm. At its lower part is arranged as a lid with folded edges. The lid is inside and is held in the box by bosses made with a pin punch.

Two lateral foldings form guides which slide along two grooves machined in aluminum shims (L = 100 mm, l = 20 mm, h = 5 mm), and fixed to the cabinet. On the upper face, holes and cutouts for rectangular windows are made.

On this box, on the front face, a piece of wood cut in the volume of wood is added, which constitutes the front face of the cabinet.

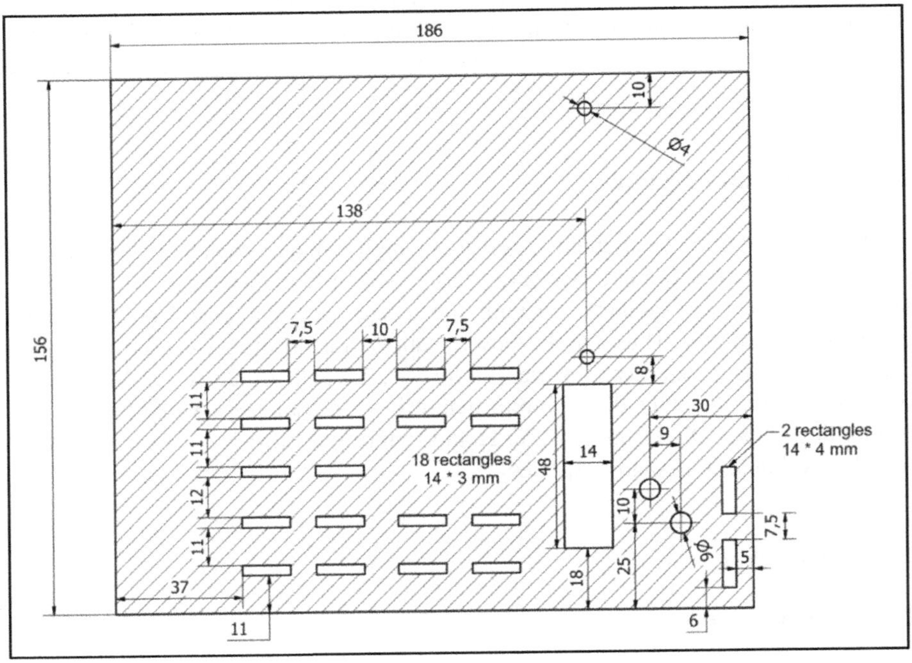

Figure 6.20. *Partial cutting plan of the upper face of the drawer (from the dimensions noted on ondes no. 208)*

Finally, a device imagined by Mr. Martenot prohibits the extraction of the drawer of the cabinet. In the first units of the 1953 model: it is simply a large metal blade used for making the springs. It has rounded edges; it is mounted on the upper face of the drawer, behind and on a bakelite shim. Angled slightly upward, it partially covers the power control knob. When the drawer is opened, it comes up against the inside face of the front panel of the cabinet; also to completely extract the drawer, it is necessary to lower this blade. This device will subsequently be replaced by a more complex system of removable stops, arranged behind and on the sides of the drawer box.

6.8.3. *Switches*

All the switches or timbre contactors, except T8, are made by Mr. Martenot. Similarly, the keyboard/ribbon control inverter is constructed from a tone switch to which an additional contact has been added. The T8 timbre switch is a classic wafer switch with rotational control via an axis. Quarter-tone adjustment control push buttons are another Mr. Martenot fabrication.

To make the timbre switches, we used Mr. Martenot's schematics (see Figure 6.21). On a block of bakelite (B) are fixed silver contacts and with cup between which can be inserted a brass rivet struck on a shim (A) in plastic material in the shape of ears. The movement of the part (A) in Plexiglas is limited by a part in bakelite (C) of polygonal form. Fixed contacts are associated in pairs. Each bakelite block can support one or two pairs of fixed contacts.

We took the elements of the silver contacts, fixed, with cup on the switches used in old radio receivers. The shim in the shape of ears was cut and engraved in a sheet of plastic material using a laser cutter.

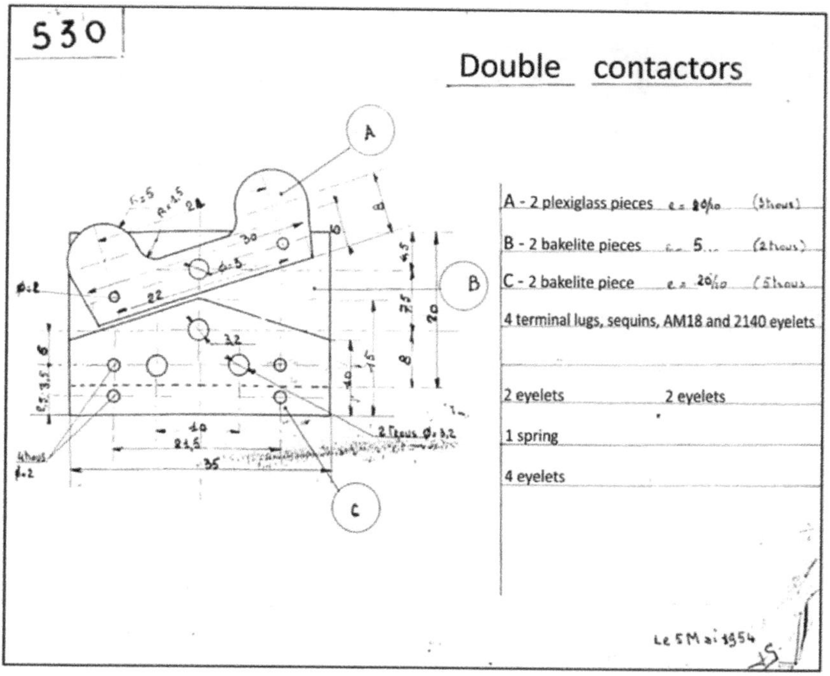

Figure 6.21. *Timbre contactor (Musée de la Musique, Paris)*

When the assembly of the contactor is finished, it is fixed inside the drawer using a threaded rod comprising a crew of two or three switches. This threaded rod is soldered with tin on metal terminal lugs. These metal terminal lugs are spot-welded to the inside of the top of the drawer.

The play selection switch on the ribbon or on the keyboard is extremely stressed, the realization is unreliable, and it will be quickly replaced by a side-operated wafer switch.

Figure 6.22. *Ondes 2208, prototype contactor, front and side views*

6.8.4. *The quarter-tone control buttons*

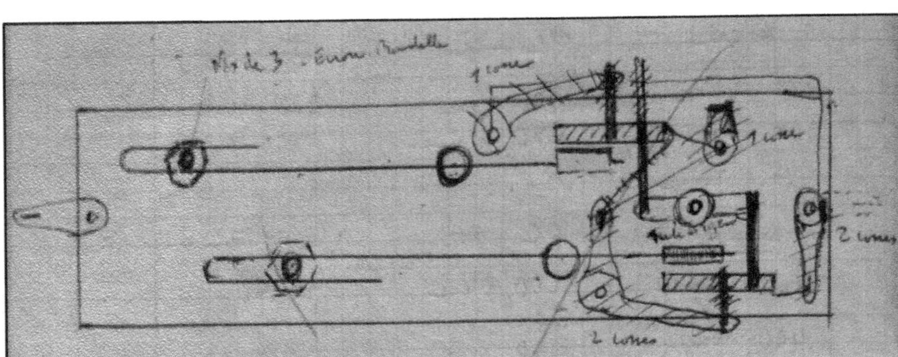

Figure 6.23. *¼ tone plate (Musée de la Musique, Paris)*

The ¼ tone controls allow us to tune the instrument while playing, especially during a concert with an orchestra. On the first instruments of the 1953 model, this device was designed and manufactured by Mr. Martenot. Thereafter, it will be replaced by an industrial device.

A cylindrical knob, made of plastic, slides vertically through a guide, driving a section of piano wire 0.6–0.8 mm in diameter. At one of its ends is welded a brass rivet (moving contact) which will establish electrical contact with an eye terminal (fixed contact). To increase or reduce the musical frequency played by a quarter tone, a capacitor (12 pF) is either connected in parallel to the tuning capacitor, or disconnected and grounded. We have taken the schematics of Mr. Martenot from his workshop notebooks (Martenot n.d.c).

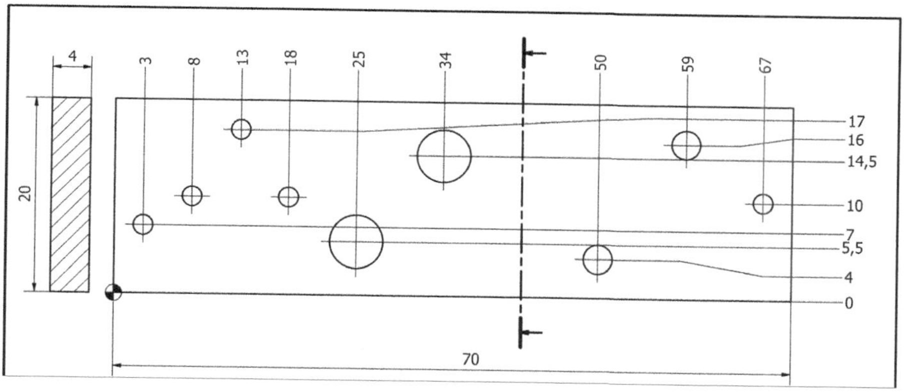

Figure 6.24. *Ondes 2208, schema of the ¼ tone plate*

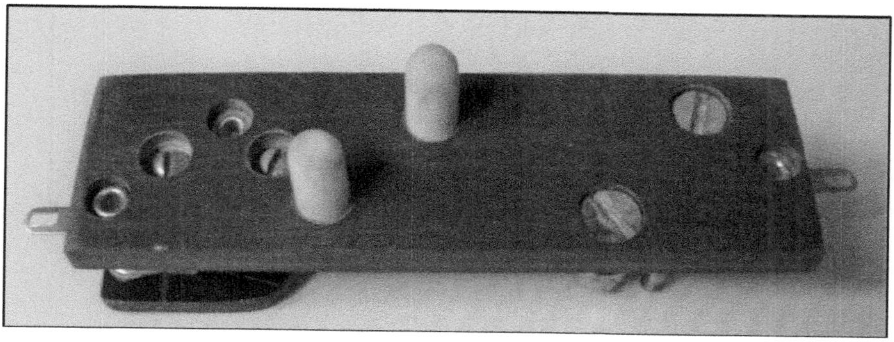

Figure 6.25. *Ondes 2208, ¼ tone plate*

6.8.5. *The T8 timbre control knob*

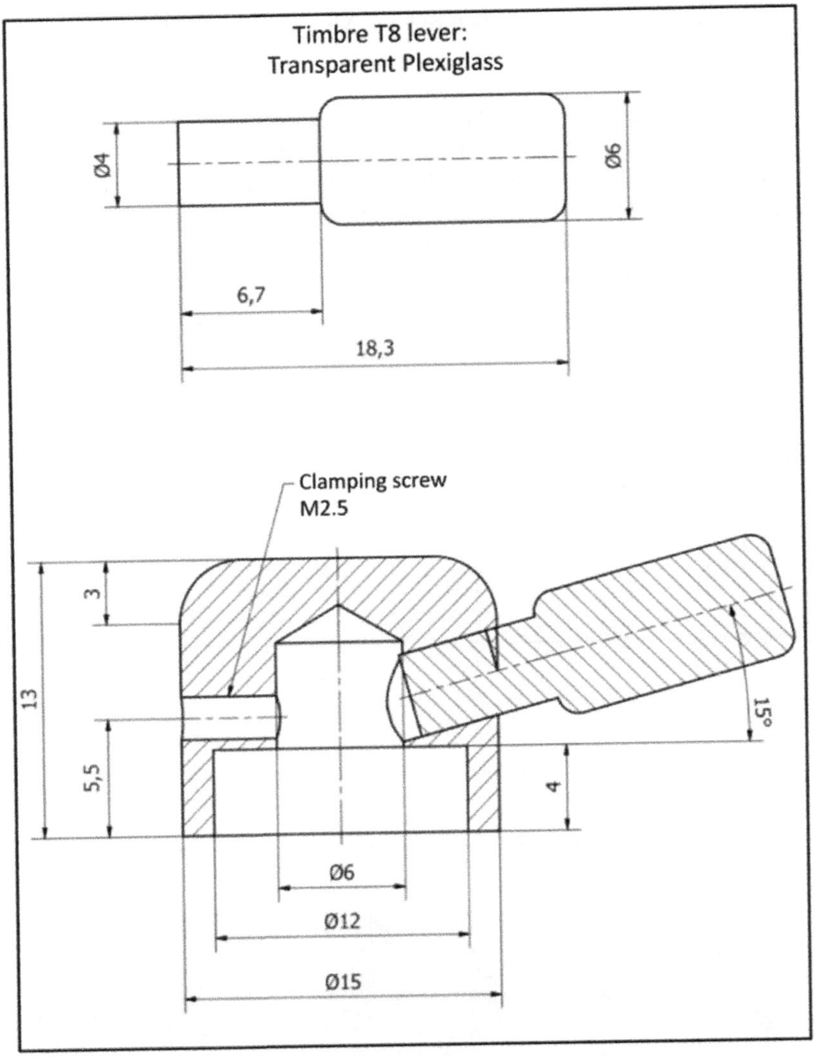

Figure 6.26. *Ondes 2208, T8 timbre control button diagram*

This control is carried out using a three circuits two positions wafer switch, although the electromechanical device deserves no particular comment, on the other hand, the control button with its inclined lever is particularly aesthetic.

6.8.6. *The power control knob*

This device only applies to the loudspeaker; it allows us to adjust the power of the signal in it. It is obtained by placing four resistors in the series (15 Ω, 27 Ω, 47 Ω, 100 Ω). The switch has 11 positions/one circuit, one position out of two is useful, and the last ones are shunted. The control is carried out by a notched ebonite button graduated in positions I, II, III, etc. Subsequently, this system will be replaced by a low power linear control rheostat (<2 W).

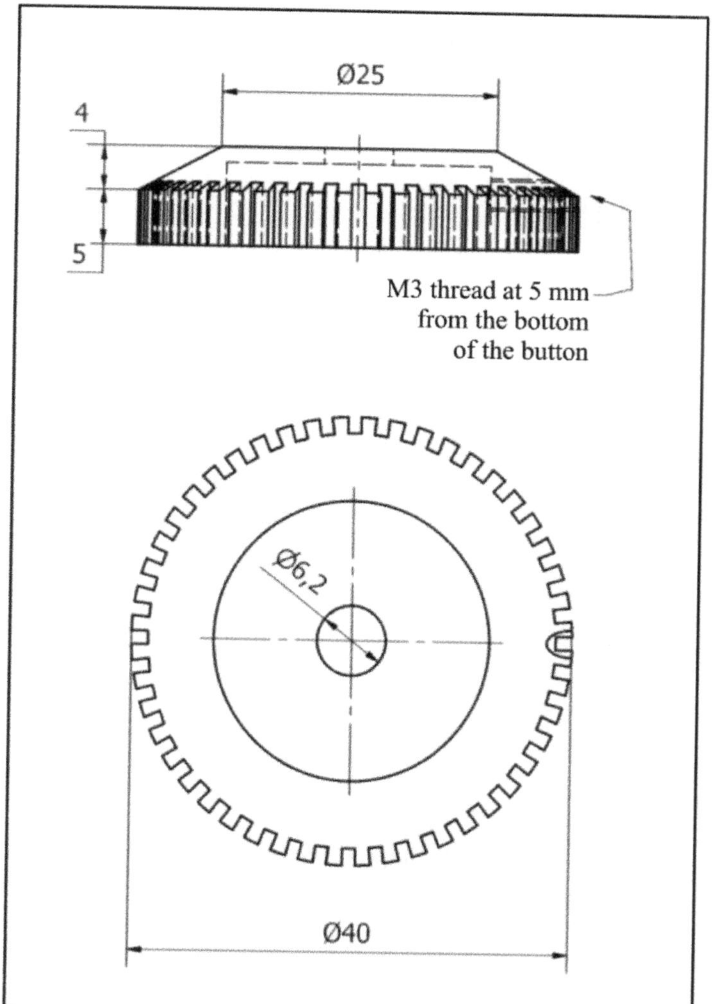

Figure 6.27. *Ondes 2208, power control knob diagram*

6.8.7. *The needle block*

When the performer puts their fingers on the contacts of the terminal of the needle block, they modify the frequency of the musical note played. This effect is achieved by circuiting low capacitance capacitors through the finger contacts and ground. The ondes chassis is itself grounded. This block is made of ebonite; it is a parallelepiped whose dimensions are: L = 30 mm, l = 8 mm, h = 10 mm. It is crossed by three brass rods 1.5 mm in diameter; these are slightly protruding from the upper face of the terminal and polished. It is fixed to the drawer by two screws.

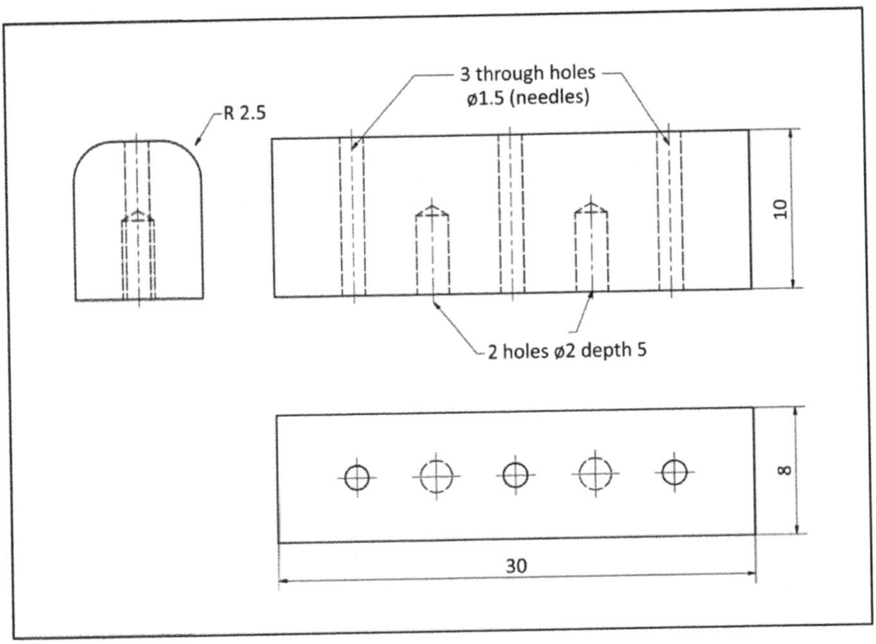

Figure 6.28. *Ondes 2208, needle block diagram*

6.8.8. *The intensity key*

Because of its importance in the operation of the ondes, the intensity key will be studied in detail in Chapter 7.

6.8.9. *The wiring*

All the components are available, the inductor (2.2 mH) of the T1 timbre, which was initially manufactured by Mr. Martenot using a ferrite pot, is currently available in the form of a small component. Similarly, the rectifier bridge of the T8 timbre, which initially is a finned selenium rectifier, is replaced by a silicon diode bridge.

The drawer wiring is mixed. A small terminal board supports a few components and allows the connections of the large wiring (10 insulated conductors + two shielded cables) between the chassis and the drawer. It should be noted that this strand is welded at both ends, which makes it difficult to extract the drawer from the cabinet, for example, in the context of a maintenance operation.

Two connectors are made by pairs of 4 mm banana plugs. The first located on the left side of the drawer allows for the connection of the pedal (it is identical from the operating point of view to the intensity key). The second pair, located on the upper side, behind and on the right, allows the links with the oscillators to establish the keyboard/ribbon control. These banana plugs will be replaced later by an industrial connector.

Another useful accessory is the lighting lamp (6.3 V/0.3 A); it is located in a groove in the front of the drawer. It produces a grazing light whose intensity is regulated by an adjustable resistance and low value collar 10 Ω/4 W.

Figure 6.29. *Ondes 2208, drawer wiring*

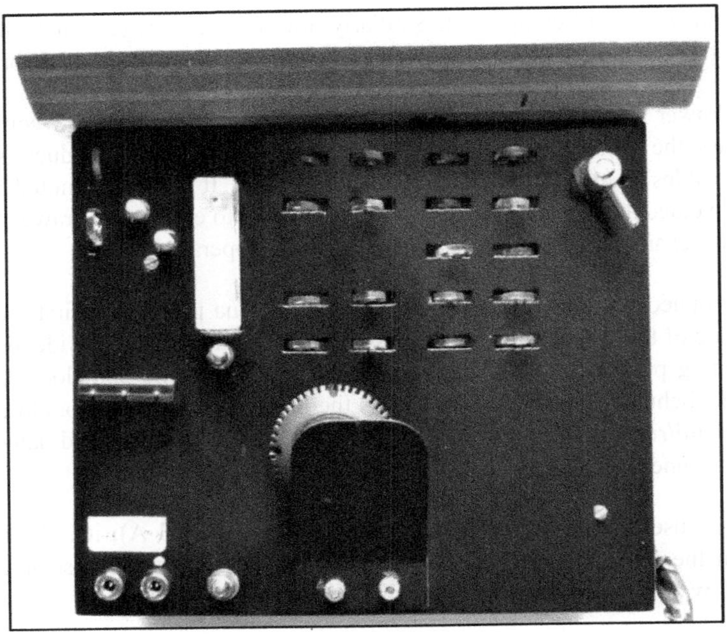

Figure 6.30. *Ondes 2208, drawer seen from above*

6.9. The keyboard/ribbon system

6.9.1. *Introduction*

The keyboard/ribbon system is a complex device whose realization was the most delicate. It consists of two interlocking rigid frame structures, one of which is suspended by nylon threads inside the other. The fixed frame supports the ribbon capacitor with the ribbon guide system. The frame, suspended, supports all the keys of the keyboard and is mobile in relation to the fixed frame. This freedom of movement allows control of the vibrato. We will only describe the manufacture of the essential elements so as not to unnecessarily weigh down this chapter.

6.9.2. *The ribbon capacitor*

Aluminum blades of 5/10 mm are arranged in pairs inside a U-shaped shield fitted with a removable cover, also in aluminum. The blades are associated in pairs; the progressive distance between each is obtained by means of plastic shims. On some of its shims are placed a guide rod and a pad to keep the movement of the ribbon parallel to the axis of symmetry of the pair of blades. All the pairs are arranged in a broken line in the middle to reduce lateral vibrations of the ribbon.

In practice:

– the shielding is a U-fold of a 10/10 mm aluminum sheet: L = 952 mm, l = 36.4 mm, h = 40 mm. The lid is made from the same sheet metal, the length is shorter so as to avoid an overlapping on the side flanges; L = 858 mm, l = 39.4 mm, h = 6 mm;

– three metal U 20/10 mm are arranged on the upper part of the shielding U to allow the fixing of the cover. At both ends: L = 35 mm, l = 20 mm, h = 15 mm; middle piece: L = 35 mm, l = 10 mm, h = 15 mm. All holes are M3;

– 12 blades 5/10 mm aluminum sheet of L = 148 mm, h = 31 mm;

– three groups of 12 plastic shims, with guide pins and pads. All mounting holes are M3.

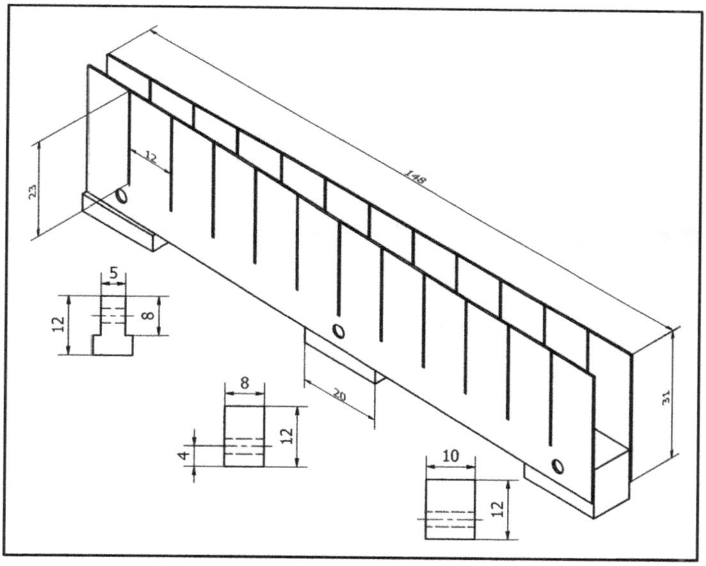

Figure 6.31. *Ondes 2208, diagram of one of the blocks of 12 blades that make up the ribbon capacitor, with its three shims*

The ribbon is stretched using a ribbon made of elastic material. It closes by hooking on a ring in which the ondist slips onto their finger. The ribbon slides around four pulleys arranged at the four corners of the fixed structure. To the left of this structure is a brass friction pin; it establishes electrical contact between the ribbon and the heterodyne circuit.

Figure 6.32. *Ondes 2208, top view, with shims and guide rods*

Figure 6.33. *Ondes 2208, ribbon tuning capacitor (with the ribbon, seen from above, indicated by the red line)*

To these devices must be added the ribbon tuning capacitor. It consists of an 8/10 mm spring blade, cut in a rectangle L = 12 mm * 30 mm. The useful wall is a square of 30 mm * 30 mm which faces one of the walls of the capacitor of the highest octave. The 90 mm * 30 mm return slat is fixed by two M2.5 screws on the wall of the shielding U. The adjustment of the proximity of this blade with one of the walls of the ribbon capacitor is carried out using an M5 threaded rod engaged in an M5/M8 threaded cylinder which is itself screwed through the wall of the U-shaped shield.

Two other accessories complete the ribbon capacitor tuning. The first is a wall made of 4/10 mm sheet metal, the dimensions of which are the same as that of a wall of the ribbon capacitor. L-shaped, it is placed parallel and close to it (top left in Figure 6.33), the sole is taken under two shims which support the pair of walls. Thus, the capacitor produced allows for the tuning adjustment when calibrating the ribbon capacitor. The second device is similar to the first but of smaller dimensions 20 mm * 30 mm, the wall is placed under the central fixing part of the cover.

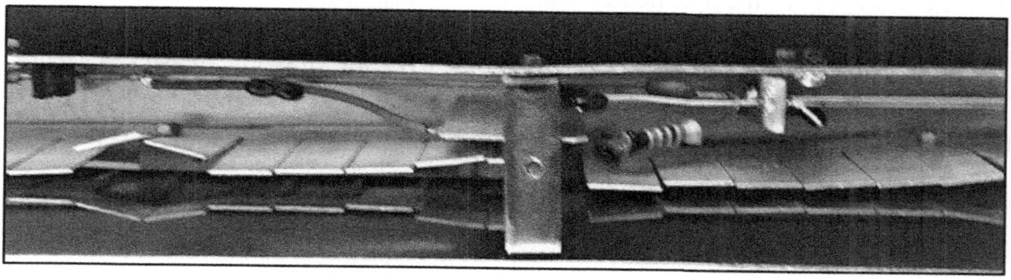

Figure 6.34. *Ondes 2208, auxiliary ribbon capacitor tuning device*

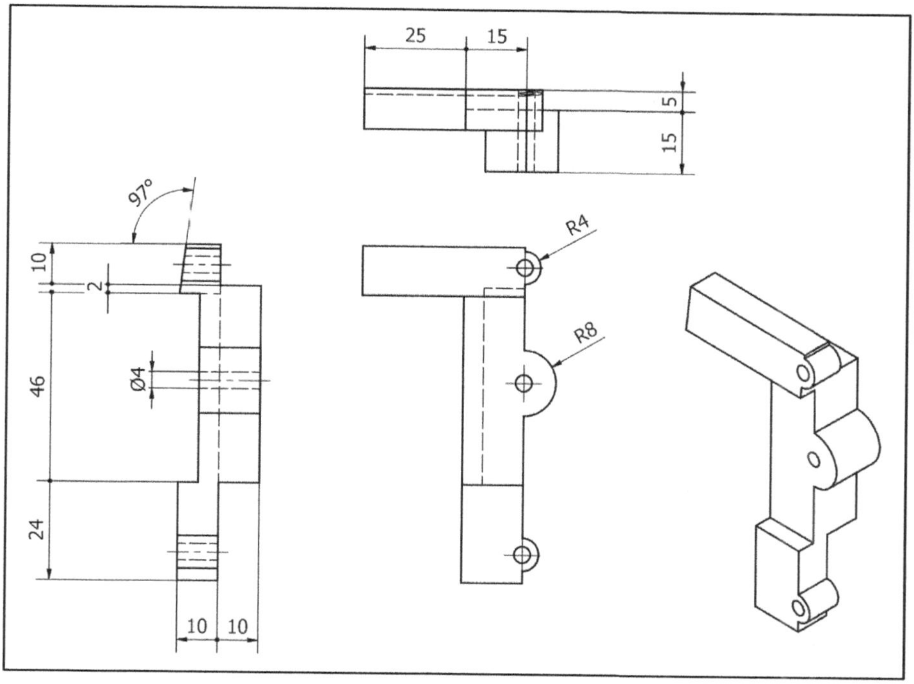

Figure 6.35. *Ondes 2208, diagram of the right side flank*

The U-shaped shield is the rear part of the fixed frame. At the front, we have an angle iron of 20 mm * 15 mm and 900 mm in length, in aluminum of 2 mm thickness. These two lengths are held at their ends by two lateral sides. These two parts with a complex design which stiffen the whole are subcontracted. They also ensure the suspension of the mobile frame which supports the keyboard. By means of guide parts, the pulleys are arranged on the upper part of the flanks. Two 50 mm long M6 threaded rods in brass, fitted with a knurled cylindrical head, ensure the reversible locking of the mobile keyboard when transporting the instrument.

For the machining of the pulleys, we took over the drawing of Mr. Martenot. Three of them are made of plastic; the fourth is made of aluminum and is placed near the friction pin.

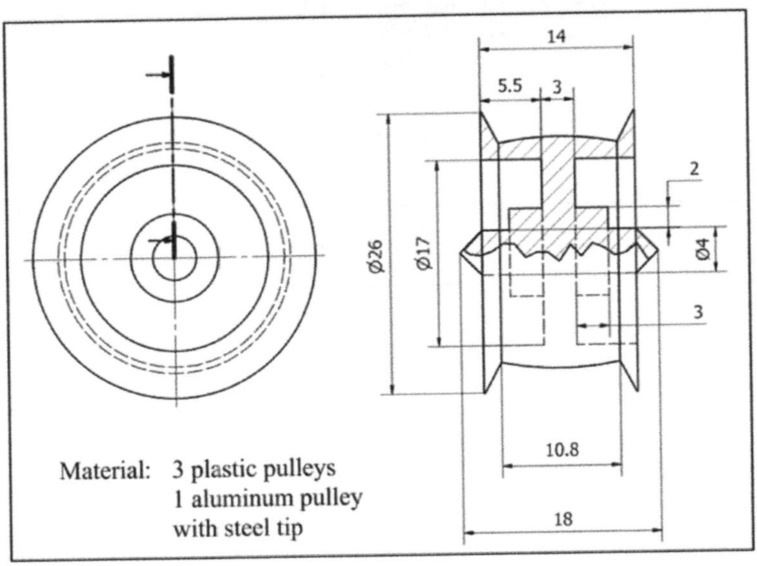

Figure 6.36. *Ondes 2208, diagram of the pulleys*

To these devices must be added the 4 mm thick bakelite shims which insulate the friction pin support plate and the guide parts of the pulleys. Similarly, a spacer is positioned at the front of the side flank, but behind the pulley guide part, it will be used for fixing the cover.

At the end of the assembly:

– On a bakelite support fixed to the outer bottom of the U-shaped shielding, on the ribbon input side, the LC system (17 mH/15 pF) for compensation of the treble register is mounted. The adjustable tubular-piston trimmer-type capacitor C is soldered inside the shield, between the rear wall and one of the input blades of the ribbon capacitor.

– Two thin sheet metal parts, formed and painted black, are arranged and screwed to the side parts of the fixed frame. They are equipped with a rectangular opening allowing the passage of the ribbon, they maintain by an M3 screw, the wooden piece in the shape of a bracket which hides the screws for tightening the keys.

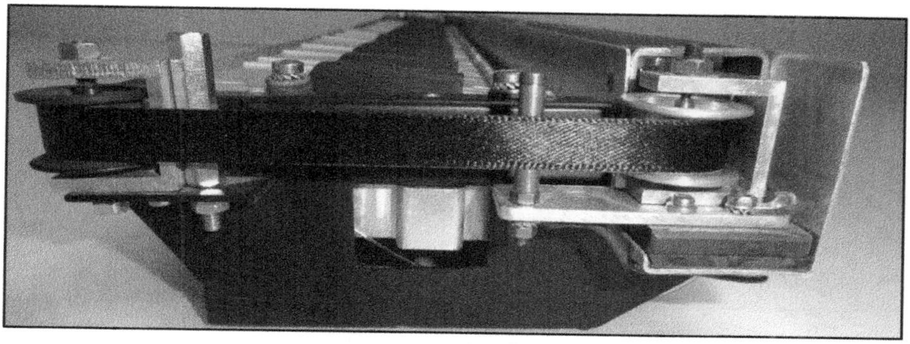

Figure 6.37. *Ondes 2208, assembly of the rod and pulleys*

Figure 6.38. *Ondes 2208, the LC compensation circuit for the high register*

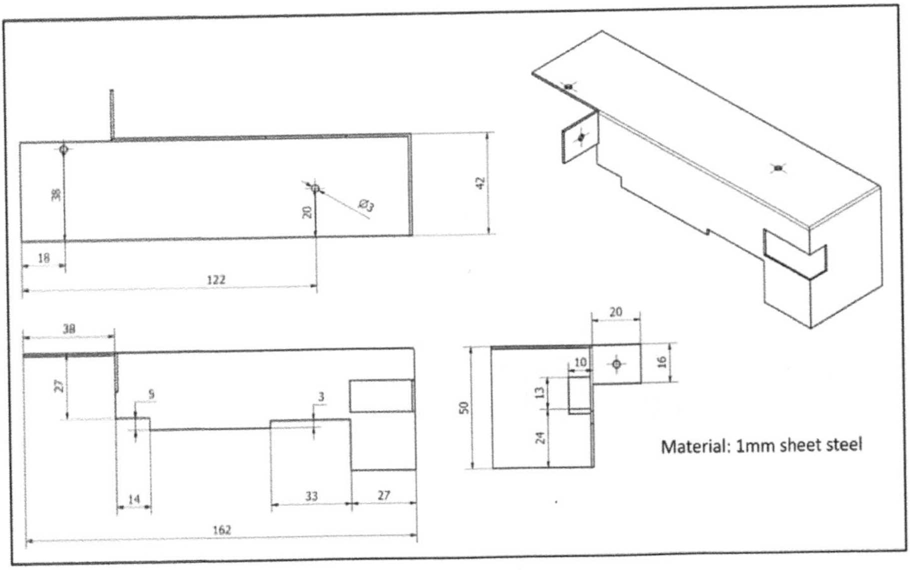

Figure 6.39. *Ondes 2208, diagram of the right side flange*

Figure 6.40. *Ondes 2208, mobile frame of the keyboard system*

6.9.3. *The keyboard*

6.9.3.1. *Introduction*

The keyboard is the most complex device to make. All 72 white and black keys rest on a mobile frame suspended by nylon threads inside a fixed frame. This set features many outsourced parts.

6.9.3.2. *The moving frame*

An angle iron (15 mm * 12 mm * 920 mm) in aluminum of thickness e = 3 mm, in the rear position, ensures the partial rigidity of the mobile frame. At the front, another U-shaped angle iron (15 mm * 25 mm * 15 mm * 920 mm) in aluminium, thickness e = 2 mm, is kept parallel to the previous one thanks to two parts (subcontracted) located at both ends and five legs (subcontracted) of particular shape. The rigid assembly obtained constitutes the mobile frame. On the back angle, the keyboard keys will be fixed.

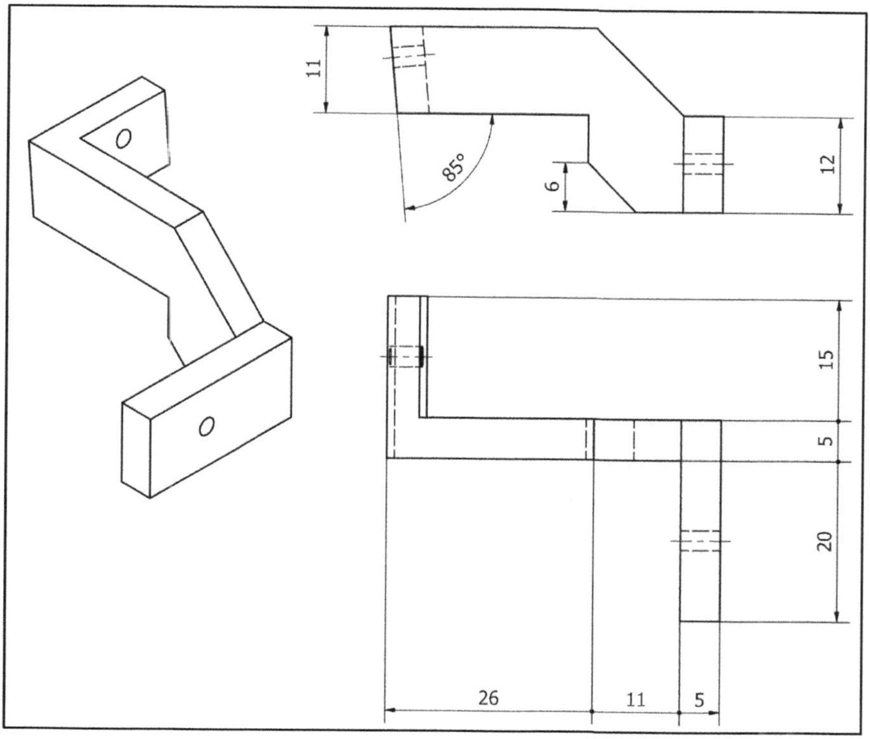

Figure 6.41. *Ondes 2208, diagram of the support part (leg) of the mobile frame*

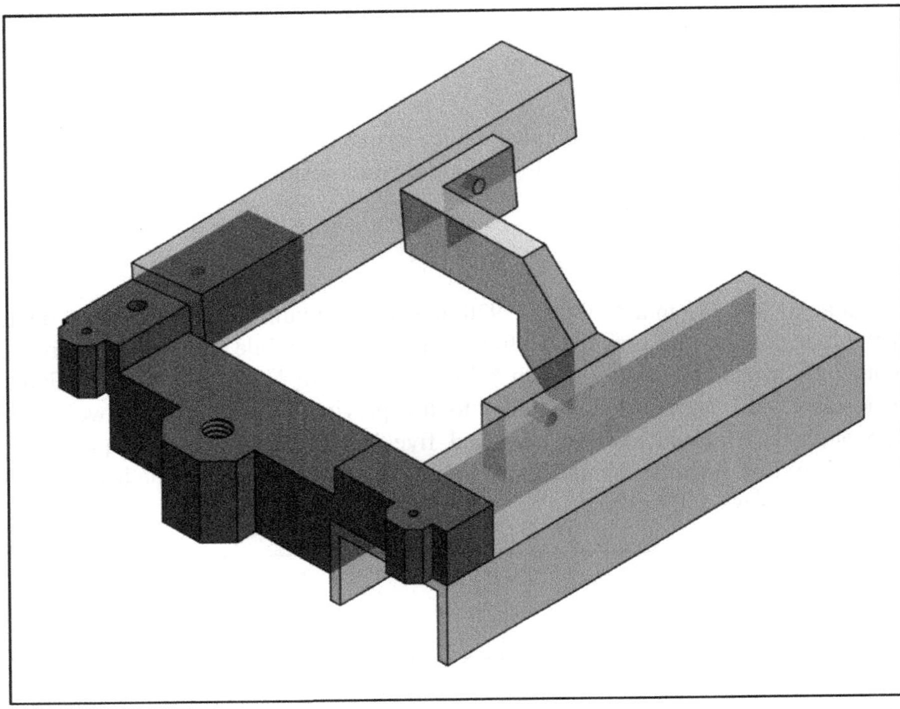

Figure 6.42. *Ondes 2208, diagram explaining the assembly of the mobile frame*

On the upper and front face of this frame, a 4 mm thick hardboard plane ("Isorel") is fixed. A length of cork 15 mm * 15 mm in the section is glued to this plane. Pins are inserted through the cork shim. They ensure the connection between the keyboard inductors and the fixed contact of the key. The keyboard coils are grouped by octave on a printed circuit; this one is fixed by two M2.5 screws on the support in "Isorel".

Figure 6.43. *Ondes 2208, cork square*

The suspension of the mobile frame is made using Nylon threads. These threads are attached to the upper part of the outer sides of the fixed frame. They pass through the end pieces of the mobile frame, and then are fixed to the ends of a leaf spring. The tensioning of this blade is carried out by a return screw which crosses the external side.

In the transport position, the brass screws (M6) block the mobile keyboard. They lift it up and press it against the two cut and folded wall elements on each of the outer sides (see Figure 6.51).

In the playing position, the mobile frame is released by unscrewing the two brass screws, the mobile frame moves from top to bottom parallel to it. Until the end, pieces of the mobile frame come up against the knots made on the Nylon thread.

In both cases, the Nylon threads are always stretched by the spring blade. But the adjustment of the tension of the Nylon threads determines the flexibility of the lateral movement of the mobile frame.

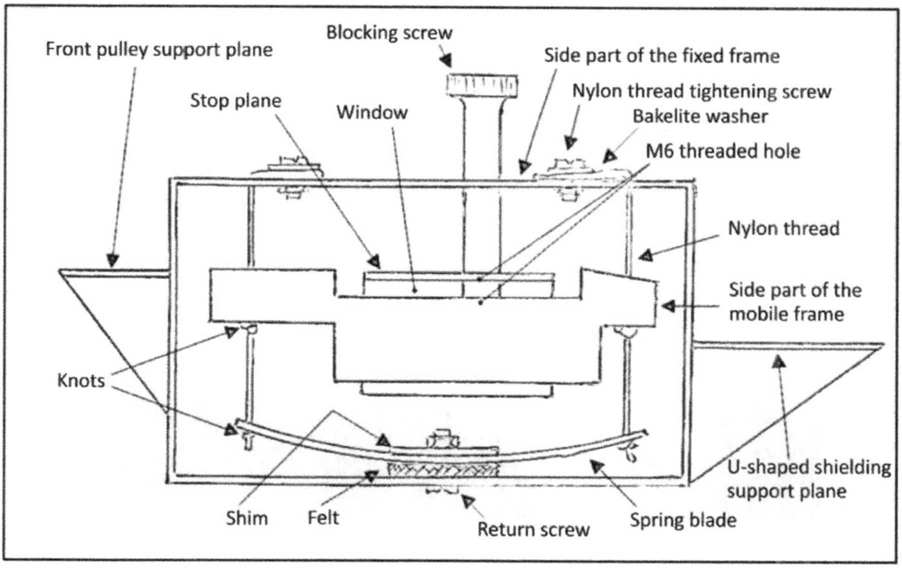

Figure 6.44. *Ondes 2208, sketch of the mobile keyboard suspension system*

6.9.3.3. *Keyboard keys*

The keyboard keys are made, on the one hand, by laser cutting in a sheet of white plastic material for the unaltered notes, and, on the other hand, in ebony wood for the altered keys.

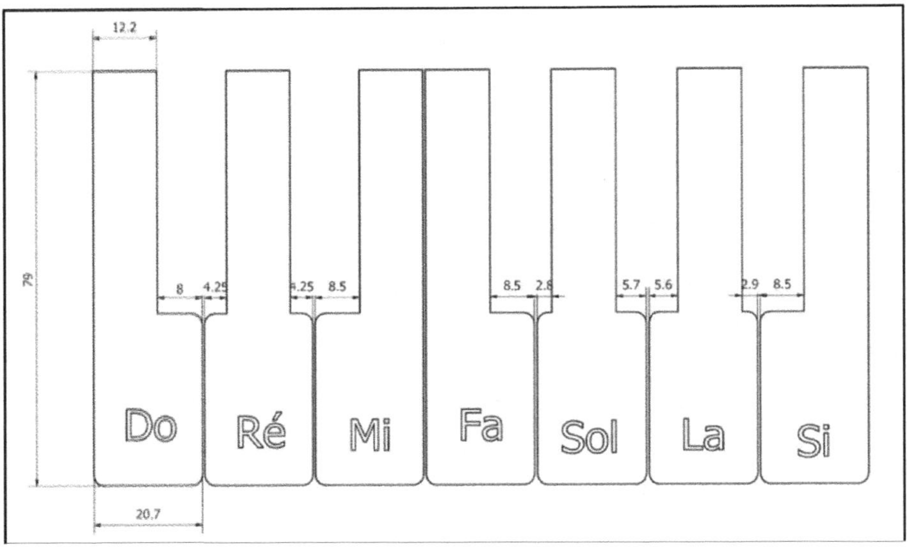

Figure 6.45. *Ondes 2208, white key diagram*

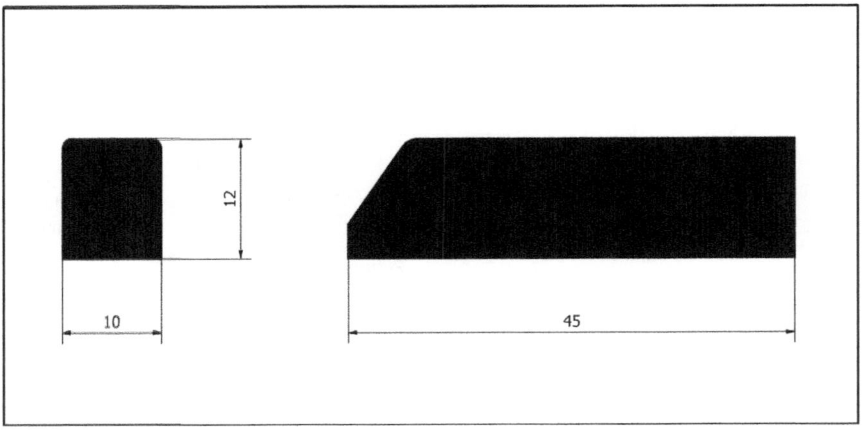

Figure 6.46. *Ondes 2208, black key diagram*

They are all glued to a plastic shim obtained by laser cutting. This shim has a determining role; it will ensure the assembly of the elastic blade (35/100 mm) in tinned phosphor bronze. A pair of moving contacts is welded with tin on this blade which is also secured using three rivets on the shim. Mr. Martenot thus avoids performing a metal/plastic bonding, which would indeed have little chance of lasting (the deformation stress applied to the blade when the finger presses the key is significant).

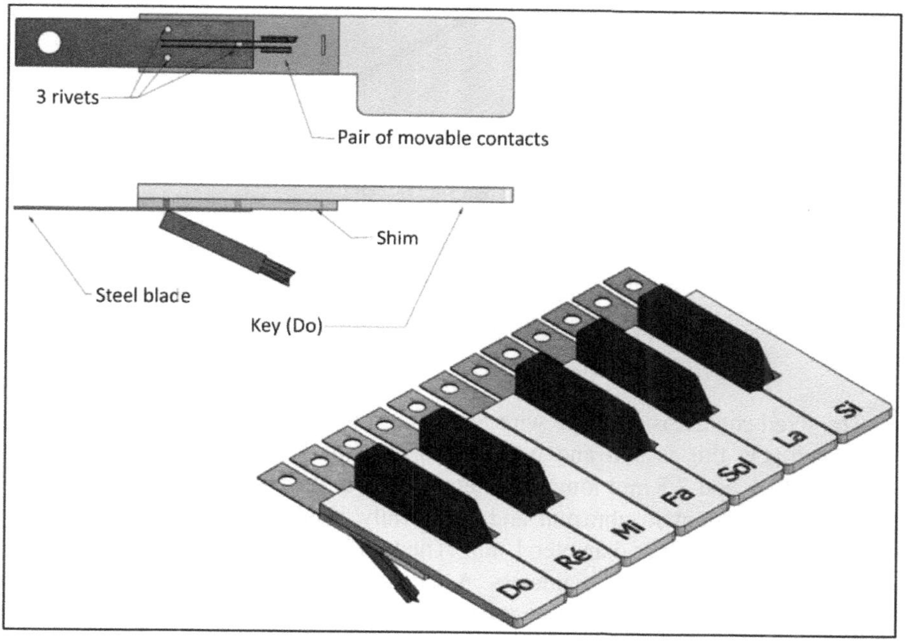

Figure 6.47. *Ondes 2208, diagram of a complete key (Do) and an octave*

The keys are distributed in an interval of 150 mm so as to best coincide with the index mark of the ribbon. They are fixed to the rear angle iron by American screws (UNC #8, 32 threads/inch) of 6.35 mm in length, in aluminium; on either side of the blade are two washers, also made of aluminum.

The moving contacts are cut from a 25/100 mm copper sheet and then formed at one end in a V shape using a template. The two blades associated in pairs are welded with tin on the spring blade of the key, before riveting on the plastic shim.

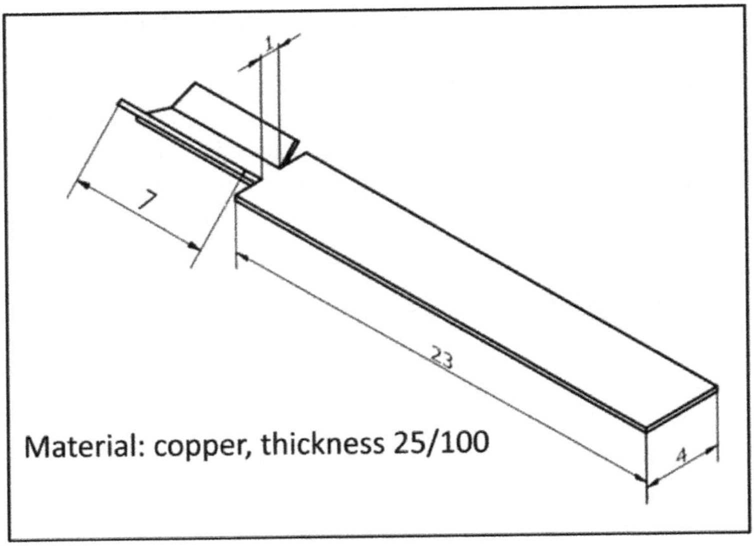

Figure 6.48. *Ondes 2208, moving contact diagram*

The fixed contact is a bronze wire 5/10 mm in diameter and of total length equal to 35 mm. On this is slid and welded a brass tube 1 mm in external diameter, 2/10 mm thick and 7.5 mm long. In extension and at the upper end is also slipped a plastic tube resistant to abrasion and electrically insulating. The dimensions are: length 7.5 mm and outer diameter 1 mm. The two tubes placed end to end have the same internal and external diameters. A small folding is made at the end of the bronze wire to avoid any displacement of the insulating sheath. The other end is bent as a pin and will be tin-soldered to the office pin. The finished part will then have a total length of 25 mm.

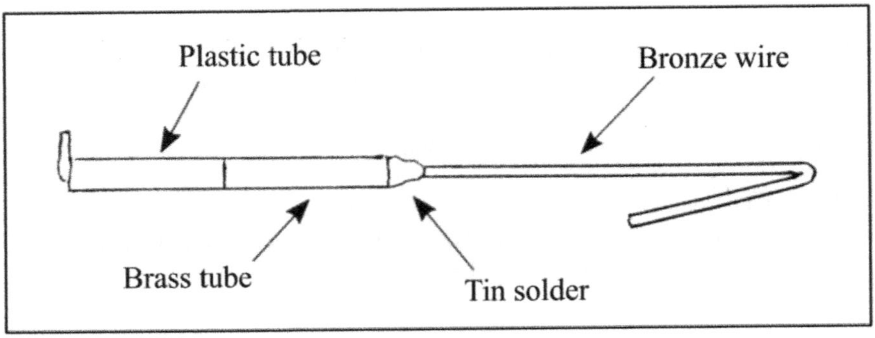

Figure 6.49. *Ondes 2208, keyboard fixed contact*

By pressing the key, electrical contact is established when the pair of V-shaped conductors grips the brass tube. The inductor that corresponds to the note played is brought into service. In the rest position, this pair of conductors presses on the insulating part of the fixed contact.

A satin ribbon 4 mm wide is immobilized by gluing to the shim at one of its ends. The other end is slipped through a strand of felt and then secured with a pin. The tension of the ribbon compensates, in the resting state, the tension of the spring blade. By adjusting the tension of this ribbon, we define the rest position of the key and its alignment with the other keys of the keyboard. The "depressed" position of the key is defined by the top surface of the felt edge.

Figure 6.50. *Ondes 2208, keyboard key tensor ribbon*

6.9.3.4. *The fixed frame*

This rigid assembly is made up of two aluminum angles secured by screwing on two end pieces (subcontracted) of complex shape. As indicated in Figure 6.44, this structure supports the mobile keyboard and the ribbon capacitor with its four pulleys.

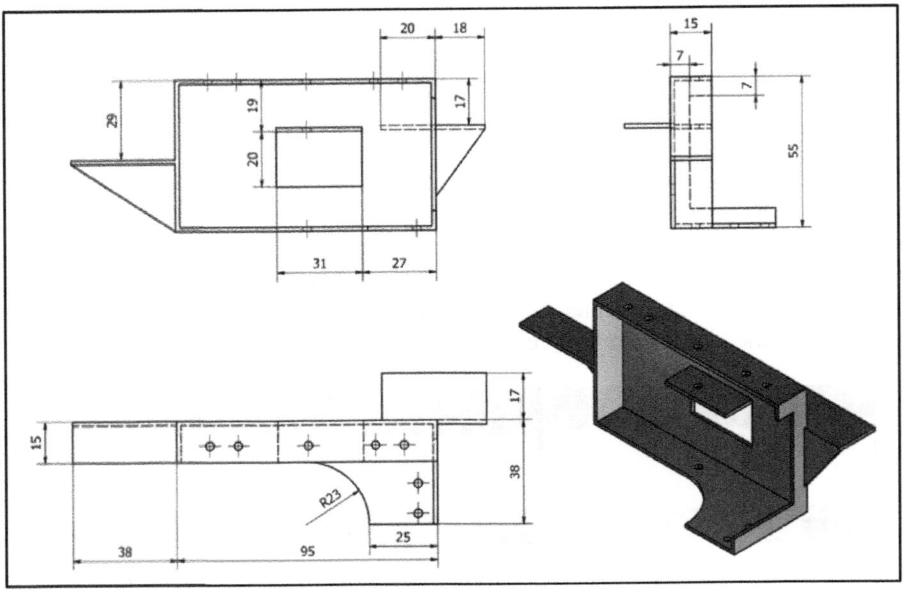

Figure 6.51. *Ondes 2208, diagram of the right outer flank*

When the system is completely assembled, it is placed in the cabinet on two pieces of wood, above the chassis. These pieces of wood make it possible to position the keyboard in height so as to ensure the closure of the lid of the cabinet. The sides of this piece of cabinet laterally immobilize the external frame.

6.10. The accessories

The accessories have a decisive importance in the identification of the ondes musicales.

6.10.1. *Command buttons*

On the drawer: the various double ear contactors (see Figure 6.22), the T8 timbre (see Figure 6.26) and the Ebonite power adjustment button (see Figure 6.27).

On the front of the instrument (see Figure 5.5), from left to right:

– ribbon tuning (on the horizontal axis to the left above the keyboard); round plexiglass diameter 18 mm;

– tuning of the low register of the keyboard (on the vertical axis to the left of the chassis); beveled round in plexiglass, diameter 20 mm;

– notched wheel for tuning the high register of the keyboard (horizontal axis to the left of the chassis near the previous one); diameter 90 mm, thickness 3 mm;

– bass register (to the left)/treble register (to the right) switching button; threaded rod, brass ring, with plastic button at the end (see Figure 6.53);

– T7 timbre and T0 timbre; threaded rod, brass ring, with a plastic knob on the horizontal axis (see Figure 6.53);

– notched diapason tuning wheel (horizontal axis approximately in the middle of the frame); diameter 90 mm, thickness 3 mm.

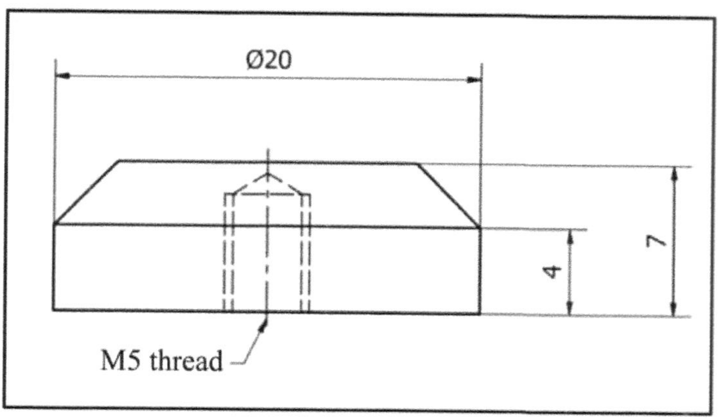

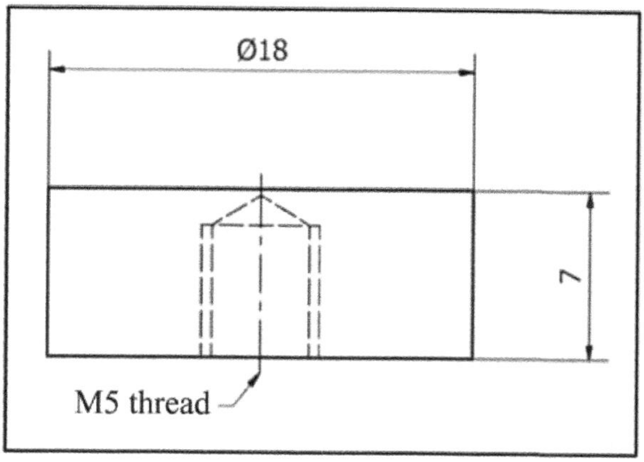

Figures 6.52. *Ondes 2208, diagrams of the tuning knobs of the ribbon (left) and the low register of the keyboard (right)*

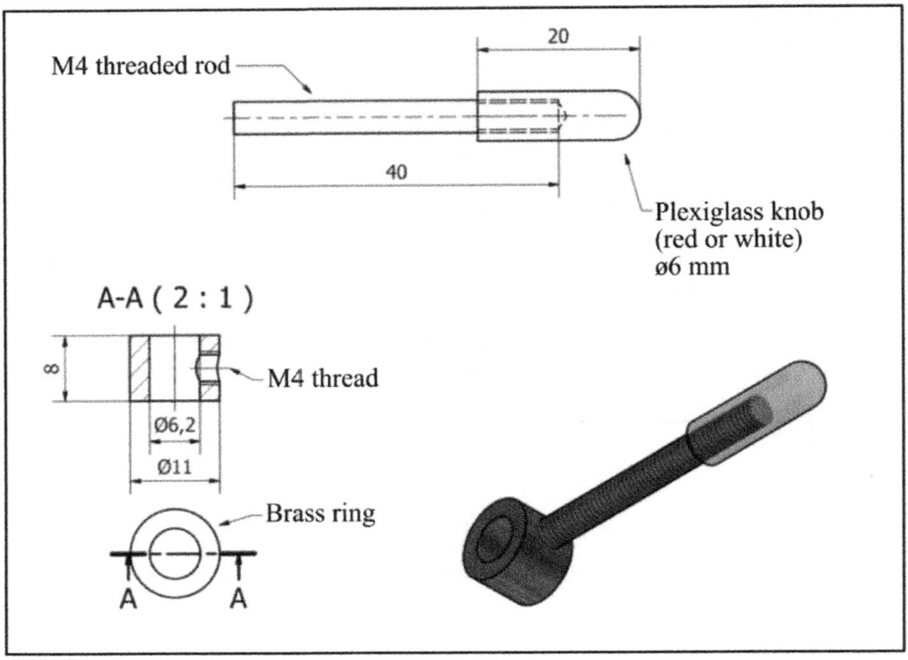

Figure 6.53. *Ondes 2208, diagram of the register change lever, and the T0 and T7 timbres*

6.10.2. *Other accessories*

The index mark:

This device is essential in the functioning of the ondes; it allows the performer to locate the musical note played by positioning the index either on a "hole" (white key) or on a "bump" (black key), but it also allows us to perform a *glissando* on all six octaves. We used a strip of PVC + wood composite material in black color. The dimensions are: L = 904 mm, l = 15 mm and e = 3 mm. With respect to the central axis, holes are drilled 1.3 mm in diameter, 12.5 mm apart along the entire length of the strip. Then, a conical hole is made (with a 6 mm drill, 6 mm depth) for the markers of the white keys. After positioning and gluing this strip in the intended location on the cabinet, brass and dome-head nails (1.3 mm * 12 mm) are driven into the remaining holes. We then find both the initial aesthetics of the device manufactured by Mr. Martenot and the quality of sliding necessary for the functionality of the device.

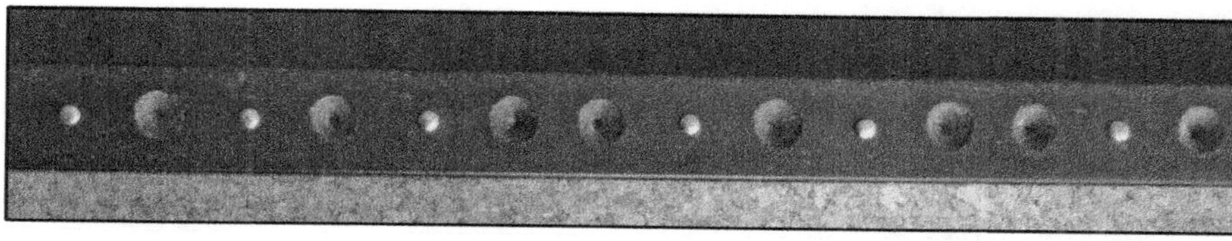

Figure 6.54. *Ondes 2208, index mark*

The ring:

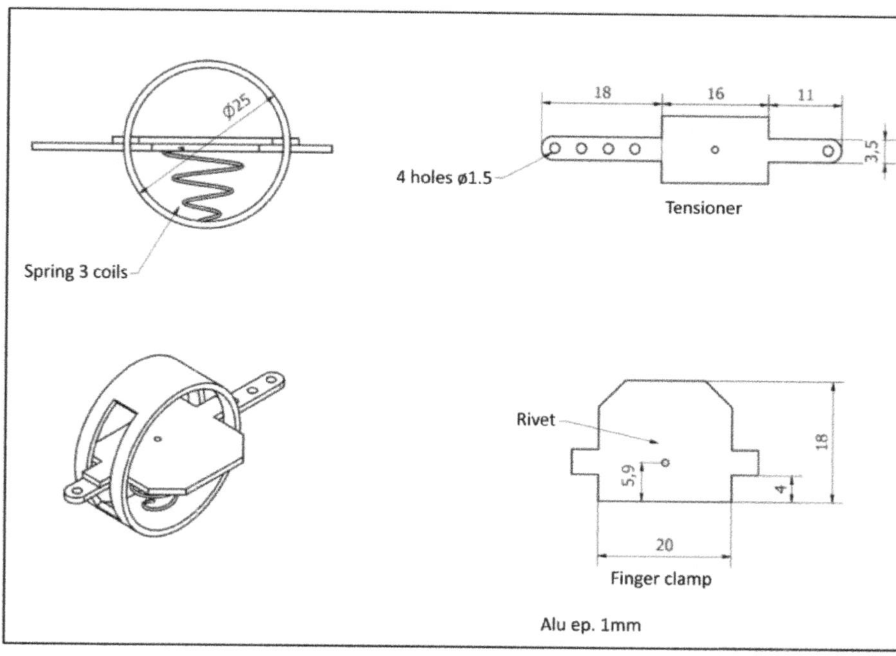

Figure 6.55. *Ondes 2208, diagram for the realization of the ring*

The music stand:

Figure 6.56. *Ondes 2208, the music stand (the brass fixing parts, the three blades stored under the Plexiglas shim) and the identification plate*

A set includes two inclined brass supports arranged on each side of the identification plate of the ondes, in which one slides a set of three transparent plastic blades of different lengths, forming a wave propagation structure. These three blades (width 15 mm, thickness 1.5 mm) have lengths of 585, 685 and 820 mm. They are superimposed and aligned on the left end with a rivet. The music stand fixing parts are cut from a rectangular brass profile (20 mm * 10 mm, thickness 1.5 mm), with an attached and welded bottom.

Identification plate:

This is a Plexiglas plate (300 mm * 20 mm) engraved with a laser cutter (see Figure 6.56).

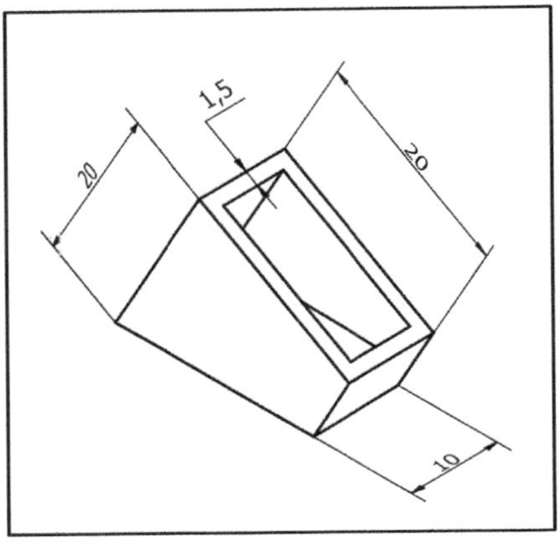

Figure 6.57. *Ondes 2208, fixing part diagram*

Hook to keep the ribbon taut:

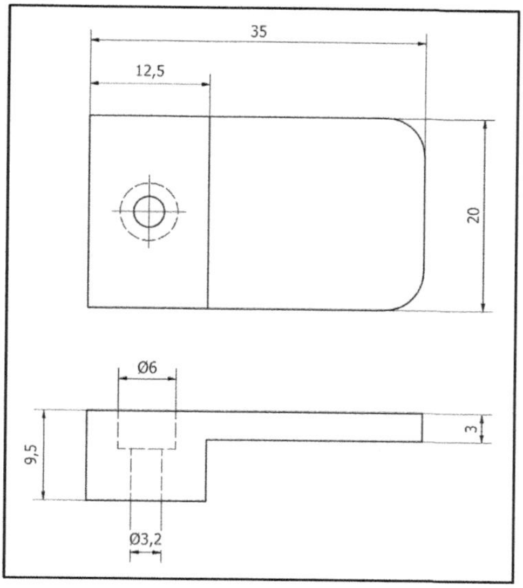

Figure 6.58. *Ondes 2208, diagram of the storage wedge of the music stand*

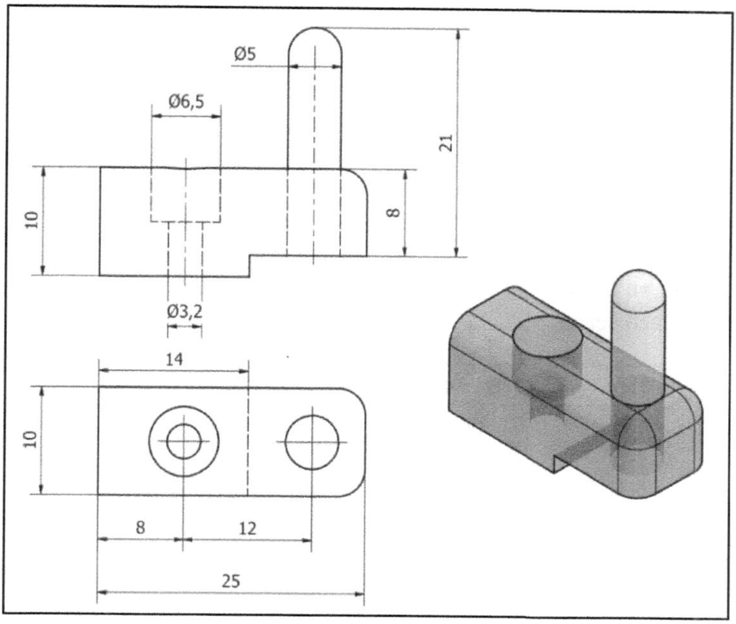

Figure 6.59. *Ondes 2208, stretched ribbon crochet diagram (plexiglass)*

6.11. Calibration and tuning

6.11.1. *Introduction*

Tuning an ondes Martenot with vacuum tubes is a long and delicate operation which requires a lot of patience; the first operation is to calibrate the ribbon capacitor, and the second is the calibration of the keyboard. Only then can the tuning procedure begin.

6.11.2. *Ribbon capacitor calibration*

The ribbon capacitor is a sub-assembly extremely sensitive to external disturbances; it is very laborious to calibrate it with the instrument itself, mainly because of the drift of the oscillators.

The first step is to form the capacitors using a second ribbon capacitor which will serve as a reference. We measure for each note the value of the capacitance on the prototype and the value of the corresponding capacitance on the reference ribbon capacitor, in this way; it is possible to calibrate the ribbon capacitor of the prototype.

The measurement technique with a balanced bridge was not used because it requires having a box of very low value capacitors. Moreover, its implementation is extremely delicate because the capacitances to be measured are of the same order of magnitude as the interfering capacitances which appear in any wiring.

The second step consists of using a very stable quartz oscillator and taking measurements with a frequency meter. We have made quartz mixing oscillator device. This device, which works with transistors, incorporates an 80 kHz crystal and an oscillator controlled by the ribbon capacitor.

Beforehand, in Table 6.1, we will give the values of the capacitors to be placed in series with each sector.

Octave 0	Octave 1	Octave 2	Octave 3	Octave 4
18 pF	22 pF	39 *pF*	56 pF	150 pF

Table 6.1. *Value of capacitors put in series for each of the sectors*

Similarly, a capacitor with a value of 82 pF is soldered across the ribbon capacitor wire terminals inside the shield. It ensures the equality of the capacitance values between the capacitor C_{57} (220 pF) put into service in keyboard mode and the capacitance value seen by the variable oscillator in ribbon mode.

The pre-tuning blade on the left end of the device is positioned midway between the ribbon capacitor wall and the shield wall. It is the same for the blade controlled by the tuning adjustment wheel.

The blades of the capacitor are first preformed by measuring the gaps (in mm) between the wall and the ribbon, at the entrance and at the exit of each section. Then, we separate the first two blades of each section (discontinuity). A first measurement of the overall capacitance variation is taken; it must be approximately 33.5 pF (the LC compensation circuit is short-circuited). The swept range extends from Do_0 to La_5.

At this stage, we can start a more precise adjustment by measuring the capacitances at the entrance of each section by positioning the ring in front of the corresponding note (Do). Again, the compensation LC circuit is short-circuited. Table 6.2 gives the values to be obtained.

Wall/ribbon capacitance measurements (in pF), after tuning and short-circuiting the corrector LC dipole						
Octave	**0**	**1**	**2**	**3**	**4**	**5**
Do	29.40	35.60	40.40	45.20	50.90	57.80
La						62.90

Table 6.2. *Values measured for the walls/ribbon capacitances (measurements made with a Peak LCR45 impedance meter, at 200 kHz)*

If the differences from one sector to another are too great, carefully observe the path of the conductive part of the ribbon and we can identify a few sensitive places where the ribbon approaches or deviates from the walls (guide rods, breakline); the blades of the sector are separated or brought closer to these places, then it is checked that a continuous increase in capacity without sudden variation is obtained. We end the calibration with a frequency measurement for each note. If for a sector, we observe a progression of the variation of the capacitances which is too fast or too weak, it is necessary to think of modifying the value of the series capacitor of the sector.

We then connect the device to the quartz measurement system. The compensation LC circuit is then connected. One carries out a round trip of the ribbon in order to check that the range of frequency variation is of the order of

2 kHz; if this is not the case, we act on the magnetic core of the inductor of the compensation circuit.

We finally proceed to the final adjustment of the ribbon capacitor with frequency measurement for each note. It is preferable to start with octaves 5, 4 and then 3; indeed, on these octaves, the values of capacities are the greatest, then we take again the adjustment on octaves 2, 1, 0. And we renew these operations until the final tuning.

At the end of this whole procedure, the ribbon capacitor is calibrated.

6.11.3. *Calibrating the keyboard coils*

The procedure is much faster because the direct measurement of each coil is possible. By adjusting the core on each inductor, we will make sure to obtain the values given in Table 6.3.

Keyboard inductor measurements after tuning (en µH)						
Octave	**0**	**1**	**2**	**3**	**4**	**5**
Do	0.31	0.40	0.99	2.45	5.19	10.51
Do#	0.30	0.64	1.31	2.81	5.83	10.99
Ré	0.37	0.67	1.54	2.78	6.56	11.20
Ré#	0.49	0.76	1.65	3.02	6.27	11.64
Mi	0.46	0.72	1.66	3.65	6.77	13.68
Fa	0.36	0.88	1.68	3.95	7.64	13.69
Fa#	0.42	0.95	1.87	3.63	8.05	15.24
Sol	0.44	1.01	1.99	3.89	7.70	17.05
Sol#	0.40	1.06	2.18	4.43	7.81	16.53
La	0.52	1.02	2.17	4.41	9.73	19.69
La#	0.66	1.06	2.29	4.75	10.19	20.03
Si	0.64	1.21	2.23	5.07	9.72	20.00

Table 6.3. *Measured values for keyboard inductors (measured with a Peak LCR45 impedance meter, at 200 kHz)*

NOTE.– These values were measured at the terminals of the inductors after having tuned the instrument, taking into account the low values of these ones, variations can appear, in particular according to the elements of the wiring used.

6.11.4. *Tuning setting*

Connect the ribbon capacitor and the keyboard to the terminals of the instrument; all the adjusting devices are placed in the middle position and the voltage is turned on waiting for thermal stabilization.

The following settings are available (see Figure 5.5):

1) Preset devices:

– the adjustable capacitor C_{30} marked L which adjusts the frequency of the fixed oscillator of the high register, F = 160 kHz;

– the adjustable capacitor C_{32} marked F which adjusts the frequency of the fixed oscillator of the low register, F = 80 kHz;

– the adjustable capacitor C_1 marked E which acts on the frequency of the variable oscillator.

2) Tuning devices:

– the ribbon capacitor tuning wheel marked R;

– the control wheel of the S_3 inductor for tuning the low register and the keyboard marked M;

– the large capacitor C_{31} control button for tuning the high register of the keyboard marked K;

– the large control button of the $C_{41A/B}$ differential capacitor known as the diapason marked G.

The adjustment procedure that we have determined is based on the following observations:

1) A calibrated ribbon capacitor adapts easily, at the level of the tuning, to a small variation of the fixed frequencies. Indeed, let us observe how the adjustment frequency of the variable oscillator evolves, when ΔC, which represents the variation in the capacitance of the ribbon, is the variable. In the absence of control by the ribbon or the keyboard, on the low register, the frequency of the fixed oscillator is equal to the frequency of the variable oscillator in order to obtain the zero beat.

This table is the application of the relation (see section 5.4):

$$1 - \frac{\Delta F}{F_0} = \frac{1}{\sqrt{1 + {\Delta C}/{C_0}}}$$

ΔC (*pF*)	F_0(*kHz*)	C_0(*pF*)	ΔF (*Hz*)
65.0	80.000	1412.0	1781.35
67.5	80.000	1412.0	1847.51
70.0	80.000	1412.0	1913.50
72.5	80.000	1412.0	1979.32
73.4	*80.000*	*1412.0*	*2000.25*
75.0	80.000	1412.0	2044.98
77.5	80.000	1412.0	2110.47
80.0	80.000	1412.0	2175.80
82.5	80.000	1412.0	2240.96

Table 6.4. *Values of ΔF, when the variable is ΔC (F0 fixed at 80 kHz, C0 at 1412 pF)*

Now, observe this same table but taking F_0 as a variable quantity; recall that C_0 and F_0 are linked by the relation:

$$F_0 = \frac{1}{2\pi\sqrt{L_0 C_0}}$$

ΔC (*pF*)	F0 (*kHz*)	C0 (*pF*)	ΔF (*Hz*)
73.4	65.000	2140	1086.19
73.4	67.500	1190	1214.03
73.4	70.000	1850	1351.24
73.4	72.500	1720	1498.1
73.4	75.000	1610	1654.93
73.4	77.500	1510	1821.94
73.4	*80.000*	*1410*	*1999.43*
73.4	82.500	1330	2187.63

Table 6.5. *Values of ΔF, when the variable is F0 (ΔC being fixed at 73.4 pF)*

On these two tables, the quantities indicated in italics are the theoretical adjustment values.

2) A calibrated keyboard only works with a pair of capacitor values $\{C_1 + C_2 + C_{57}\}$ and $\{C_3\}$ which guarantees the frequency variation range to be obtained (i.e. approximately 2 kHz).

It follows the idea that it is necessary to seek the optimum setting in keyboard mode, which is compatible with operation in ribbon mode. It is rare that at the end of the adjustment, the fixed oscillators are aligned on 160 kHz and 80 kHz.

6.11.5. *The tuning procedure*

1) Adjust fixed oscillator frequencies, high register (L) 160 kHz, low register (F) 80 kHz.

2) Put yourself in ribbon mode and position the ring at the left end of the index mark, adjust (E) so as to obtain F = 80 kHz, on the variable oscillator.

3) Place the ring in front of the La_{440} and tune the note using the tuning wheel (R). If adjustment is not possible, action is taken on the pre-tuning blade of the ribbon capacitor.

4) We switch, in the low register in keyboard mode, and we play the note La_{440} we tune with the wheel (M). Then, we play all the notes on the keyboard and we first check that the range of frequencies to be covered is obtained. If this is not the case, the setting of the capacitor (E) is modified to obtain the range of frequencies to be reached.

5) We go back to the La_{440} and we seek the agreement with the capacitor (F), then with the tuning wheel (M) of the keyboard.

6) We go back to ribbon mode and tune the La_{440} with the tuning wheel (R) of the ribbon.

7) The operations 4-5-6 are repeated until complete tuning is obtained at the level of the ribbon and at the level of the keyboard. Adjusting the diapason capacitor (G) using the wheel can make tuning easier in some cases. We measure the frequency of the fixed oscillator of the low register which, at the end of the adjustment, has varied, and we note its value (e.g. 79.857 Hz).

8) We switch to the high register, and we adjust by (L) the frequency of the oscillator to twice that measured previously (in our example, F = 2 * 79857 = 159714 Hz).

9) In keyboard mode and on the high register, we check that we can obtain the tuning of the La_{440} by acting on the tuning wheel (K) of the high register.

10) From this step, we can tune the keyboard note by note to the high register and check that the low register is working correctly.

11) Fine-tune the tuning of the ribbon capacitor. Tunings on the lowest notes (Do_0, $Do\#_0$, $Ré_0$) are very difficult to obtain.

When all of these operations ("factory settings") are carried out, the procedure becomes simpler: with each new use, the tunings are obtained only by using the settings by the commands R, M, K.

In Table 6.6, we summarize the final values of the essential quantities.

F_0 high register	F_0 low register	C_{34}	C_{29}	C_2	C_3
159.146 kHz	79.578 kHz	1383pF	313pF	498pF	573 pF

Table 6.6. *Values obtained at the end of the tuning of ondes 2208*

6.12. Some photographs of ondes 2208

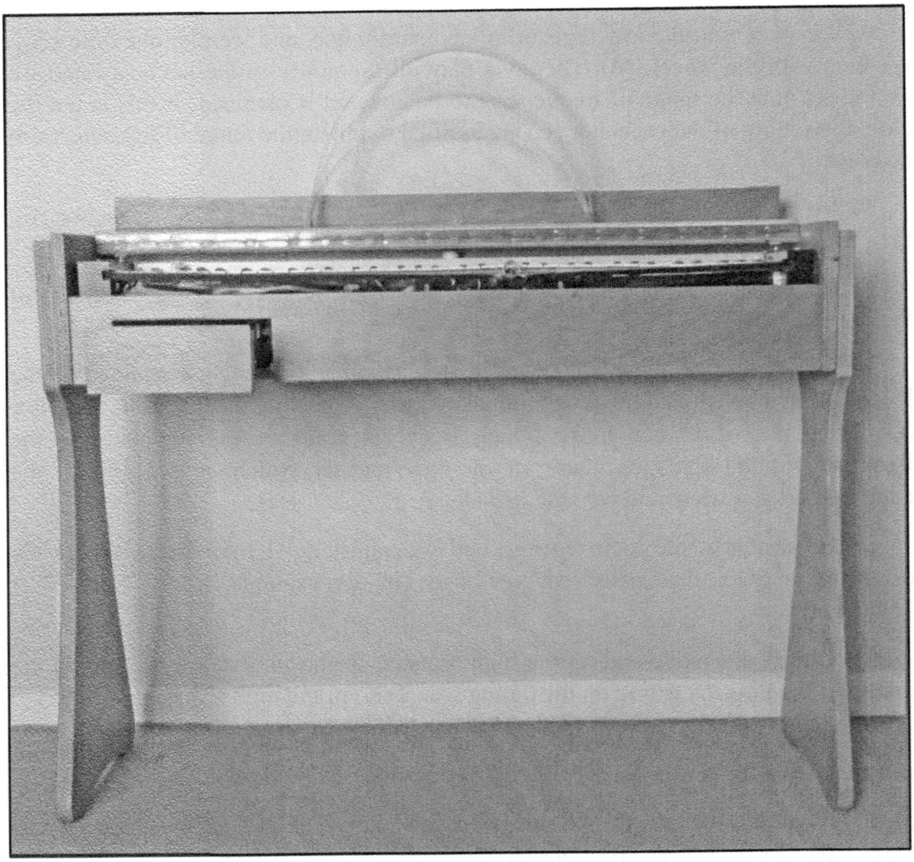

Figure 6.60. *Ondes 2208, front view*

Figure 6.61. *Ondes 2208, the ring*

Figure 6.62. *Ondes 2208, the drawer*

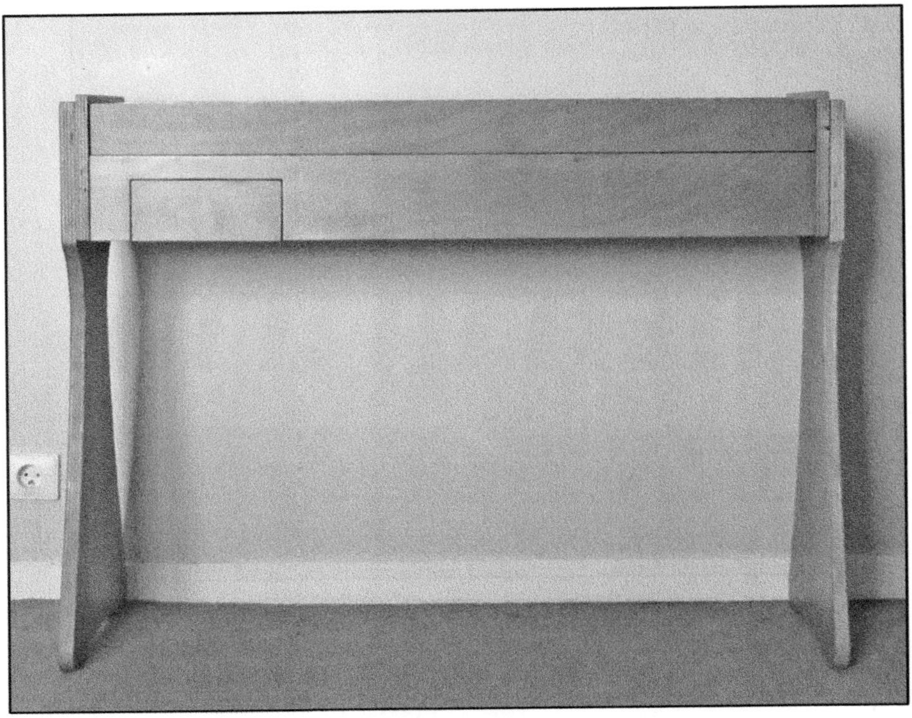

Figure 6.63. *Ondes 2208, cabinet closed*

7

Manufacture of the Leather Bag of the Intensity Key and the Ribbon

7.1. The intensity key

Before approaching the technique of manufacturing the intensity key, we will describe its evolution over time and according to the ondes models. This work is necessary because this device, which may seem simple in its design, is actually very difficult to describe. There is currently no physical model that fully describes its properties. It remains a mystery, and its manufacture is partly based on a recipe that only Mr. Martenot knew.

The intensity key is an essential device of the ondes-type musical instruments built by Maurice Martenot from 1928 until the 1980s. It allows the sound intensity delivered by the instrument to be controlled.

The purpose of the following study is to analyze the different solutions that Mr. Martenot has considered to control the sound level. Currently, we have identified five types of keys, which chronologically follow each other, and best meet the constraints imposed by both the user and the instrument:

Around 1925–1928: the mercury key.

1928–1937: first version of the powder key.

1937–1953: second version of the powder key.

1953–1973: third version of the powder key.

For the color versions of the figures in this chapter, see www.iste.co.uk/courrier/ondesmartenot.zip.

1973–1980: fourth version of the powder key adapted to transistor-based ondes.

We are going to describe these different keys as precisely as possible, as well as those which have been developed by Mr. Martenot for specific uses: pedal, knee lever.

We will end this part with a detailed description of the manufacture of the key powder bag.

7.2. The mercury key

7.2.1. *Description*

To date, there is no copy of this type of key built by Mr. Martenot for his first ondes models. On the other hand, Mr. Martenot provides us with some information reported in the book by Laurendeau (1990, p. 72):

> Well! Thanks to mercury, a small piece of frosted glass, and a lead pencil [...]. On this piece of glass, we have drawn a few lines [...]. These are the electrically conductive elements [...]. By pressing the manipulator, that is to say the intensity key, this piece of lead-penciled glass sinks itself into a small vat filled with mercury [...].

Later in this chapter, Mr. Martenot specifies that for this command must be as fast as possible, the device must therefore be as light as possible, and the stroke will be limited to 15 mm.

Figure 1.2 gives us an idea of how this key looked; it is built in duplicate: we see the two levers which are relatively long compared to the levers of the keys of the last generations.

7.2.2. *Operating constraints*

Around the years 1925–1928, in wireless radio receivers, the power amplifier was made from a triode. At the time, the most widely used model in France was the B406 triode. It made it possible to obtain sufficient electrical power to excite a diffuser.

The diffusers used by Mr. Martenot were of the electromagnetic type with a cardboard cone. The impedance of the order of 2,000 Ω allowed a direct connection between the anode and the high voltage. To obtain an adjustment of the sound intensity, Mr. Martenot places a "variable" resistor in series with the diffuser. The diagram is given in Figure 7.1.

Here is how it works:

– when the key is inactive or raised, there is no electrical contact, the resistor is an open circuit, and the diffuser is silent;

– when the key lever is lowered, the electrical contact is established and then the resistance (R_T) in series decreases to a zero value; in this position, the power delivered to the diffuser is maximum.

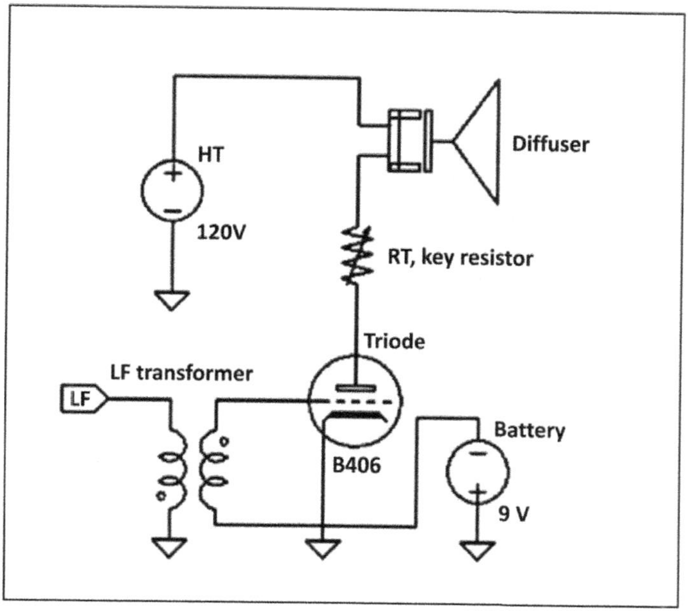

Figure 7.1. *Insertion of the mercury key in the circuit of the power triode*

This mode of operation has drawbacks: when the contact is established, the electrical circuit is suddenly supplied with high voltage, which generates contact noise. It would therefore be necessary, ideally, to start from an infinite resistance (open circuit) to gradually evolve towards a zero resistance (closed circuit). There was no such device at the time. The other constraint comes from the progressiveness of the resistance values to obtain sound volume control. In 1925–1928, variable resistors were rheostats (see section 3.4.4). These devices were incompatible with the need for very rapid variation required by musical performance.

Mr. Martenot realized a very judicious compromise by designing the mercury key (we remind you that mercury (Hg) is a liquid metal at room temperature and is a conductor of electricity). The resistor is made by sketching resistive lines on a small piece of glass. In the inactive position, the piece of glass is out of contact with the mercury; the circuit is open. When the latter comes into contact with the mercury, the current passes through the resistance formed by the pencil lines. Finally, in the fully depressed position, the brass strapping electrically connected to the external circuit, and which holds the glass slide, comes into contact with the mercury. In this position, the resistance is then zero.

With one of the poles being electrically connected to the high voltage via the diffuser, the lever must be insulating.

7.2.3. *Realization*

Analysis of the diagram shows us that a minimum resistance value of 400 kΩ is necessary to reduce contact noise as much as possible; at this operating point, the current is established at a value lower than 1 mA. The other constraint to respect is the stroke of 15 mm. The mobile contact device is actuated from a lever whose length can be estimated at 20 cm. The fulcrum is located around two-thirds of the length of the lever (see Figure 1.2), so a useful stroke of 10 mm is obtained. By providing the fixing system (5 mm) of the glass plate, it measures 15 mm * 10 mm. The pencil lines, arranged vertically and parallel to the movement axis, are drawn until the value of 400 kΩ is obtained. In order to minimize the volume of mercury, we use a test tube of 15 mm in diameter and 50 mm in length. Figure 7.2 gives an overview of this realization.

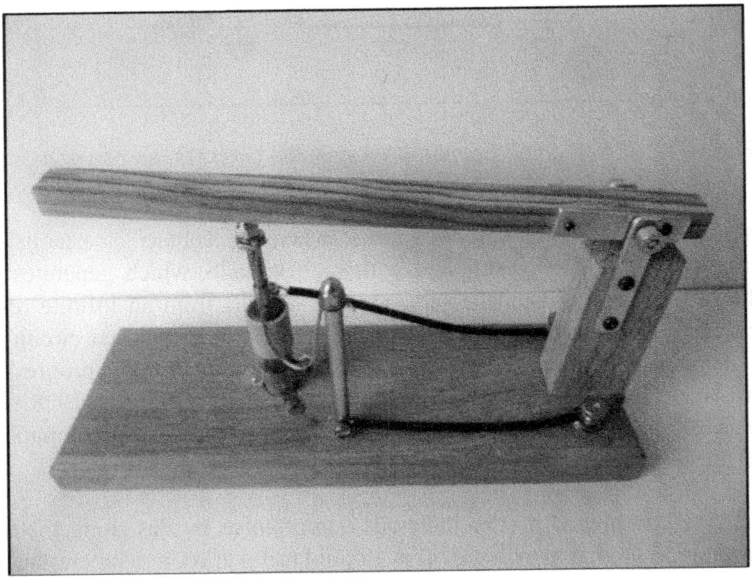

Figure 7.2. *Mercury key mockup*

In use, we observe a very rapid passivation of the penciled surfaces on the glass plate, which requires frequent disassembling of the key in order to reactivate the penciled surface. In the work of Laurendeau (1990), Mr. Martenot indicates that he used a lead pencil. In our case, it is graphite, which could explain the premature aging of the resistive tracks.

We proceeded in another way, which is totally equivalent to the previous one: we used a carbon resistance body, as they were manufactured during the first years of wireless radio technology. We stripped it of its insulating paint and filed the end to obtain approximately 400 kΩ, in contact with mercury. Then, we mounted it on the sliding axle. The results are conclusive:

– useful stroke: 14 mm;

– resistance to contact with mercury: 436 kΩ;

– resistance at the end of the resistive body: 9 kΩ;

– resistance at the end of the stroke: $R < 0.1\ \Omega$;

– total absence of noise.

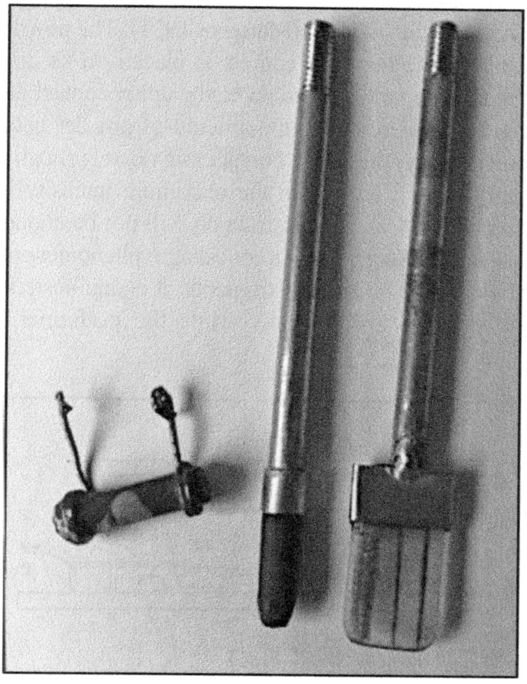

Figure 7.3. *On the left, a resistance manufactured in the years 1928–1930; on the right, the two types of moving contact*

These results are sufficient to demonstrate the feasibility of the device as Mr. Martenot imagined it. With mercury being a very toxic product and difficult to handle, we end the experiment at this stage. Figure 7.3 shows the realization of the moving contact for the two techniques, as well as a resistor as it was manufactured in the 1930s.

7.3. The powder key, first version

7.3.1. *Abandonment of the mercury key*

The operation of this key was quite sufficient for Mr. Martenot, but unfortunately during the year 1928, when he was preparing to give his first concert, the box that contained the intensity key was accidentally knocked over and the mercury spread over the floor in innumerable little pearls. Mercury was no longer the ideal product; Mr. Martenot considered replacing it.

7.3.2. *Description*

This key works with a powder consisting of a mixture of graphite and mica. The device is described in patent no. 703.923 (Martenot 1931). The powder is contained in a small bag of fine leather. The movable contact is placed on its upper part; the fixed contact is placed on the lower part. Using a lever, the upper contact is lowered vertically towards the fixed contact, thus reducing the amount of powder held between the two contacts. To function correctly, the upper contact must barely touch the powder at rest so as to obtain an open circuit, and when the maximum intensity is desired, the two electrodes must be in contact. Careful construction will not be enough to eliminate the phenomenon of "leakage". Recall that the leakage phenomenon is a defect that appeared on the first electronic musical instruments; it is manifested by the continuous emission of a sound of very low intensity while the performer has stopped their performance.

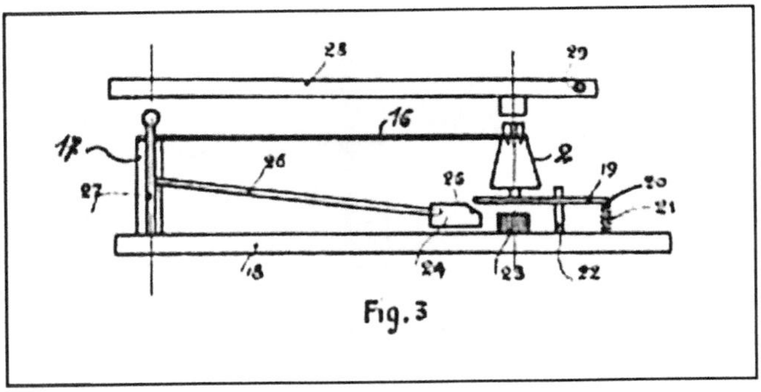

Figure 7.4. *Design of the key shown in the patent*

Mr. Martenot specifies in his patent that the upper electrode is placed in an ebonite washer in the shape of a bowl; this is to protect the upper electrode against the agglomeration of powder. Experience shows that this solution is applicable if the powder has a high grain size; when the powder is made from materials with low grain size, the mixture becomes very adhesive and the bowl fills up. Mr. Martenot will replace this part with a cone with, however, a slight shrinkage of the surface of the electrode, probably to avoid premature wear due to repeated docking between the metal surfaces of the two electrodes. On the other hand, the principle of the electrode embedded in a bowl of insulating material will be taken up for the powder resistances with foot control.

7.3.3. *Operating constraints*

The power circuit diagram has not been modified; the control of the sound level is carried out in the same way as in the previous case: one of the electrodes is connected to the anode of the triode; the other is connected to the diffuser. The constraints are the same.

7.3.4. *Realization*

As shown in Figures 7.5 and 7.6, the key is controlled by a lever (wood + ivory siding) similar to that used for the mercury key.

Figure 7.5. *Ondes no. 15 bag (Musée de la Musique, Paris)*

A movable wedge made it possible to rise more or less the sole on which the bag rests, and thus to shorten the total stroke of the lever, according to the tastes of the performer. The control is indirect since the control lever acts by pressure on the knurled nut arranged on the upper part of the bag. The bag tension spring is of the "blade" type.

Figure 7.6. *View of the control lever (underside); the ivory siding is not visible here. Ondes no. 15 (Musée de la Musique, Paris)*

Note the relatively small size of the leather bag, as well as the position (at the bottom) of the string closing the bag, which makes its manufacture delicate. We do not have any information relating to the composition of the mixture, as well as the particle size of the components used.

7.4. The powder key, second version

7.4.1. *Description*

This key will be implemented on the '37 models. In principle, it is identical to the previous one, but significant modifications are made (see Figure 3.30): increase in the size of the bag, hairpin return spring, lever in direct contact with the upper contact, reduction in the length of the lever.

Adding a mobile keyboard poses a lot of problems for Mr. Martenot, particularly in terms of contact noise each time a new note is played. Having done a lot of research (Martenot n.d.a), he filed a patent (Martenot 1931) where he indicated that with associations of capacitors and resistors, it is possible to eliminate the "snap". It is ultimately the intensity key that will also control the playing on the keyboard.

7.4.2. *Operating constraints*

The variable resistor is always in series with the diffuser(s). The 2A3 triode will be used instead of the B406, which will provide a sufficient power reserve to simultaneously power several diffusers (Palme, Gong, etc.). The anode voltage is increased to 230 V, the current is in tens of milliamps and it is probably for these reasons that Mr. Martenot will adopt larger dimensions for the bag. The use of an insulating element, which eliminates the risk of electrical contact with the performer, is essential. Mr. Martenot also tried to modify the electrical diagram to reduce the leak (Martenot n.d.c). It seems that this modification has not been applied to the instruments in service. The key is placed in a much shallower drawer; the sole has the dimension: 130 mm * 20 mm.

7.4.3. *Realization*

Figures 7.9–7.11 suffice to describe the realization of this key.

In his workshop notebook 21 (Martenot n.d.c), Mr. Martenot indicates the composition of the powder. Two dosages are offered:

– first composition: *graphite 90/99% 425 sieve after sedimentation + mica 150 + MB, 1g + 1g +1g*;

– second composition: *graphite 5M + mica 150 + MB, 2g + 3g + 3g*.

In both cases, he uses two micas of different grain sizes, commercially available fine mica (150[1]) and mica (MB), which he obtains by grinding an ore in a coffee grinder. The difficulty in reconstituting this mixture is to determine the duration of the grinding. Indeed, this duration determines the size of the grains obtained. We will return to this aspect of the problem in section 7.6.3.

The mixture will evolve in terms of the choice of graphite, Mr. Martenot specifies:

> The 425 mesh graphite must contain too large grains, which cause an audible "friction". This graphite can only be used after sedimentation.

---

1 These numbers correspond to grain sizes: 150↔150 μm; 425↔425 μm; 5M↔5 μm.

By using 5M graphite, there seem to be too many differences between this fineness and the much coarser grain of mica MB, hence the need to use very fine mica in equal measure.

7.5. The powder key, third version

7.5.1. *Description*

This is the latest version of the powder key for electronic tubes for ondes. It is the result of a long process of development by Mr. Martenot. It works like the previous ones, but it is more compact.

7.5.2. *Operating constraints*

In this model, the intensity key is placed in series with the cathode of the final pentode and allows the amplifier to be controlled (instead of a variable resistor in series with the diffuser).

An anti-leakage circuit is placed as follows: a high value resistor (>10 MΩ) is connected between the high voltage (+ 330 V) and one of the two poles of the bag, (point C), the lever must be carefully insulated.

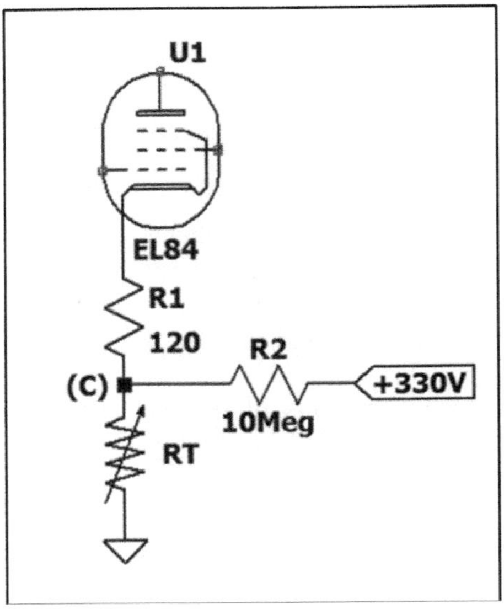

Figure 7.7. *Partial diagram of the output amplifier of a 1953 ondes model*

7.5.3. *Realization*

In this new manufacture, it seems that Mr. Martenot sought to increase the efficiency of the mixing of the powder with the movement of the upper pole. The leather bag reaches its final dimensions: a diameter of 15–16 mm for a height of 11 mm, from the upper washer to the tightening wire. Its shape at rest is initially a right cylinder. The upper electrode is placed inside a cone of insulating material whose base diameter is 15–16 mm and height is 6 mm; the contact surface is slightly recessed. It is the mobile pole. A brass washer in the low position ensures the maintenance of the leather bag and the electrical connection with the external circuit. This is the fixed pole.

The intended operation is as follows:

When the key is at rest, the cylindrical wall of the leather bag holds the powder, the upper electrode is just above the powder, the contact is open.

When the key is lowered, contact is made with the powder and simultaneously the leather bag takes the form of a cushion; the powder opposes a resistance to the electric current; the more the cone goes down inside the powder, the more the resistance decreases.

In the low position, the resistance is minimal; the key is then equivalent to a closed switch.

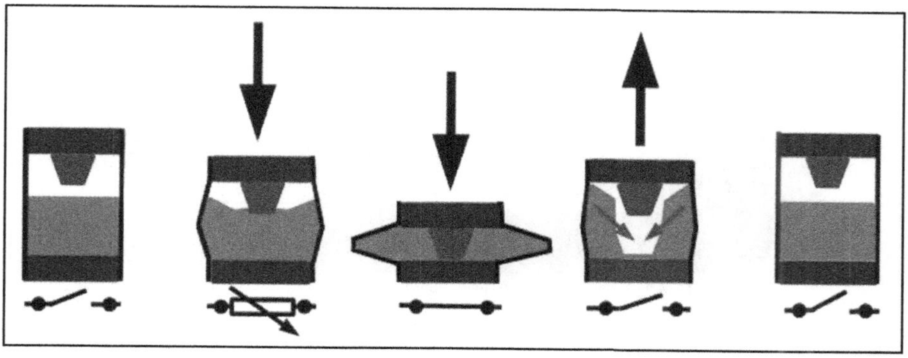

Figure 7.8. *Powder bag working principle*

When the lever rises under the effect of the return spring, the cone releases space in the volume of the powder, but this is immediately filled under the combined effects of the movement of the wall of the leather bag, which regains its initial cylindrical shape and gravity. This return spring has a rectangular blade shape of 44 mm * 18 mm and a thickness of 5/10 mm. A 6 mm to 9 mm hole allows the stiffness to be adjusted.

Figures 7.9–7.11 show different aspects of the key of ondes no. 257.

Figure 7.9. *Intensity key, side view. Ondes no. 257*

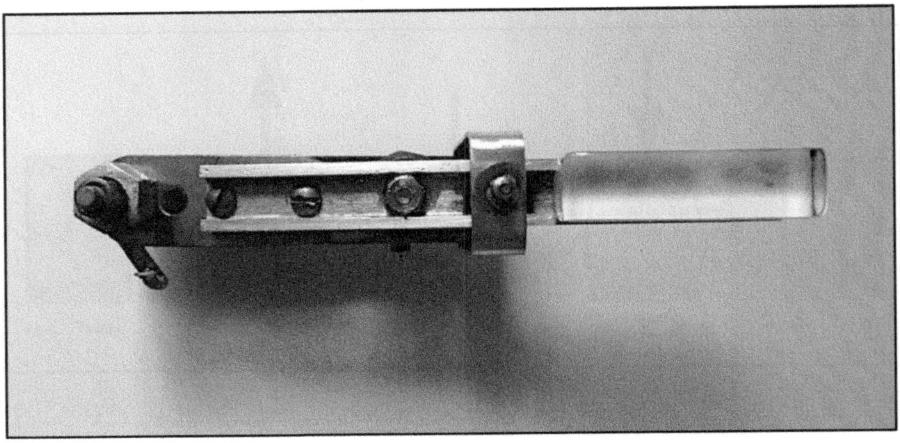

Figure 7.10. *Intensity key, top view. Ondes no. 257*

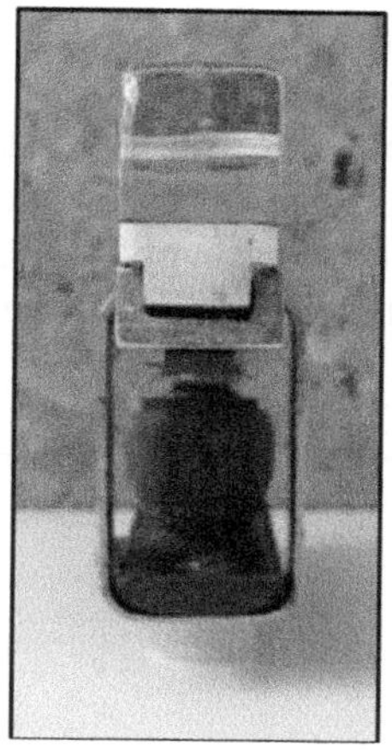

Figure 7.11. *Front view*

The dimensions are as follows:

– sole: 85 mm * 20 mm * 4 mm;

– lever length: 126–130 mm;

– key height at rest: 34–35 mm;

– lowered key height: 25–26 mm;

– lever tilt: 3°;

– total stroke: 9.5–10.5 mm (lever end);

– upper pole stroke (inside the bag): 4–5 mm.

On some models, a stiffness adjuster, consisting of an aluminum blade applied against the spring blade at its fixed point, allows the key to be finely adjusted according to the demands of the performers.

Given the remark made by Mr. Martenot (section 7.4.3), the most probable composition is as follows: graphite (5 μm): 2 g; fine mica (155 μm) 3 g; MB mica: 3 g.

7.6. The powder key, fourth version

7.6.1. *Description*

This is overall the same model as the previous key, but adapted for the new type of ondes, which is now equipped with transistors.

7.6.2. *Operating constraints*

In terms of electrical diagram, the key now works as a potentiometric divider, so it is located on the path of the LF signal. The two poles are isolated and the energy levels are much lower. To solve the problem of leakage, Mr. Martenot adds a micro-cut contact (circuit breaker) at the end of the stroke, to ensure the certain opening of the two poles.

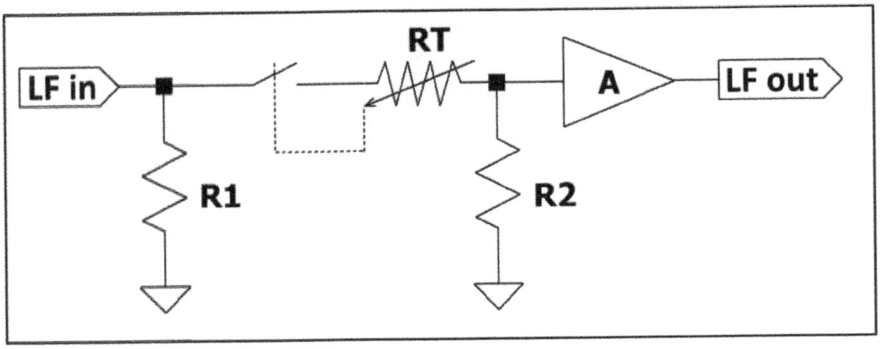

Figure 7.12. *Partial diagram: inserting the intensity key*

7.6.3. *Realization*

Another modification is the elimination of the small recess of the electrode relative to the bearing surface of the insulating cone. The contact then takes place on a residue of adherent powder. Incidentally, Mr. Martenot has placed a connector to facilitate maintenance operations.

Figure 7.13. *Modifications made to the intensity key: the micro-cut contact, the connector. Photograph taken from Meurisse (2011)*

In terms of powder dosage, we did not find any written trace; however, a study was carried out (Jossic 2011). The dosage conclusion is as follows:

Graphite: 3 g; mica (3’30 min. grinding in a coffee grinder): 9 g.

Remember that Mr. Martenot prepared his powders with two types of mica, mica 155, which is a commercially available product, and a mica ground in a coffee grinder, the grain size of which depends on the grinding time.

7.7. Other powder bags

7.7.1. *Introduction*

From the 1953 model, to expand the possibilities of the instrument, Mr. Martenot will introduce two additional controls, the pedal and the knee lever. These two devices use the powder bags.

7.7.2. *The pedal*

The pedal works on the same principle as the key, but the bag is placed on a bakelite structure which supports a mobile plane, on which the performer’s foot rests; this mobile plane is equivalent to the lever of the key. Electrically, it is connected in parallel with the intensity key.

Figure 7.14. *Pedal, ondes no. 257*

The moving contact is embedded in a bowl made of cylindrically shaped insulating material; the countersunk and grooved screw head come to the end of its stroke, in contact with a residue of compressed powder, itself in contact with the fixed pole. There is therefore a residual resistance at closing. It is likely that the choice of this arrangement was made to give priority to the manual key, particularly at the time of *staccato*.

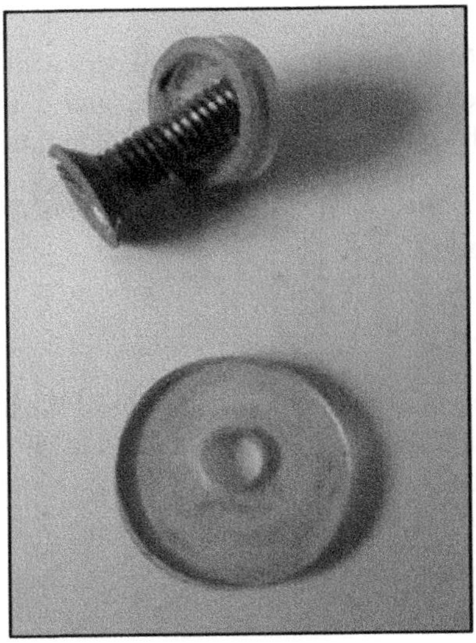

Figure 7.15. *The moving contact cup and its insulating washer*

In accordance with the descriptions that Mr. Martenot formulated in his patent, this bowl must be emptied at the opening; it is therefore necessary to use a powder that is not very adherent (i.e. which will contain high particle size mica).

7.7.3. *The knee lever*

This device also uses a powder bag (see Figure 5.58). A lever placed under the lower part of the instrument cabinet is operated by the performer's knee. This lever is connected to a mechanism by a nylon thread. This mechanism acts on the moving contact in the same way as on the intensity key, but the desired effect is different. Indeed, the "resistor" of the powder bag (R_G) is in series with a capacitor (C19), which forms a harmonic filter. It is a first-order low-pass filter whose cut-off frequency is between 200 Hz and 400 Hz. Thus, the effect of this device is to carry out a filtering of the harmonic components of the sound, which modifies its timbre.

The moving contact is slightly set back in the insulating cone; the assembly (bag + poles) is strictly identical to that of the intensity key. But the desired effect requires finer control of the resistance variation. The bag is placed in an inclined way with respect to the vertical, playing with the knee lever is clearly slower than the *staccato* played on the intensity key, the mixture is no longer stirred energetically but moderately compressed. In this way, the key could best fulfill the role of a variable resistor.

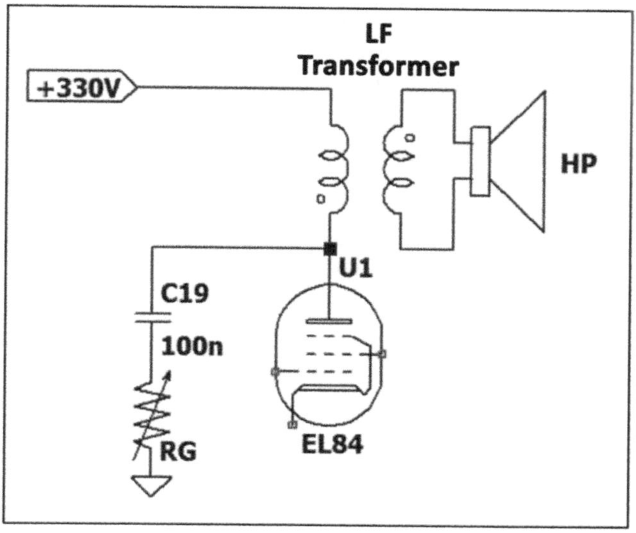

Figure 7.16. *Partial diagram, the knee lever filter*

Figure 7.17 shows the different aspects of this device.

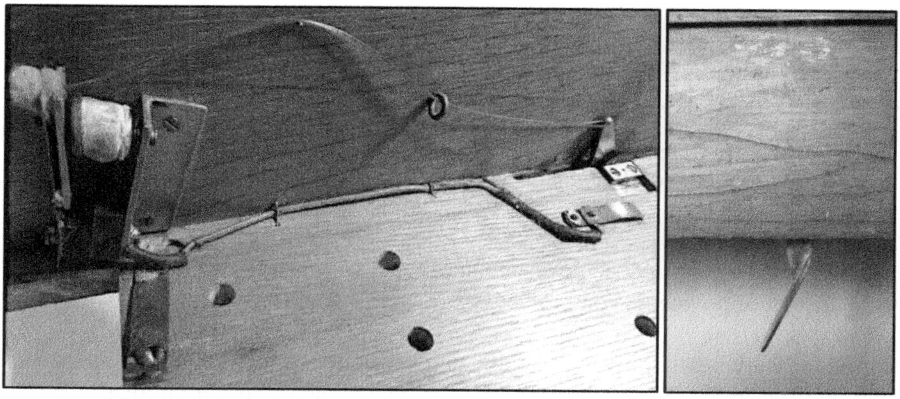

Figure 7.17. *Knee lever: (left) bag control mechanism, (right) knee lever (under the instrument)*

7.8. Intensity key and gesture control

In all of the above, the focus has been more on the electronic aspects of the intensity key. In this chapter, the focus is on the mechanical characteristics.

In 2011, a study was carried out (Meurisse 2011) to characterize the intensity key of a transistor ondes from the CNSM (ondes no. 320). It is a question of measuring the force with which the musician presses the key, as well as its displacement, and the sound produced. All these measurements were carried out simultaneously, and in playing conditions.

Figure 7.18 shows the time signals of the electrical voltages upstream and downstream of the bag of powder, as well as the force applied to the key, while a note (*Do*$_0$, 32 Hz) is played twice for one second.

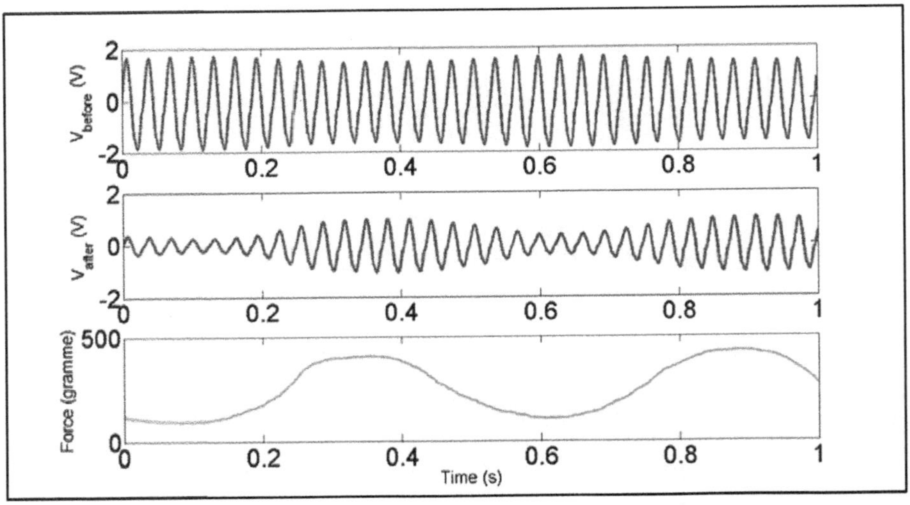

Figure 7.18. *Electrical voltage measured upstream (top) and downstream (middle) of the bag, and force applied to the key (bottom)*

Thus, the intensity key achieves an amplitude modulation of the electrical signal when the bag of powder is compressed.

The measurements show that the dynamics (difference between the lowest sound level and the highest) is around 50 dB; for comparison, most traditional acoustic instruments rarely exceed 20 dB.

All of these sound dynamics correspond to an instrumental gesture of low amplitude (the pressing of the key varying by a few millimeters). Finally, the variation in sound level according to the displacement or the force with which the finger presses does not seem to depend on the speed of execution of the notes by the musician.

The force exerted by the musician and the stroke of the key are linked by the mechanical response of the key. Figure 7.19 shows the variation of this force according to the displacement of the key. Musical nuances (sound level intensity) are plotted in Figure 7.19.

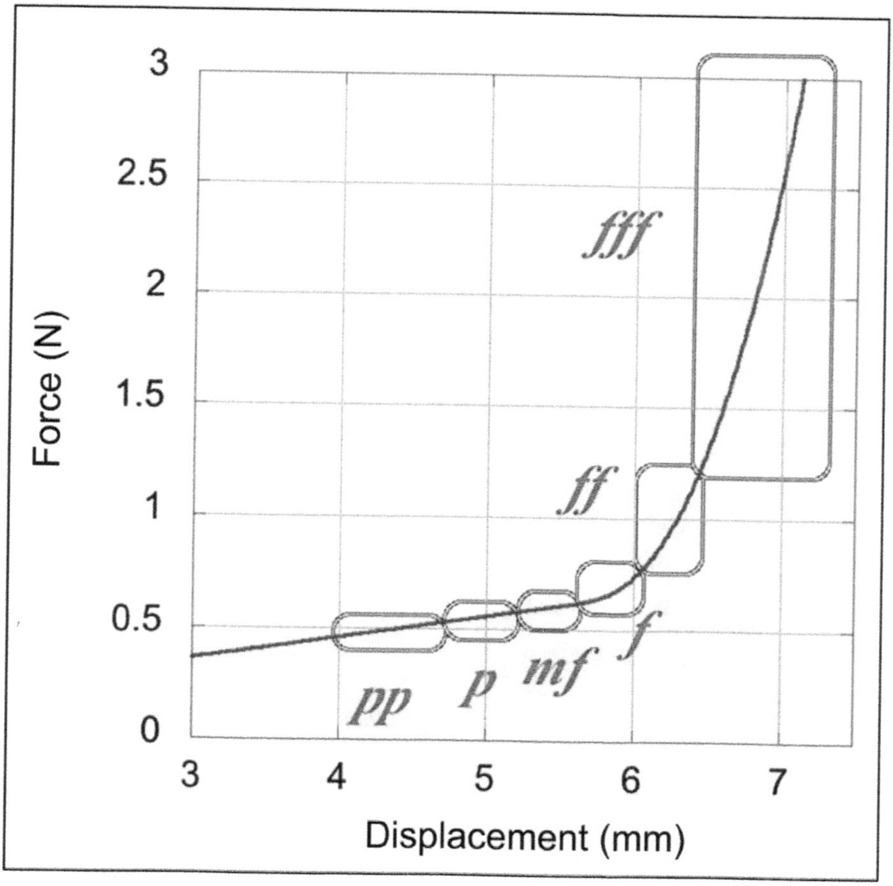

Figure 7.19. *Mean mechanical response of intensity key and associated musical nuances*

Quantitatively (for this intensity key), the musician's finger only needs to move less than 4 mm to go from an inaudible sound to maximum sound intensity. This can be done fast enough to allow for the sharpest attack transients (*percussive* sounds). Thus, when the musician uses the intensity key, it is in fact all the temporal dynamics of the notes (including the attacks and the extinctions) that they control. As it is written in Loriod (1987): "*this infinitely sensitive key enables the sound to be shaped in the same way as a sculptor moulds his clay*".

Along with the increase in sound intensity, the force required to actuate the key increases sharply, in a nonlinear fashion with displacement. This particular mechanical behavior seems to play a role in the finesse of control exploitable by Ondists (Quartier and Meurisse 2014).

7.9. Manufacturing a powder bag

7.9.1. *Introduction*

This key is subject to severe constraints. The most frequent breakdown is the rupture of the leather bag, with its unpleasant consequences: complete stoppage of the play, dispersion of the powder in and out of the drawer. But another type of wear also appears frequently. It does not suspend the play, and in the absence of intervention, the Ondistes often continue the play as it is. It is about the wear of the powder by the innumerable crushings which it undergoes; the key then loses its initial dynamics. It is therefore necessary to be able to manufacture powder bags.

Figure 7.20. *Breakage of a powder bag*

7.9.2. *Problem*

We have seen in this study how much the formulation of the powder mixture has evolved over time. Mr. Martenot makes modifications to his mixtures according to the materials available, the instrument model and the type of function that the powder bag must perform. The difficulty comes from the fact that these materials are not easily available today or too expensive. It is therefore necessary to establish a formulation with the current materials. For this, we will rely on the following observations:

1) The first materials used had a large grain size of 100–500 µm. But Mr. Martenot quickly used finer materials (graphite 5 µm).

2) The mixture may contain one or two grain sizes of mica. But the proportion of mica to graphite remains constant 3:1.

3) The thin mica allows more extensive control of the resistance variation (knee lever). Coarse mica allows faster switch operation (*staccato*) but brings noise.

4) The mass of coarse mica must not exceed that of fine mica.

It is the proportion between mica (electrical insulator) and graphite (conductive) which will define the overall resistivity of the mixture. Graphite powder alone is very cohesive and quickly turns into a very fine plastic paste (which allows its use as a "pencil" as well as a lubricant in the mechanical industry). Hence, the need to combine it with mica to ensure a physical separation between the graphite grains, which allows the mixture to remain a powder.

The noise generated by the key is caused by electric micro-arcs which occur in the air spaces when the powder is in motion. If the particle sizes of graphite and mica are close, as the dielectric strength of mica is much higher than that of air, the noise is reduced in this way. We would then be tempted to use the same particle size for all the components; we then obtain a powder which agglomerates very quickly and which leads, according to remark 3, to a key of the "knee lever" type.

7.9.3. *Manufacturing the mixture*

Two products are readily available: white mica for cosmetics: 8–42 µm, and graphite powder for drawing or lubricant: 5–7 µm. In the absence of being able to use ingredients of equivalent particle size for the graphite/fine mica mixture, we have observed empirically that the ratio between the particle sizes of one component compared to the median particle sizes of another component must be between 4 and 7.

By closely analyzing the technical instructions, we note that the grains are far from all having the same size; the distribution is statistical between the smallest size and the largest. The majority of the grains on a sample have the average size, i.e. 25 μm for mica and 6 μm for graphite. It remains to find coarse mica whose size will be between 100 μm and 175 μm. We have selected a product which comes closest in terms of grain size to coarse mica which corresponds to our needs; it is mica for decoration material: 300–600 μm (median 450 μm). Like Mr. Martenot, we will reduce its size using a coffee grinder. The grinding time is around 2 min 30 s (10 passes of 15 s); this time is only an indication, because too many parameters come into play to determine the exact value.

Figure 7.21 shows how heterogeneous the powder obtained is, we will then use a sieve of 150–160 μm.

Figure 7.21. *Photograph of raw crushed mica obtained on an X 40 binocular loupe; the reference element is the thickness of the copper 35 μm*

For the powder bags intended for tube ondes, the proportions are as follows:

– intensity key: graphite: 2 g; fine mica: 3 g; coarse mica: 3 g;

– pedal: graphite: 2 g; fine mica: 3 g; coarse mica: 3 g;

– knee lever: graphite: 2 g; fine mica: 6 g;

– total mass of the mixture for filling the bag: 0.77 g.

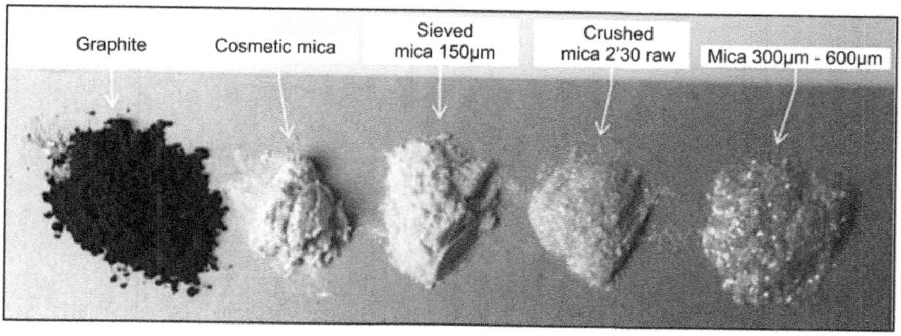

Figure 7.22. *The constituents of the powder at different stages. Note the difference in shine of the mica grains according to their size*

REMARKS.–

1) The dosage of the powder for the intensity key can be modulated according to the desired sensitivity by adjusting the masses of the two types of mica and taking care to respect conditions 2 and 4 defined in section 7.9.2.

2) Strictly speaking, a double sieving should be applied to the crushed mica, in order to eliminate the smallest grains and respect the recommended proportions. In practice, reducing the supply of fine mica from 5% to 10% with an increase of the same order of the supply of crushed and sieved mica is sufficient to restore the proportions of the mixture.

7.9.4. *Manufacturing the powder bag*

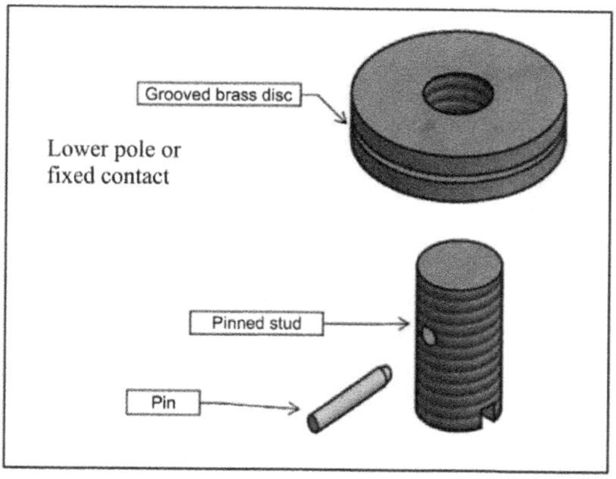

Figure 7.23. *The different components of a Martenot bag: the fixed contact*

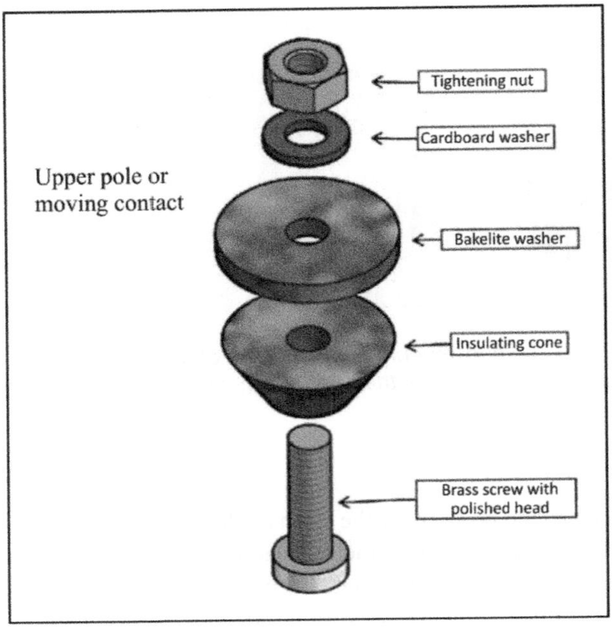

Figure 7.24. *The different components of a Martenot bag: the mobile contact*

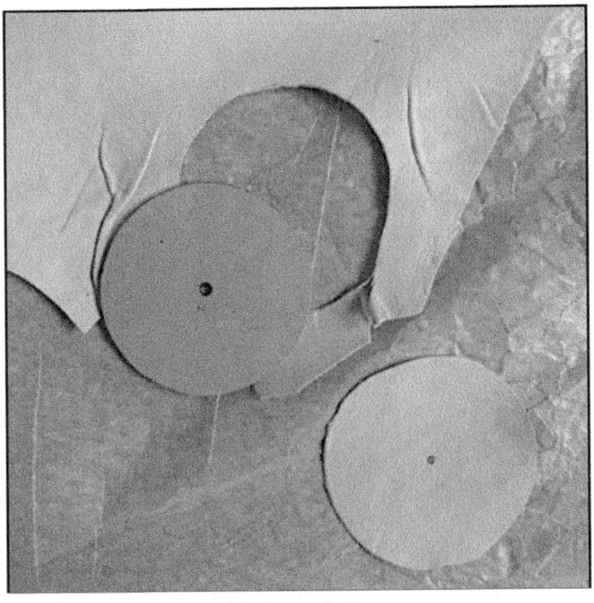

Figure 7.25. *Bag making. Leather cutting*

We start from a goatskin for glove-making; we determine by transparency an area where the skin is thin but not pierced. Using a cardboard template (diameter 90 mm, pierced in the center), we transfer the drawing of the cutting circle and its center. Then, we cut out the piece and pierce the center using a punch. The hole must have a diameter equal to the clamping screw minus 1 mm.

A shim is prepared consisting of a threaded bolt A (6 mm * 40 mm) with a nut B.

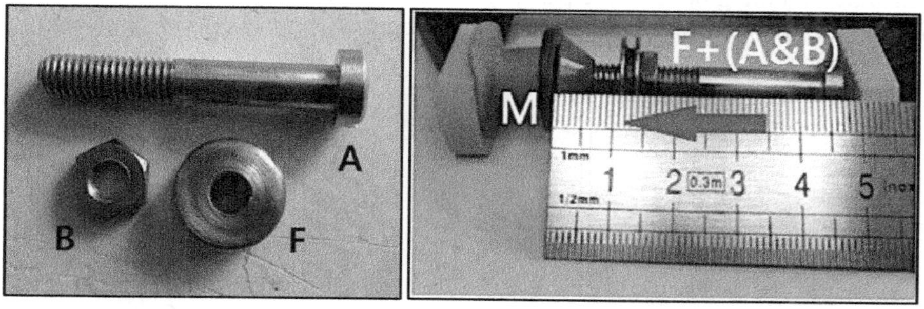

Figures 7.26. *Bag making. The wedge and its adjustment*

The part constituting the fixed pole (F) is inserted on the threaded rod (A) resting on the nut (B). The end of the threaded rod (A) is resting on the mobile pole (M) (the contact surface is slightly set back from the end of the cone). We measure very precisely 12 mm with the groove machined on the part constituting the fixed pole (F).

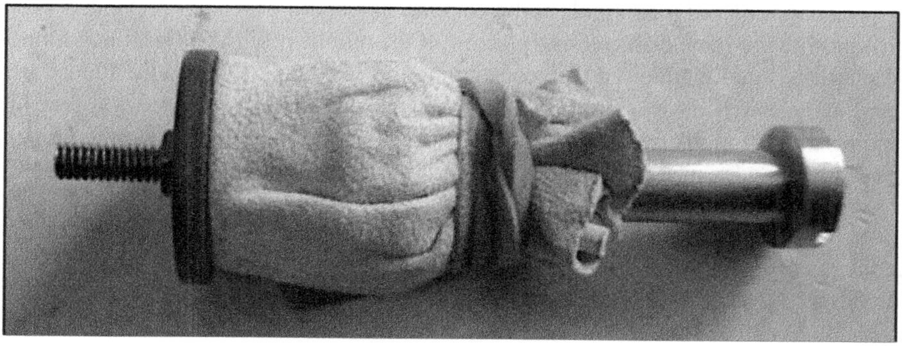

Figure 7.27. *Bag making. Setting up the two poles*

Then, we mount the piece of leather between the rear face of the cone and the bakelite washer, and tighten it with the nut. The tanned side of the leather is on the inside, the raw side is on the outside. We take the wedge (A&B), press it against the contact surface of the mobile pole (M) and carefully fold the piece of leather by stretching it behind the nut (B) of the bolt (A). The folded parts are immobilized with elastic.

The leather bag has the shape of a cylinder but is irregularly pleated. At this stage, distribute the folds evenly by pulling the leather towards the back of the nut (B). We finish by tightening very tightly two turns of a linen thread; the leather must penetrate into the groove of the part (F) in order to ensure the tightness of the bag. We can then remove the elastic and then the wedge.

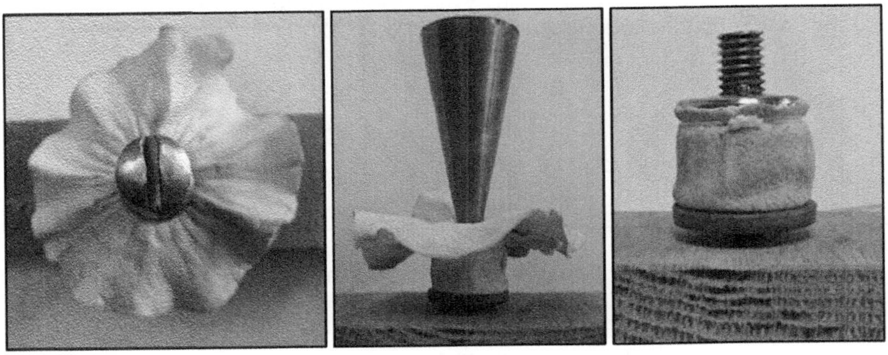

Figure 7.28. *Bag manufacturing. Spread the folds (left). Powder filling (center). Finished bag (right)*

The bag is placed (upper pole at the bottom) on a board which has previously been pierced with the same diameter as the screw of the mobile pole (M). The filling funnel is introduced into the pierced disc constituting the fixed pole (F), and the funnel and the slightly stretched bag are held by hand, then the exact mass of mixture is poured in, tapping the bag regularly. After cleaning with a soft brush and removing the funnel, the threaded and pinned rod is screwed in until it stops. End of filling operation.

It is necessary to check that the distance between the bakelite washer and the groove is of the order of 11 mm. If so, cut the excess leather as close as possible to the grooved disc and mount the bag on the intensity key.

7.10. Ribbon manufacturing

7.10.1. *Introduction*

The ribbon is an indispensable element of the ondes. This is the device that allows the ondist to perform glissando. It essentially consists of a textile ribbon, on which a weave with gold thread is added. This golden weaving constitutes the mobile armature of a capacitor which defines the frequency of oscillation of the variable-frequency heterodyne. In the first ondes models, the electrical contact is obtained by adhesion on one of the four pulleys which guide the ribbon, the pulley which ensures the electrical contact is made of aluminum. Later, the conductive part of the tape is brought into contact with a brass rod, so there is abrasion. To ensure the tension of the ribbon, on the left end, a connection is made by stitching a ribbon of elastic material. The two remaining ends are connected on each side of the ring by hooking. As for the intensity key, we observe different patterns of ribbon over time.

7.10.1.1. *The policeman's stripe*

This type of ribbon is found in the first ondes models, including no. 15. We use this denomination because to make the distinctive signs of rank in the gendarmerie, the seamstresses used this stripe to adorn the uniforms. This type of ribbon was made in trimmings workshops; in general, they were intended for clothing, as well as for furnishing. At the time, wearing a uniform was widespread in many state bodies, including the army and the gendarmerie. There was therefore a great diversity of ribbon, of all sizes and colors.

Figure 7.29. *Ribbon woven with braid-type gold threads. Ondes no. 15 (Musée de la Musique, Paris)*

This ribbon has the advantage of presenting these two golden faces to the fixed walls of the capacitor; on the other hand, it is relatively deformable by stretching. When the performer places the ring in front of the Do_0, a large part of the conductive surface of the ribbon is apparent; electrical contact is possible. Finally, the connection by a seam of the elastic ribbon to the golden ribbon is difficult to achieve, and an extra thickness remains, which causes a lateral displacement of the ribbon on the guide rods or the pulleys.

7.10.1.2. *The ribbon gilded by the weft thread*

This is the ribbon described by Mr. Martenot in his patent (Martenot 1947). The weft thread is gold thread, the warp thread is silk. This tape is very resistant to traction, and deforms little. Connections by sewing with elastic tape are easier to make, because the thickness of the covering can be reduced by ironing. This ribbon has a black side and a golden side, because the weaving technique used is more complex than simply replacing the initial weft yarn with an electrically conductive yarn.

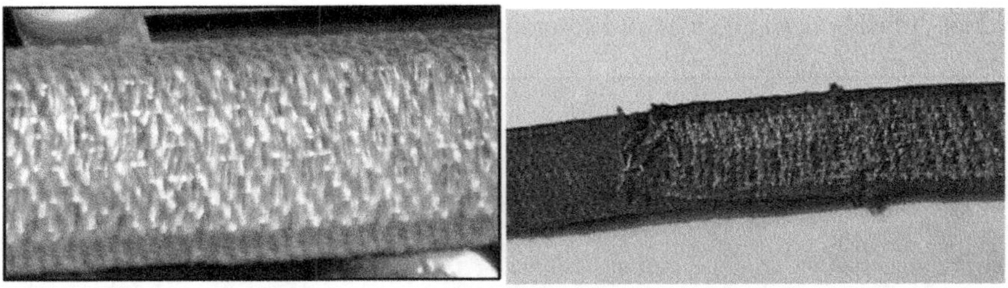

Figure 7.30. *(Left) Golden ribbon by the weft thread (ondes no. 180). (Right) Junction by sewing between the elastic tape and conductive tape*

7.10.1.3. *Ribbon obtained by gluing*

Figure 7.31. *Ribbon obtained by gluing*

This is the last ribbon model that Mr. Martenot used for the last ondes models with electronic tubes. It is made by gluing a very thin ribbon of gold thread on a ribbon of black cotton (?). This tape is stiffer and thicker than the previous two. It has a black side and a golden side; it is much more sensitive to wear.

7.10.2. *Problem*

During our research, we have observed on many instruments the degradation of the ribbon, which very often leads to the stoppage of play. The most frequent are:

– wear of the gold wire by abrasion;

– fraying gold threads;

– local conductive surface losses;

– in the absence of tension, the elastic tape is distended.

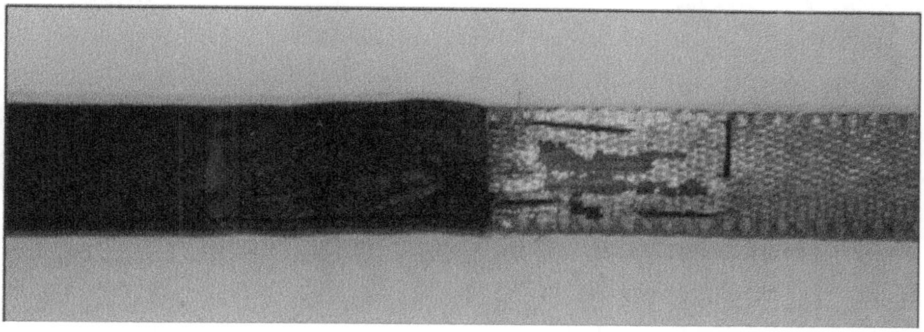

Figure 7.32. *Ribbon local wear*

We therefore looked for solutions to repair these ribbons and thus give them a second life. However, this was not possible and new ribbons had to be made. The search for gold ribbon-type policeman's braid was unsuccessful, because currently this type of product is made from synthetic materials. For model 15-169, we used a policeman's braid-type ribbon available from an old stock supplier of trimmings. Eventually, we will make the ribbons from new materials.

7.10.3. *Ribbon manufacturing*

7.10.3.1. *Raw materials*

To make a ribbon, we supply the following materials:

– one black ribbon of width equal to 10 mm, in polyester or viscose;

– one adhesive conductive tape 10 mm wide;

– one ribbon of elastic material 10 mm wide;

– one brass square 6 mm * 6 mm * 4/10 mm and drilled with a hole of 2.1 mm in diameter;

– two rivets 2 mm * 5 mm;

– two pieces of piano strings 5/10 mm in diameter for the hooks;

– fabric glue.

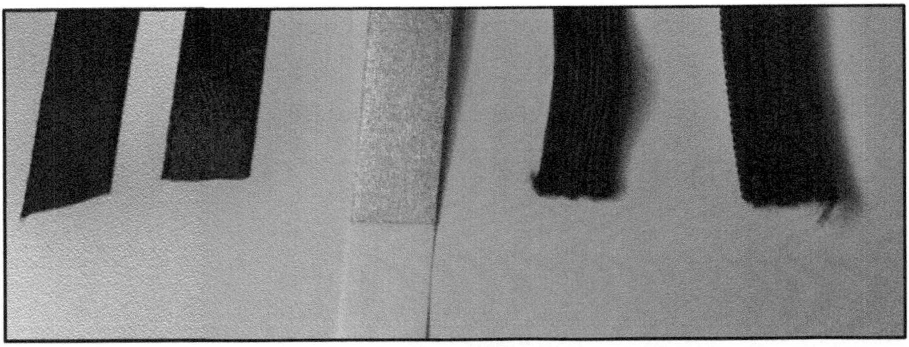

Figure 7.33. *Various ribbons useful for making the Martenot ribbon*

In Figure 7.33, from left to right: the polyester tape, the viscose tape, the conductive and adhesive tape, the tape of the elastic material of the current manufacturer, and the tape of the thin elastic material.

It should be noted that the ribbons of fabric or elastic material are not easy to obtain with a width of 10 mm. Thus, the viscose ribbon and the ribbon of elastic material are commonly offered in 9 mm or 12 mm widths; moreover, the ribbon of elastic material is too thick; it causes jumps when it comes to move on the pulleys. Thin elastic material ribbon (intended for making corsetry) and 10 mm viscose ribbon are only available from specialist suppliers.

The conductive and adhesive tape is available from suppliers for the aeronautical and space industry; this tape is normally used to make shields on devices of complex shapes, in order to shield them from the action of magnetic and electrostatic fields. It consists of a deposit of copper and nickel on a very tight mesh polymer fabric. The conductivity is excellent, and the tensile and abrasion resistances are sufficient for our use.

7.10.3.2. *Mounting*

Figure 7.34 will help to understand how the assembly of the tape is carried out.

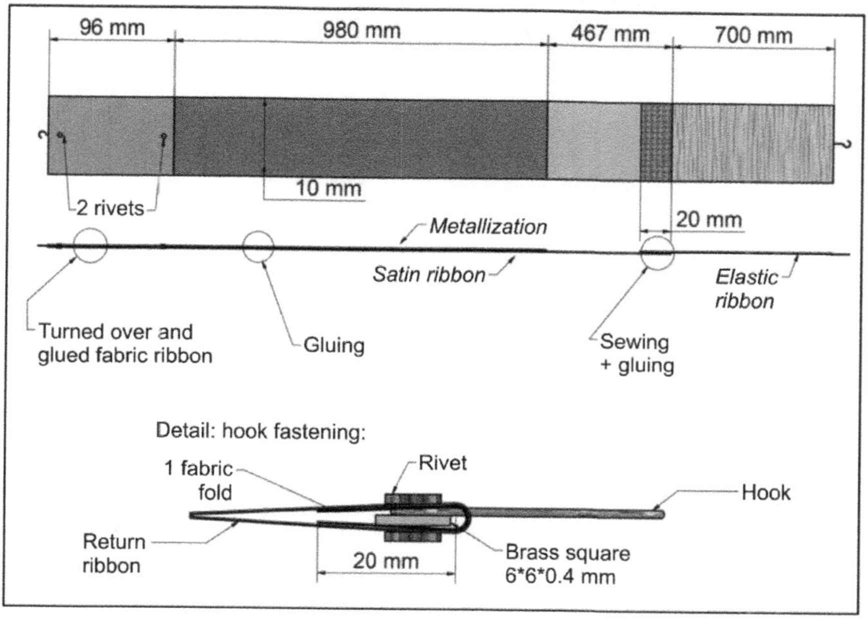

Figure 7.34. *Assembly sketch of the new ribbon*

The ribbon of fabric (~1.64 m) is stretched on a support at right angles by a mass of 200–400 g; we mark the beginning (467 mm) of the bonding, and we cut the conductive tape to length (980 mm). We glue it on the previous one by aligning edge to edge along the length, one of the sides of each ribbon thanks to the square support. We pass several times on the conductive side of a wooden wedge to perfect the bonding.

On the right end, we glue and we sew, with an overlap of 2 cm, the ribbon of elastic material; this ribbon is glued on the side opposite to the conductive face of the conductive ribbon. After drying, we iron to reduce the thickness.

Then, on the space remaining between the conductive strip and the strip of elastic material, a second thickness of fabric strip can be glued. This makes it possible to obtain an assembly whose thickness is approximately the same over the entire length of the assembly, but this gluing also stiffens the assembly.

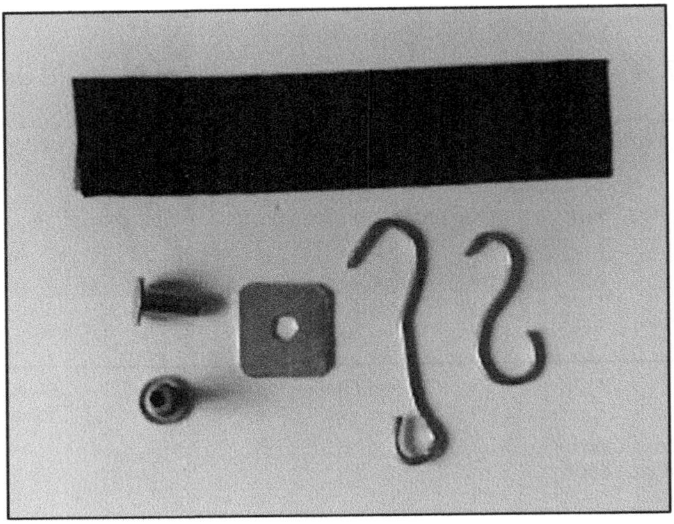

Figure 7.35. *Ribbon mounting accessories*

After having prepared the hook with an eye of 2.1 mm in diameter and a brass square of 6 mm * 6 mm * 4/10 mm, we fold and glue the other end of the ribbon, placing around the brass wedge a second layer of fabric over a length of 20 mm. The purpose of this assembly is to reduce the shearing stress exerted on the tape by the brass spacer and the hook, as well as to keep the tape flat, which is subjected to localized traction at the point of attachment of the hook. On some ribbons, a piece of folded cardboard was also added around the brass wedge. The rivet[2] is struck to join these parts together. The flap is made on the same side as the conductive tape and is glued edge to edge with it. The thread of the hook crosses the assembly in the middle of its height. Then, you can put the second rivet near the conductive tape, it is not always present, and if the bonding is well done, it is not essential. Ironing is carried out at moderate temperature, excluding the part that supports the conductive tape. And to finish, we place the other hook in the shape of an S on the ribbon in elastic material.

Finally, it should be noted that the lengths given in Figure 7.34 are indicative, variations exist between the different instruments; it will be prudent to make exact measurements on the ribbon to be replaced before proceeding with manufacturing.

---

2 To prevent the rivet from splitting during installation, it may be necessary to anneal it in bright red for a few seconds to give it some plasticity, and also to reduce its length by 1 mm using a file.

We have by this method manufactured several ribbons; most of the time, the insertion and the replacement on the instrument are carried out without any adjustment of the tuning. The ribbon, when it is made with a viscose support, is a fragile device, from its manufacture, it must be installed on the instrument and stretched, and otherwise it must be kept vertically. Ribbon made from polyester tape is less sensitive to traction. In any case, avoid folding or storing with too small a rolling radius. The mounting strength is much better, using nonadhesive conductive tape and gluing it yourself, which requires a good mastery of the gluing technique.

8
Transistorization

8.1. Introduction

From 1925 to 1973, Mr. Martenot did not stop perfecting his instrument. The first model works with triode tubes and a wire, the ultimate development will be obtained with the last model with pentode tubes, which works with a keyboard and a ribbon. The common denominator is the intensity key. From the 1960s, the transistor replaced the electronic tube, first in industrial and then in consumer equipment. Sensitive to this technological change, Mr. Martenot attempted a transistorization of the latest model with pentode tubes. This ondes bears the number 270 A; this model will never leave the workshop.

8.2. The support

Logically, Mr. Martenot reuses existing elements in the latest models of ondes with electronic tubes: the cabinet, the keyboard/ribbon system, the drawer box and the intensity key.

Figure 8.1. *Ondes no. 270 A, open cabinet (Musée de la Musique, Paris)*

For the color versions of the figures in this chapter, see www.iste.co.uk/courrier/ondesmartenot.zip.

The first step is to produce transistor oscillators that can operate with the keyboard or ribbon interfaces initially designed for electronic tubes. Due to the lack of availability, we have not carried out live tests of this ondes.

Figure 8.2. *Ondes no. 270 A, the drawer seen from above (Musée de la Musique, Paris)*

Figure 8.3. *Ondes no. 270 A, the drawer seen from below (Musée de la Musique, Paris)*

At the level of the amplification chain and the sound effects, Figures 8.2 and 8.3 show that the timbre keys are generally preserved, but a closer analysis shows that a reworking of the drawer has been made. We are moving towards the design of the drawer used for transistor-based ondes, with the use of integrated circuits. We also observe that the wiring technique used suggests that it is a prototype under development. We also note the presence of a power rheostat, which is probably used to adjust the sound level of the loudspeaker.

8.3. The problems of transistorization

Several difficulties arise at the level of the amplification chain. The intensity key, which in the model with electronic tubes is inserted in the power circuit, either in series with the plate circuit when using a triode tube, or in series with the cathode circuit when using a pentode, cannot be inserted in this way on a power amplifier with transistors. Indeed, in the case of electronic tube models, the amplifier operates in class A (with a low efficiency). Figure 8.4 of the upper face of the ondes no. 270 chassis shows two complementary power transistors. These are probably the active elements of an amplifier that operates in class B. We find this amplifier structure on the ondes with transistors. Mr. Martenot noted that making a class A power amplifier with a single transistor was not the optimal solution.

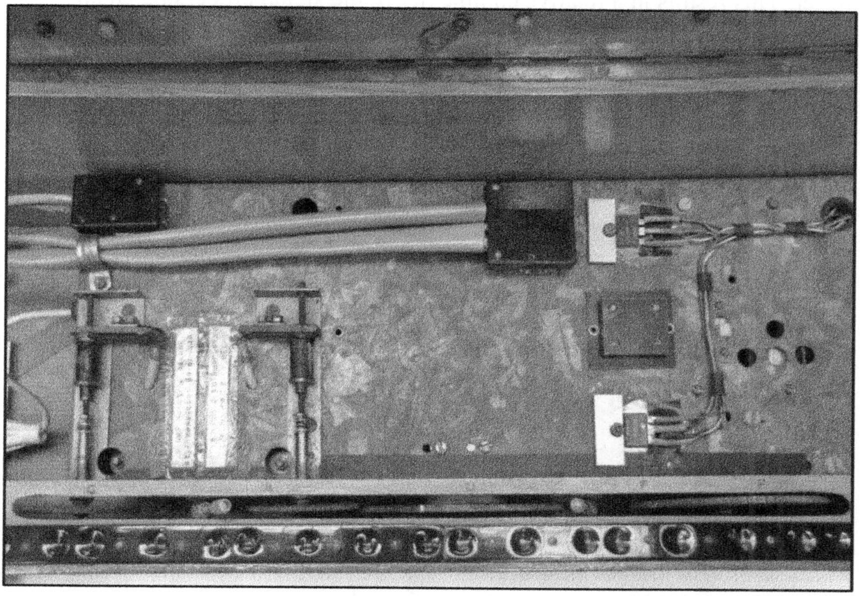

Figure 8.4. *Ondes no. 270 A, upper face of the chassis (Musée de la Musique, Paris)*

Therefore, if we want to keep the intensity key, it must be inserted in a series between the amplifier and the loudspeaker, but the impedances of these circuits are low and the currents that cross them are of great intensity, incompatible with the use of the powder bag. Therefore, it is necessary to insert the intensity key between two signal processing blocks. This is the solution that Mr. Martenot adopts and uses for the ondes with transistors. However, the intensity key has to change. Indeed, experience shows that the cut in the signal path must be clean; otherwise, the phenomenon of leakage will occur. Mr. Martenot is forced to add a *pre-cut* contact, which will break down the travel of the key into two zones: the pre-cut zone and the active sound intensity control zone (see the partial signal control diagram, figure 7.12).

The adoption of a class B amplifier with transistors does not make it possible to carry out a simple control of timbres. Indeed, for the T2 and T3 timbre filters, it is the internal resistance of the amplifier associated with a capacitor that defines the bandwidth. With a transistor amplifier, this internal resistance is low, which leads to the use of capacitors of several tens of μF. They are therefore electrochemical capacitors with large tolerances on the values; the definition frequencies will vary depending on the capacitor wired to the amplifier. Finally, the breaking currents during the choice of timbres can cause malfunctions in the amplifier. These devices must therefore be placed in the signal processing chain, upstream of the power amplifier. This results in a relative complexity of the signal processing chain.

Figure 8.5. *Ondes no. 270 A, the multiplication of electronic sub-assemblies (Music Museum, Paris)*

But one problem persists: frequency drift. To obtain good stability, the heterodynes must be very efficient. For a fixed frequency oscillator, a quartz system can be used (this solution was never implemented by Mr. Martenot). But for the variable frequency oscillator, the problem remains.

For his ondioline, Jenny (1938) used an electronic tube relaxation oscillator: the frequency stability can be improved by using resistors and capacitors with low tolerances and by stabilizing the supply voltage of the oscillator. Later on, Moore invented the synthesizer (1964) that uses the voltage-controlled oscillator whose design was relatively complex. To obtain a stable frequency and linearity in voltage/frequency control, it was necessary to design high-performance current sources, but the results were conclusive.

Mr. Martenot uses a voltage-controlled oscillator in his transistor model. The ribbon is replaced by a steel wire that rotates a potentiometer, and the numerous inductances of the keyboard are replaced by resistors.

8.4. Compatible transistor experimental ondes

8.4.1. *Introduction*

We have developed and experimented with a transistor-based ondes on an experimental basis, which operates by using the intensity key, the ribbon and the keyboard of an electronic tube model. We will not develop in detail the diagrams and the development for lack of space.

8.4.2. *Diagram of the HF sub-assembly*

This diagram is structured in the same way as the diagram of the HF part of an electronic tube model; however, for the fixed frequency oscillator, we have adopted operation with a quartz (f = 8 MHz). Using a counter and some logic gates, it is easy to generate the two fixed frequencies of 160 kHz (treble register) and 80 kHz (low register). Shaping is achieved by a classic LC resonant circuit.

The variable oscillator has a structure very similar to that established for the Ondes model with electron tubes, i.e. a Hartley structure built around a JFET transistor. But in our case, the signal obtained is purely sinusoidal, which is not suitable for switching the high register, which requires operating with the second harmonic of the HF signal. The oscillator is therefore followed by a shaping circuit with an NPN transistor that delivers a signal whose duty cycle is chosen in such a way that the amplitudes of the fundamental and the second harmonic are almost the same.

Finally, the two HF signals are applied to a Gilbert cell (NE602 from Signetics) that performs the multiplicative mixing. At output, we obtain an LF signal that varies approximately from 30 Hz to 4 kHz.

The different switchings (80 kHz/160 kHz, ribbon/keyboard) are carried out with relays.

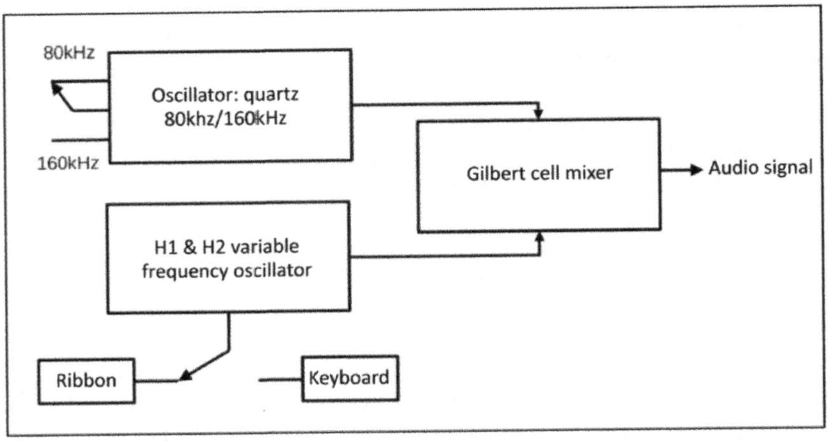

Figure 8.6. *Synoptic diagram of the HF sub-assembly*

8.4.3. *The LF subset*

This sub-assembly naturally fits between the HF part and the diffusers. The LF signal from the mixer is of low amplitude and contains many harmonics. We then use an amplification cell whose gain is around 22 and whose bandwidth is limited from 30 Hz to 4 kHz. The next stage is a comparator, we can choose two threshold values, in this way we obtain a rectangular signal whose duty cycles depend on the thresholds (50%, 70%), and this is the T5 timbre control. The signal processing that follows is a succession of filters that strictly respects the effects obtained by the action of the timbres produced on the ondes with electronic tubes. The amplification or filtering cells are built from operational amplifiers; the on/off switching of the filters is carried out by relay commands. Note that in synoptic diagram 8.8, for the RC filter of timbre T2-T3-T4, we have not included the relay contacts. Indeed, the path of the LF signal is not modified.

Power amplification is achieved by three independent class D amplifiers, each of which drives a diffuser: HP, Gong and Palme. Note that timbres T1 and T8 apply only to the signal delivered to the loudspeaker, in accordance with the operation of the Electronic Tube Ondes.

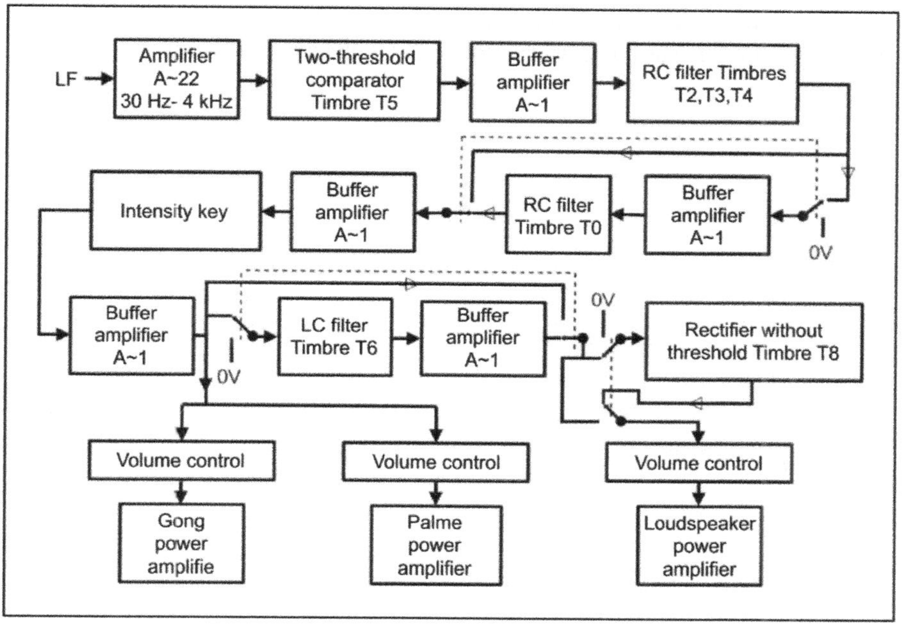

Figure 8.7. *Synoptic diagram of the LF sub-assembly*

The power amplifiers are modular systems built around Texas Instrument's TPA3110 integrated circuit, powered at 9 V DC. It makes it possible to obtain an output power of the order of 6 W under 4–5 Ω. Although operating in class D, this chip contains an amplifier with high-performance audio properties. We used industrially manufactured boards that support this chip. The footprint is extremely small. Each amplifier has a volume control and switching on/off is ensured by a "mute" input that is connected to the 0 V potential for passivation.

8.4.4. *Power supplies*

The different sub-assemblies are supplied with different voltages (9 V, 12 V, ± 12 V). We used power supply devices available from suppliers.

8.5. Tests

At the prototype stage, we carried out tests that turned out to be very promising. Indeed we find all the subtleties of the "Martenot" sound, even if the bandwidth is more extensive towards the low frequencies. The design of the amplification chain, which uses independent LF filters, allows for easy adaptations. Musician gestures are of course the same as those used in the ondes with electronic tubes.

References

Alindret, R. (1925). *Les C119*, 1st edition. Publications Henry Etienne, Paris.

British, P. (1934). Music from the ether [Online]. Available at: https://www.youtube.com/watch?v=f4phKWMaOr0 [Accessed 6 March 2022].

Castellengo, M. (2015). *Ecoute musicale et acoustique*, 1st edition. Eyrolles, Paris.

Jossic, M. (2011). Caractérisation et étude paramétrique de la poudre de l'Onde Martenot. L3 internship report, Université Pierre et Marie Curie, Paris.

Lampkin, A.F. (1939). An improvement in constant frequency oscillator. *Proc. I.R.E*, 27(3), 199–201.

Laurendeau, J. (1990). *Maurice Martenot, Luthier de l'électronique*, 2nd edition. Beauchesne, Paris.

Loriod, J. (1987). *Technique de l'Onde électronique*. A. Leduc, Paris.

Martenot (n.d.a). Fonds "Martenot" du Musée de la Musique. Notebook 15, Paris.

Martenot (n.d.b). Fonds "Martenot" du Musée de la Musique. Notebook 18, Bloc d'Alimentation, Paris.

Martenot (n.d.c). Fonds "Martenot" du Musée de la Musique. Workshop notebook 21, Note, Modification Instrument 1937, Paris.

Martenot (1929). Perfectionnement aux instruments de musique électriques. INPI Patent, 666807.

Martenot (1931a). Dispositif de jeu pour les instruments de musique électriques et radioélectriques. Patent, FR 708073.

Martenot (1931b). Résistance variable pour instruments de musique électriques et radioélectriques. Patent, FR 703923.

Martenot (1939). Perfectionnement aux instruments de musique électriques. Patent, FR 841866.

Martenot (1947). Perfectionnements aux dispositifs de variations de fréquence dans les instruments de musique électriques ou électroniques. Patent, FR 1004406.

Martenot (1954). Sons physiques – Sons musicaux. Conference given in Gravesano, 12 August, verbatim published in G.A.M. Notebook no. 60, "Les Ondes Martenot", 26 April 1972 [Online]. Available at: http://www.lam.jussieu.fr/index.php?page=BulletinsGAM [Accessed 6 March 2022].

Meurisse, T. (2011). Etude d'un système à retour d'effort simulant la touche d'expression des Ondes Martenot. Master's Thesis, Université Pierre et Marie Curie, Paris.

Quartier, L. and Meurisse, T. (2014). La "Touche d'Intensité" des Ondes Martenot : un contrôleur physiologiquement adapté à la main. *Congrès Français d'Acoustique*, Le Mans.

Weiss, E.H. (1930). *Phonographes et musiques mécaniques*, 1st edition, Hachette, Paris.

Willems, J. (1980). Etude de trois ancêtres de la musique électronique. *Technologia*, 3(1) [Online]. Available at: https://www.astrolabium.be/spip.php?article164&artpage=9-17 [Accessed 6 March 2022].

Index

Other titles from

iSTE

in

Electronics Engineering

2022

METHAPETTYPARAMBU PURUSHOTHAMA Jayakrishnan, PERRET Etienne, VENA Arnaud
Non-Volatile CBRAM/MIM Switching Technology for Electronically Reconfigurable Passive Microwave Devices: Theory and Methods for Application in Rewritable Chipless RFID

2021

NGUYEN Thien-Phap
Organic Electronics 1: Materials and Physical Processes
Organic Electronics 2: Applications and Marketing

2020

CAPPY Alain
Neuro-inspired Information Processing

FOUQUET François
Noise in Radio-Frequency Electronics and its Measurement

MAHDOUM Ali
CAD of Circuits and Integrated Systems

2018

MURET Pierre
Fundamentals of Electronics 2: Continuous-time Signals and Systems
Fundamentals of Electronics 3: Discrete-time Signals and Systems, and Quantized Level Systems

2017

MURET Pierre
Fundamentals of Electronics 1: Electronic Components and Elementary Functions

BUCCI Davide
Analog Electronics for Measuring Systems

2016

BAUDRAND Henri, TITAOUINE Mohammed, RAVEU Nathalie
The Wave Concept in Electromagnetism and Circuits: Theory and Applications

FANET Hervé
Ultra Low Power Electronics and Adiabatic Solutions

NDJOUNTCHE Tertulien
Digital Electronics 1: Combinational Logic Circuits
Digital Electronics 2: Sequential and Arithmetic Logic Circuits
Digital Electronics 3: Finite-state Machines

2015

DURAFFOURG Laurent, ARCAMONE Julien
Nanoelectromechanical Systems

2014

APPRIOU Alain
Uncertainty Theories and Multisensor Data Fusion

CONSONNI Vincent, FEUILLET Guy
Wide Band Gap Semiconductor Nanowires 1: Low-Dimensionality Effects and Growth
Wide Band Gap Semiconductor Nanowires 2: Heterostructures and Optoelectronic Devices

GAUTIER Jean-Luc
Design of Microwave Active Devices

LACAZE Pierre Camille, LACROIX Jean-Christophe
Non-volatile Memories

TEMPLIER François
OLED Microdisplays: Technology and Applications

THOMAS Jean-Hugh, YAAKOUBI Nourdin
New Sensors and Processing Chain

2013

COSTA François, GAUTIER Cyrille, LABOURE Eric, REVOL Bertrand
Electromagnetic Compatibility in Power Electronics

KORDON Fabrice, HUGUES Jérôme, CANALS Agusti, DOHET Alain
Embedded Systems: Analysis and Modeling with SysML, UML and AADL

LE TIEC Yannick
Chemistry in Microelectronics

2012

BECHERRAWY Tamer
Electromagnetism: Maxwell Equations, Wave Propagation and Emission

LALAUZE René
Chemical Sensors and Biosensors

LE MENN Marc
Instrumentation and Metrology in Oceanography

SAGUET Pierre
Numerical Analysis in Electromagnetics: The TLM Method

2011

ALGANI Catherine, RUMELHARD Christian, BILLABERT Anne-Laure
Microwaves Photonic Links: Components and Circuits

BAUDRANT Annie
Silicon Technologies: Ion Implantation and Thermal Treatment

DEFAY Emmanuel
Integration of Ferroelectric and Piezoelectric Thin Films: Concepts and Applications for Microsystems

DEFAY Emmanuel
Ferroelectric Dielectrics Integrated on Silicon

BESNIER Philippe, DÉMOULIN Bernard
Electromagnetic Reverberation Chambers

LANDIS Stefan
Nano-lithography

2010

LANDIS Stefan
Lithography

PIETTE Bernard
VHF / UHF Filters and Multicouplers

2009

DE SALVO Barbara
Silicon Non-volatile Memories / Paths of Innovation

DECOSTER Didier, HARARI Joseph
Optoelectronic Sensors

FABRY Pierre, FOULETIER Jacques
Chemical and Biological Microsensors / Applications in Fluid Media

GAUTIER Jacques
Physics and Operation of Silicon Devices in Integrated Circuits

MOLITON André
Solid-State Physics for Electronics

PERRET Robert
Power Electronics Semiconductor Devices

SAGUET Pierre
Passive RF Integrated Circuits

2008

CHARRUAU Stéphane
Electromagnetism and Interconnections

2007

RIPKA Pavel, TIPEK Alois
Modern Sensors Handbook

Printed and bound by CPI Group (UK) Ltd, Croydon, CR0 4YY

13/09/2026

14961047-0006